VOLUME NINETY FIVE

ADVANCES IN PROTEIN CHEMISTRY AND STRUCTURAL BIOLOGY

Proteomics in Biomedicine and Pharmacology

VOLUME NINETY FIVE

ADVANCES IN
PROTEIN CHEMISTRY AND STRUCTURAL BIOLOGY

Proteomics in Biomedicine and Pharmacology

Edited by

ROSSEN DONEV

Singleton Park
Swansea University
Swansea, UK

ELSEVIER

AMSTERDAM • BOSTON • HEIDELBERG • LONDON
NEW YORK • OXFORD • PARIS • SAN DIEGO
SAN FRANCISCO • SINGAPORE • SYDNEY • TOKYO

Academic Press is an imprint of Elsevier

Academic Press is an imprint of Elsevier
225 Wyman Street, Waltham, MA 02451, USA
525 B Street, Suite 1800, San Diego, CA 92101-4495, USA
The Boulevard, Langford Lane, Kidlington, Oxford, OX5 1GB, UK
32 Jamestown Road, London NW1 7BY, UK

First edition 2014

ISBN: 978-0-12-800453-1
ISSN: 1876-1623

For information on all Academic Press publications
visit our website at store.elsevier.com

14 15 16 17 10 9 8 7 6 5 4 3 2 1

CONTENTS

CONTRIBUTORS

Khaled Alawam
Forensic Medicine Department, Ministry of Interior, Kuwait City, Kuwait

Hossein Baharvand
Department of Developmental Biology, University of Science and Culture, and Department of Stem Cells and Developmental Biology at Cell Science Research Center, Royan Institute for Stem Cell Biology and Technology, ACECR, Tehran, Iran

A. Elizabeth Bond
Institute of Mass Spectrometry, College of Medicine, Swansea University, Swansea, United Kingdom

Christoph H. Borchers
University of Victoria—Genome British Columbia Proteomics Centre, and Department of Biochemistry and Microbiology, University of Victoria, Petch Building Room 207, Victoria, British Columbia, Canada

Nicola Luigi Bragazzi
Nanobiotechnology and Biophysics Laboratories (NBL), Department of Experimental Medicine (DIMES); Nanoworld Institute Fondazione ELBA Nicolini (FEN), Pradalunga, Bergamo, and School of Public Health, Department of Health Sciences (DISSAL), University of Genoa, Genoa, Italy

Juan Casado-Vela
Centro Nacional de Biotecnología, Spanish National Research Council (CSIC), Madrid, Spain

Ed Dudley
Institute of Mass Spectrometry, College of Medicine, Swansea University, Swansea, United Kingdom

Octavio Luiz Franco
Centro de Análises Proteômicas e Bioquímicas, Programa de Pós-Graduação em Ciências Genômicas e Biotecnologia, Universidade Católica de Brasília, Brasília, Brazil

José Manuel Franco-Zorrilla
Centro Nacional de Biotecnología, Spanish National Research Council (CSIC), Madrid, Spain

Dustin C. Frost
School of Pharmacy, University of Wisconsin, Madison, Wisconsin, USA

Manuel Fuentes
Centro de Investigación del Cáncer/IBMCC (USAL/CSIC), IBSAL, Departamento de Medicina, Unidad de Proteomics & Servicio General de Citometría, University of Salamanca, Salamanca, Spain

Lingjun Li
School of Pharmacy, and Department of Chemistry, University of Wisconsin, Madison, Wisconsin, USA

Claudio Nicolini
Nanobiotechnology and Biophysics Laboratories (NBL), Department of Experimental Medicine (DIMES), University of Genoa, Genoa; Nanoworld Institute Fondazione ELBA Nicolini (FEN), Pradalunga, Bergamo, Italy, and Biodesign Institute, Arizona State University, Tempe, Arizona, USA

Eugenia Pechkova
Nanobiotechnology and Biophysics Laboratories (NBL), Department of Experimental Medicine (DIMES), University of Genoa, Genoa, and Nanoworld Institute Fondazione ELBA Nicolini (FEN), Pradalunga, Bergamo, Italy

Bernardo A. Petriz
Centro de Análises Proteômicas e Bioquímicas, Programa de Pós-Graduação em Ciências Genômicas e Biotecnologia, Universidade Católica de Brasília, Brasília, Brazil

Evgeniy V. Petrotchenko
University of Victoria—Genome British Columbia Proteomics Centre, Victoria, British Columbia, Canada

Ghasem Hosseini Salekdeh
Department of Molecular Systems Biology at Cell Science Research Center, Royan Institute for Stem Cell Biology and Technology, ACECR, Tehran, and Department of Systems Biology, Agricultural Biotechnology Research Institute of Iran, Karaj, Iran

Faezeh Shekari
Department of Molecular Systems Biology at Cell Science Research Center, Royan Institute for Stem Cell Biology and Technology, and Department of Developmental Biology, University of Science and Culture, ACECR, Tehran, Iran

PREFACE

In the last decade, proteomics emerged as a very valuable tool in biomedical and pharmacological research. Different proteomic techniques have been employed in the screening for biomarkers for different disorders and diseases, in understanding molecular mechanisms underlying pathological alterations in humans, in studying protein structures, design of potential therapeutics, etc. Considering the wide application of proteomics in biomedicine and pharmacology and the increasing number of specialists from different fields employing proteomic techniques, we focused this volume of the *Advances in Protein Chemistry and Structural Biology* on Proteomics in Biomedicine and Pharmacology.

Chapter 1 in this volume presents the main classical gel-based methods and the advances of gel-free quantitative proteomic techniques. The application of these proteomic methods in elucidation of host–bacteria interactions and design of treatment for a number of infectious diseases is reviewed.

Protein phosphorylation and glycosylation play fundamental roles in many biological processes as one of the most common, and the most complex, posttranslational modification. Alterations in these posttranslational modifications are now known to be associated with many diseases. As a result, the discovery and detailed characterization of phosphoprotein and glycoprotein disease biomarkers is a primary interest of biomedical research. There have been many advances in this area and these are detailed in Chapters 2 and 3, both in relation to available protocols for phospho/glycoproteomic analysis and to the widening range of biomedical fields in which such approaches are being commonly applied. A special emphasis is given to their application to cancer biomarkers and neurodegenerative diseases.

Next five chapters review in details the use of different proteomic techniques in studying oral diseases (Chapter 4), alterations in protein structure and design of personalized treatment (Chapters 5 and 6), stem cells organelle proteomics research and challenges in subcellular proteomics (Chapter 7), and screening of protein–protein and protein–DNA interactions and its application in biomedicine (Chapter 8).

The final chapter (Chapter 9) in this volume focuses on the application of different proteomic techniques in diagnosis and treatment of psychiatric disorders such as major depression, suicidal behavior, schizophrenia, and attention deficit/hyperactivity disorder. The potential of specific biomarkers determined by proteomic tools for distinguishing between comorbid psychiatric disorders in clinical setup as well as their potential for understanding mechanisms underlying the disorders and in discovery of new treatment strategies is also discussed.

The aim of this volume is to promote further proteomic-based studies in biomedicine and pharmacology in order to discover reliable tools for early diagnosis and treatment/management of different diseases and disorders.

Rossen Donev
Singleton Park
Swansea University
Swansea, UK

CHAPTER ONE

Application of Cutting-Edge Proteomics Technologies for Elucidating Host–Bacteria Interactions

Bernardo A. Petriz, Octavio Luiz Franco[1]

Centro de Análises Proteômicas e Bioquímicas, Programa de Pós-Graduação em Ciências Genômicas e Biotecnologia, Universidade Católica de Brasília, Brasília, Brazil

[1]Corresponding author: e-mail address: ocfranco@pos.ucb.br; ocfranco@gmail.com

Contents

Abstract

Advanced techniques and high-throughput protein analysis have led proteomics to substantive progress in the understanding of bacterial–host interactions. Mass spectrometry (MS)-based proteomics have been a central methodology in the discovery of new protein involved in the infectious process that leads to thousands of deaths every year. The discovery of novel protein targets, together with *de novo* drug design, improves the accuracy of early diagnosis, leading to improved new treatments. MS-based proteomics has also been widely applied to structural biology, where proteomic investigation is being used to improve knowledge on the relationship between protein sequence, structure, and function. Thus, the search for therapeutic targets for infectious diseases using these cutting-edge technologies represents the new frontiers for proteomics applications in biomedicine and pharmacology. In this review, the main classical gel-based methods (2-DE, DIGE) are discussed, as well as the advances of gel-free quantitative proteomic techniques, from metabolic and chemical labeling (SILAC, iTRAQ, ICAT, $^{16}O/^{18}O$, QconCAT) to nonlabeling (MS spectra counting and peak integration) strategies. Together, these proteomic methods are currently being used in the

Advances in Protein Chemistry and Structural Biology, Volume 95
ISSN 1876-1623
http://dx.doi.org/10.1016/B978-0-12-800453-1.00001-4

quest for tailor-made pharmaceutical and biomedical research for bacterial control, where advances in these analytical methods may represent greater improvements in the treatment of a number of infectious diseases.

1. INTRODUCTION

In recent decades, infectious diseases caused by microorganisms have become an increasing health problem. Bacterial infections caused by resistant strains are of grave concern in numerous hospitals around the world, especially in elderly patients, those compromised by illness and those receiving immunity-suppressant drugs (Grundmann et al., 2011). In this context, it is essential to improve the understanding of infectious mechanisms and the host response in order to develop drugs with potential activity against these pathogens. To fill the manifold gaps that remain in our understanding of bacterial infectious processes, proteomics has been widely used (Cox et al., 2012). In recent years, proteomic tools have accomplished significant advances in the characterization of proteins involved in the mechanism of infectious pathogens and also in the patient's response (Lima et al., 2013). In this context, this review focuses on proteomics tools used in the better understanding of proteins involved in infectious processes in microorganism and mammals, providing a broad overview of proteins possibly related to the resistance process.

2. CLASSICAL PROTEOMICS STRATEGIES FOR BIOMEDICAL RESEARCH IN GENERAL

The prominent role of proteins in all biological process, here with special attention to pathogenesis and pathophysiology, has made the study of proteins become widely incorporated into a number of fields in biomedical research, which include biomarker discovery and novel drug design (De Masi, Pasca, Scarpello, Idolo, & De Donno, 2013; Oswald, Groer, Drozdzik, & Siegmund, 2013; Parguina, Rosa, & Garcia, 2012). In this context, the discovery of protein targets associated with infectious pathologic development represents an advance in early diagnosis and drug development (Bougnoux & Solassol, 2013; Ghafourian, Sekawi, Raftari, & Ali, 2013; Konvalinka, Scholey, & Diamandis, 2012; Oswald et al., 2013). In recent years, this objective has produced a substantial amount of proteomic data, especially associated with phenotypes derived from abnormal protein profiling and biomarker discovery (Banks et al., 2000; Castagna, Polati, Bossi, & Girelli, 2012).

Proteomics is an ensemble of tools used to reveal a static profile of proteins expressed in a complex system resulting from dynamic biological signaling and gene regulation. In this way, proteomic analysis aims to identify and verify the role of a given protein or more precisely, a collection of gene products in biological processes (e.g., in pathology) (Domon & Aebersold, 2006). This process is often challenging, due to the complex dynamic range and heterogeneity of several proteome samples (e.g., plasma and tissue if much greater than genome size) (Anderson & Anderson, 2002; Harrison, Kumar, Lang, Snyder, & Gerstein, 2002). Moreover, proteomic analysis may also be laborious and time-consuming, sometimes involving several sets of biologic and technical analysis to overcome possible failures in technical reproducibility. Despite these challenges, as a great advance, some proteomic analysis may resolve thousands of proteins/peptides simultaneously (Tang, Beer, & Speicher, 2011).

Proteomic techniques may be divided into gel-based and gel-free methods, but this division does not limit the interaction of both methodologies, frequently seen in several studies as complementary strategies (Charro et al., 2011; Jungblut et al., 2010; Selvaraju & El Rassi, 2011; Thierolf et al., 2008). Gel-based techniques are represented mainly by the classic 1D, 2-DE, and 2D-DIGE, the ultimate evolution of classic two-dimensional electrophoresis (Minden, 2012). Otherwise, gel-free analyses are conducted by a wide range of liquid chromatography strategies (e.g., HPLC, UPLC, nanoLC, MudPIT) which are often directly coupled to automated mass spectrometry (MS) apparatus (e.g., LC/MS) (Franzel & Wolters, 2011; Mitulovic & Mechtler, 2006; Nagele, Vollmer, Horth, & Vad, 2004). In addition, protein/peptides may be labeled in advance to LC/MS for absolute and/or relative proteome quantitation, enhancing quantitative proteomic analysis (May et al., 2011). Independently of the chosen strategy, MS analysis is a central technology and key step for simple and high-throughput proteomic characterization and analysis (protein/peptide identification) (Domon & Aebersold, 2006; May et al., 2011). Moreover, MS is fundamental for identifying posttranslational modifications (PTMs), a key molecular signaling process, highly investigated by MS-based proteomics (Cravatt, Simon, & Yates, 2007; Zhao & Jensen, 2009), since some PTMs such as phosphorylation are associated with the development of clinical conditions (e.g., Alzheimer's, cardiovascular disease, cancer) (Kolarova, Garcia-Sierra, Bartos, Ricny, & Ripova, 2012;

Thakur et al., 2008; Toepfer et al., 2013; Trombino et al., 2004; Walker, Fullerton, & Buttrick, 2013).

Hence, biomedical and pharmacology fields have benefited from the great advances in MS-based proteomics, fundamental for high-throughput biomarker screening and development of novel pharmacologic strategies (Berna et al., 2008; Thierolf et al., 2008; Vasudev et al., 2008; Yang et al., 2011). In this review, the application of MS-based proteomics tools and strategies in biomedical and pharmacologic fields will be addressed for bacterial control and the treatment of a number of infectious diseases. Section 1 focuses on the proteomic tools used for quantitative and qualitative analysis followed by their application in the research of host–bacterial interactions.

2.1. Gel-based methods

Two-dimensional gel electrophoresis (2-DE) is still the most widely used method in quantitative and qualitative proteomic studies and is the only technique that can resolve up to 10,000 protein species from large sets of complex protein mixtures (May et al., 2011; Wittmann-Liebold, Graack, & Pohl, 2006). This technology separates the samples by two consecutive techniques: isoelectric focusing, which discriminates proteins based on their isoelectric point, followed by sodium dodecyl sulfate polyacrylamide gel electrophoresis (SDS PAGE), which discriminates proteins based on their molecular weight (Gorg, Weiss, & Dunn, 2004).

Despite the amplitude of 2-DE application, the technique is extremely laborious, time-consuming, and more sensitive to technical reproducibility error, since large sets of gel repetition and sample are usually needed (Petriz , Gomes, Rocha, Rezende, & Franco, 2012). Thus, limited sample availability is an issue in 2-DE analysis, especially concerning poor protein extraction. Moreover, 2-DE technique also fails to resolve low abundant and hydrophobic proteins as well as those with molecular size out of the range of 5–150 kDa or with extreme pH range (<3.5 and >10) (May et al., 2011). The majority of these limitations were overcome by the development of differential gel electrophoresis (DIGE) (Unlu, Morgan, & Minden, 1997). DIGE is the ultimate evolution of 2-DE technique, which significantly improved the analytical power of gel-based methods in proteome research (Minden, 2012). These improvements are based on significantly enhancing technical reproducibility and quantification over different proteome samples, previously labeled with spectrally resolvable fluorophore agents

(CyDyes™: Cy2, Cy3, and Cy5; GE Healthcare Europe GmbH). After labeling, samples are pooled together with an internal standard consisting of the mixture of all samples, which leads to a more accurate normalization of protein spots from all sample gels. When the gel is digitalized using different wavelengths, each particular fluorophore is excited, generating distinct gel images corresponding to each prelabeled proteome sample. The abundance of each protein spot is measured as a ratio to its corresponding spot present in the internal standard, by software programs, and the statistically significant changes in spots are marked for further MS analysis (Scherp, Ku, Coleman, & Kheterpal, 2011). In this way, DIGE followed by MS is extremely useful for characterizing differential proteome expression (Winnik et al., 2012). This gel-based technique may also be used together with gel-free methods (e.g., LC/MS, LC-MS/MS), improving its analytical power (Frohlich et al., 2006; Lee et al., 2012; Raggiaschi et al., 2006; Weeks, 2010).

3. GEL-FREE METHODS

As mentioned before, a limited sample is a common situation in several biomedical fields (e.g., rare cancer and invasive procedures), sometimes becoming a restrictive issue for some proteomic techniques such as huge sets of 2-DE gels, which usually require a high amount of sample. Therefore, the option for gel-free methods is often applied as an alternative to gel-based techniques, since a low amount of sample is required and because peptide mixture is less complex to analyze compared to proteins (May et al., 2011). The direct connection of LC to MS analyzers leads these methods to be referred to as MS-based methods.

A typical LC-MS workflow begins with the protein sample being enzymatically digested (e.g., trypsin, chymotrypsin, Lys-C, Glu-C, Asp-N), with the resultant peptide mixture being separated by 1D or multidimensional chromatography (HPLC or UPLC, also performed in nanoscale) depending on its complexity. Eluted peptides are then loaded directly (on line) to electrospray ionization (ESI) for further MS analysis (e.g., LC-ESI-MS, LC-ESI-MS/MS). After LC separation, digested peptides may also be connected indirectly (off line) to MS by automated loading of eluted fractions into MALDI steel plates for solid ionization process and subsequent MS analysis (e.g., LC-MALDI-MS, LC-MALDI-MS/MS) (Bodnar, Blackburn, Krise, & Moseley, 2003). After MS analysis, data on peptide masses and fragmentation of ion masses are researched against a protein database for peptide and protein identification. These research engines are well reviewed elsewhere (Eng, Jahan, & Hoopmann, 2013; Perkins, Pappin, Creasy, & Cottrell, 1999; Sharma, Eng, Maccoss, & Riffle, 2012).

As mentioned before, the use of multidimensional chromatography is an accurate strategy to reduce highly complex peptide mixtures as well as improving MS analysis. Moreover, multidimensional protein identification technology (MudPIT) is a successful multi LC-MS strategy (Elschenbroich et al., 2009) used in several types of biomarker research (Gonzalez-Begne et al., 2009, 2011). The basis for MudPIT is the combination of different chromatographic columns (e.g., RF-IEX-RF, IEX-RF, SCX-nanoRF) prior to MS (Mitulovic & Mechtler, 2006; Nagele et al., 2004). Additional procedures, such as depletion of high-abundance proteins (e.g., ionic columns) and the enrichment of low-abundance peptides, improve MudPIT sensitivity, resulting in enhanced MS quantity and quality data (Fonslow et al., 2011), but the availability of a greater amount of sample must be considered (Aebersold & Mann, 2003). Other strategies, such as enrichment methods, have been frequently implemented in order to enhance the identification of low-abundance proteins, PTMs (e.g., glycopeptides, phosphopeptides) (Fonslow et al., 2012), as well as membrane or other specific organelle proteins (Elschenbroich et al., 2009). Current enrichment strategies are reviewed elsewhere (Wu, Shakey, Liu, Schuller, & Follettie, 2007; Zhao & Jensen, 2009).

3.1. Gel-free-labeling methods

In classic gel-based methods, protein expression is detected and quantified by the spot staining intensity, which was highly improved with fluorescent dyes used in the DIGE method. Still, a limited dynamic range in protein identification is a limiting issue. However, in gel-free analysis, a series of labeling methods were developed to perform relative and absolute MS-based quantitative analysis (DeSouza & Siu, 2013). These methods are based on chemical, metabolic, synthetic, and proteolytic protein/peptide labeling in order to separate distinct proteomes or proteome states (e.g., healthy vs. pathological tissue) through MS analysis. In relative quantification analysis, the most common methods are ICAT/cICAT, iTRAQ, TMT, ICPL, $^{16}O/^{18}O$, or metabolic (SILAC and ^{15}N) isotope labeling (Julka & Regnier, 2005), while absolute quantitation is often performed by iTRAQ/MTM.

3.1.1 Chemical labeling

The isotope-coded affinity tag (ICAT) labels cysteine side chains with "heavy" or "light" isotope tags (Gygi et al., 1999). Distinctly labeled samples are then digested together, separated by LC and submitted to MS analysis (LC-ESI-MS/MS or LC-MALDI-MS/MS) (Griffin et al., 2001). Ion intensity from both labeled peptides is pair contrasted. However, peptides lacking cysteine residues are a gap in the ICAT method. Other popular isotope labeling methods, iTRAQ (isobaric tag for relative and absolute quantitation) and TMT (tandem mass tagging), present up to eight different isobaric tags that bond to N-terminus, lysine residues, and side chain amine peptides (Ross et al., 2004; Thompson et al., 2003). In these methods, protein samples are labeled, digested, and submitted to nanoLC-MS/MS, which will result in a unique peak for each individual sample within the MS/MS spectrum (Ross et al., 2004). The fragmentation of the tags will also lead to a low molecular mass reporter ion, which is used to perform the relative quantification between the samples. This approach has simplified MS analysis compared to the ICAT method. Similar to ICAT, isotope-coded protein labeling (ICPL) labels proteins within all free amino groups. In this way, two proteome samples are labeled with a "light" or "heavy" ICPL tag, pooled together, and submitted to proteolysis for further MS analysis (Schmidt, Kellermann, & Lottspeich, 2005). The difference in this method from the other chemical labeling approaches is the labeling at the protein level. However, all of the aforementioned methods may fail to fully label the entire proteome. Proteome sample may also be digested in the presence of normal $H_2{}^{16}O$ water and with heavy water ($H_2{}^{18}O$), which will result in a mass shift of 4 Da between peptides. This happens because, when digested with trypsin, the Glu-C and Lys-C peptides are incorporated with up to two oxygen atoms, but a sample with heavy water will present the extra mass shift. Digested peptides are pooled together and fractionated by LC for MS quantification and identification process.

3.1.2 Metabolic labeling

Stable isotope labeling by amino acids in cell culture (SILAC) is a proteomic technique based on the incorporation of normal essential amino acids (light label) and isotopic modified amino acids (heavy label) into cell culture, leading to light- and heavy-marked protein synthesis (Ong et al., 2002). After protein extraction and digestion, peptides are analyzed by MS. The difference between the peaks from eluted peptides is used to verify the abundance of each analyzed peptide. Compared to the other LC-MS labeling methods,

SILAC is considered the technique that achieves the most accurate results. Alternative metabolic labeling is also performed by ^{15}N labeling method, where all nitrogen atoms in peptides are labeled (DeSouza & Siu, 2013). This method is preferred for quantifying autotrophic cells, thus being more efficient in quantifying plant proteome (Oeljeklaus, Meyer, & Warscheid, 2009).

3.2. Label-free and absolute quantification

Label-free methods arise from the necessity to overcome some prime limitations of labeling methods, such as noncomplete labeling and even the high cost of some of these methods. Besides requiring a smaller amount of sample, label-free methods do not permit multiproteome analysis within the same experiment (May et al., 2011). One of the main label-free methods is based on spectra counting (Old et al., 2005). The amount of mass spectra from a protein is used as a parameter value for quantifying this ion. In this case, the ion quantification is proportional to its peptide concentration within the analyzed sample. A new variant of label-free quantification is known as LC/MS^E. This method alternates scans of low collision energy and elevated collision energy during LC/MS analysis. This method permits one to obtain both protein quantification and protein identification data in a single sample. Its advantages include a reduction in the sample consumption, an improvement in detection sensitivity, and an enhancement in data quality for proteomic studies (Silva et al., 2005).

The absolute quantification (AQUA) of proteins is one of the focuses of proteome research, where the AQUA method elicits a straight quantification from proteins and also PTMs (Gerber, Rush, Stemman, Kirschner, & Gygi, 2003). This is performed with a chemically synthesized isotope peptide, which is used as an internal standard, corresponding to a specific target protein. The ratio between the endogenous and the synthetic peptide is used to calculate the absolute amount of protein within the sample and the identification of PTM after LC-ESI-MS (Gerber et al., 2003). The use of recombinant expression of peptide concatemers (QconCATs) has recently been shown to display similar fidelity and quantitative accuracy, being also highly suitable for multiplex quantitative proteomic analysis (Austin et al., 2012; Russell et al., 2013).

Comparative investigation of gel-based and gel-free proteomic technologies has verified that both methods have advantages for investigating complex protein samples and should consequently be seen as complementary

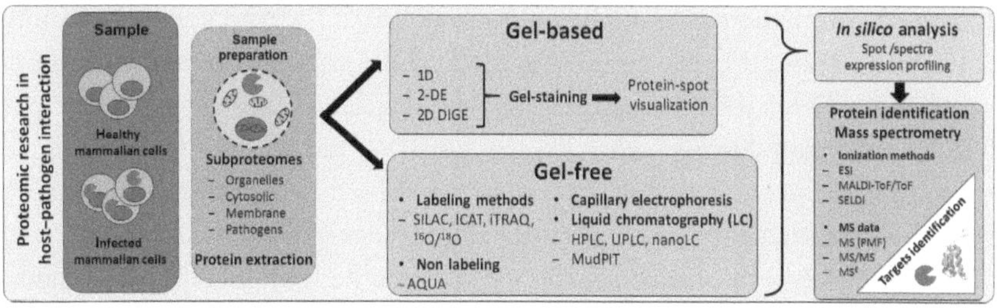

Figure 1.1 Proteomics workflow for application in host–pathogen experiments.

rather than exclusive (Wu, Wang, Baek, & Shen, 2006). Figure 1.1 summarizes the wide range of proteomic tools described in this review that are used in biomedical and pharmacologic research.

4. NEW PROTEOMIC METHODS IN LOOKING FOR BACTERIAL PATHOGENS

The use of the extraordinary tools described above has opened new perspectives in understanding bacterial pathogens during the infectious process, in spite of their multiple properties, strategies of infections, and levels of lethality (Lima et al., 2013). Moreover, not only has the whole proteome been determined for such dangerous bacteria but also numerous subproteomes from cell walls, outer membranes, cytosol, or exoproteome, which help us to better understand the proteins secreted during the infectious process (Mahdavi et al., 2014; Maria-Neto et al., 2012; Papasergi et al., 2013). These approaches have been extremely useful in clarifying the mechanisms of interaction between host and pathogens, improving knowledge of these relationships, which is essential for the development of new pharmaceuticals. In this context, novel techniques have been used to shed some light on secretomics, also known as the study of the exoproteome from pathogenic bacteria. Pathogenic microbes have been able to synthesize complex secretion systems to deliver virulence factors into host mammalian cells. The elucidation of such factors seems to be critical for understanding the progression of infection. Mahdavi et al. (2014) described a very interesting approach to labeling and identifying pathogen proteinaceous secreted compounds. Selective labeling of microbial proteins is carried out via translational incorporation of azidonorleucine, a methionine surrogate that requires a methionyl-tRNA synthetase variant form for activation. After being thus labeled, secreted pathogen proteins containing Anl can be marked by azide–alkyne cycloaddition and enriched by affinity purification. This pioneering method was applied to understand better the human pathogen *Yersinia enterocolitica* type III secretion system, enabling an efficient identification of distinct proteins involved in the infectious process, as well as the identification of distinct secretion profiles for intracellular and extracellular bacteria. Finally, the best order in which to inject substrates into host cells was also evaluated.

From all these studies, it can be seen that labeling and enrichment appear to be a good way to understand the infectious process. Nevertheless, the labeling of pathogens with green fluorescent protein or the generation of a reliable antibody does not always succeed due to multiple difficulties and pitfalls. So, a different approach was also applied to *Staphylococcus aureus*, in which the pathogen traffic was monitored after internalization into host mammalian cells by fluorescence microscopy and for separation of bacteria from host–pathogen interaction assays using iron or gold oxide core, poly(vinyl alcohol)-coated, and fluorescence-labeled nanoparticles (Depke et al., 2014). This incubation was associated with quantitative proteome analysis after enabling researchers to investigate the bacterial behavior during infection of human epithelial cells by fluorescence microscopy and proteomics using magnetic separation or cell sorting (Depke et al., 2014).

Additionally, not only have labeling techniques been used in an unusual way to elucidate bacterial and host interactions, but also Gault, Malosse, Dumenil, and Chamot-Rooke (2013) described a novel approach based on the combination of mass profiling and tandem MS in order to localize all PTMs on the major pilin protein PilE expressed by the pathogenic *Neisseria* species. This very precise work focuses on PilE, one of the main pili components, which is a filamentous organelle located at the surface of many bacterial pathogens and considered extremely important for virulence factors. Previous reports have shown that PilE can harbor various combinations of PTMs and have established clear links between such modifications and pathogenesis (Aas et al., 2006; Anonsen et al., 2012; Giltner, Habash, & Burrows, 2010). In this context, a complete PTM mapping of proteins involved in bacterial infection could be considered a main target. Such alterations, identified with high resolution by a combination of mass profiling and tandem MS, included a processed and methylated N-terminus, disulfide bonds, glycosylation, and glycerophosphorylation at two different sites (Gault et al., 2013).

Furthermore, a clear combination between MS and labeling could improve bacterial protein detection in some cases. The expansion of a mass spectrometric workflow for quantitation and identification of kinetic modifications in metaproteomes or, more especially, for bacterial pathogens after internalization by host cells was also described (Pfortner et al., 2013). This specific procedure, applied to *S. aureus* after internalization by human bronchial epithelial cells of S9 type employs three different stages, comprising SILAC pulse-chase labeling and infection assay followed by isolation of bacteria by using GFP-assisted cell sorting and further MS-based proteome analysis. This combined approach displays higher sensitivity in comparison to techniques that used conventional cell sorting and protein separation, since it employed an efficient arrangement of a filtration-based purification with on-membrane digestion. With this same approach, bacteria with high resistance to antibiotics have also been evaluated. Multidrug-resistant strains are a significant cause of hospital-acquired infections, being related to increased mortality and length of stay in hospital (Lima et al., 2013). In order to obtain a better understanding of antimicrobial resistance mechanisms in *Acinetobacter baumannii*, large-scale 2-D LC/MS/MS-based quantitative proteomics was used to compare drug-sensitive and -resistant strains. An impressive 20% of the expressed proteome modified twofold or more among the compared strains, including proteins related to resistance mechanisms, xenobiotic modification, or drug transportation (Chopra, Ramkissoon, & Anderson, 2013).

Another important contribution of proteomics to the infectious disease field consists of rapid bacterial identification by MS using ion patterns. Furthermore, these same protocols could be applied for resistance detection (Chang et al., 2013; Hoyos-Mallecot et al., 2014; Jung et al., 2013). In this context, a rapid shotgun proteomics method was applied for *A. baumannii* β-lactam-resistant identification. In this study (Chang et al., 2013), automated data-dependent scanning on a nano-LC/ion trap MS was used to characterize proteotypic peptides from the pathogen. Furthermore, SEQUEST software was applied to search a specific database named BRPDAB, the β-lactam-resistance protein database of *A. baumannii*. In this case, the authors positively found a large number of associated antibiotic-resistant proteins including AmpC, β-lactamase, and carO in resistant strains, being able to differentiate them from wild-type bacteria. In this case, the authors were able to combine the MS analyses with classical genomic bioinformatics tools including Uniprot annotations, Gene Ontology, and BLAST bioinformatics tools (Chang et al., 2013). In fact, this proteomic study provides us with a nice combination of bench and *in silico* techniques to create a platform for the rapid diagnosis of resistant bacteria.

With the emergence and rising complexity of bacterial resistance to medication, speedy and consistent susceptibility testing has become a newsworthy issue. While the uses of MALDI-ToF MS for the rapid detection of antibiotic resistance are a striking option, the current methods for MALDI-ToF MS susceptibility testing are restricted to very limited conditions. Given this, an option that may allow for rapid susceptibility testing could be based on a SILAC-like approach (Jung et al., 2013). This technique was used to visualize bacterial growth in the presence of antibiotic as observed for *Pseudomonas aeruginosa*. In this case, bacterial strains were incubated in normal broth, broth supplemented with $^{13}C_6$–$^{15}N_2$-labeled lysine, and broth supplemented with labeled lysine and antibiotic. Peak shifts arising due to the incorporation of the labeled amino acids were MALDI-ToF MS-detected. In this specific report, three antibiotics with different mechanisms of action, meropenem, tobramycin and ciprofloxacin, were evaluated. A semi-automated algorithm was yielded to empower fast and unbiased data analyses, making a clear distinction between resistant and susceptible isolates possible for three antibiotics.

Finally and no less importantly, there are some unusual multidisciplinary procedures involving proteomic techniques that have been proposed in order to clarify the infectious processes. According to Seibel et al. (2013), the current developments in apparently independent fields of the natural sciences, biophysical visualization, bioorganic synthesis, and bioanalysis open the doors for a promising interdisciplinary tactic to study human infection processes. For example, the use of special synthesized carbohydrate labels, in combination with new super-resolution imaging approaches, could allow access to mapping and identification of glycoproteins from the cell surface well below the diffraction bounds. Moreover, a similar approach could also be developed for lipids in the case of lipoproteins or nucleic acids in order to better understand the genetic processes of infectious bacteria. Such methodologies could clarify which surface or deep molecules are involved in bacterial adherence, among other biological mechanisms with potential implications for bacterial infection prevention (Seibel et al., 2013).

In summary, all procedures here described can be applied for the characterization of other host–pathogen pairs, allowing identification and quantitation of thousands of bacterial proteins in several hours postinfection, leading to a better understanding of the proteins involved in bacterial predation on mammal cells.

5. PROTEOMIC ADVANCES IN LOOKING FOR HOST ORGANISMS

If, on the one hand, it is important to deepen understanding of the bacterial nuances for elucidating infection processes, on the other, it is also clearly essential to improve patient response to the presence of such pathogens (Fig. 1.2). For that, proteomics also seems to be extremely important in a number of other cases, contributing to elucidate not only the immune response but also the deleterious effects caused by bacteria inside a mammal organism.

It is also important to remember that most of the time, animal and cell models have been used for such trials, and although they did not perfectly match a human organism, they provided an essential contribution to clarifying the host cellular response. Thus, by using an animal fish model (*Paralichthys olivaceus*), the proteomic response to bacterium *Streptococcus parauberis* infection was analyzed by using label-free protein quantitation associated with LC-MS(E) tandem MS (Cha et al., 2012). A total of 82 proteins from fish kidney were found to be differentially expressed between healthy and infected conditions. Between the differentially expressed proteins, those involved in immune responses including cathepsins, goose-type lysozyme, heat-shock proteins, and complement components were significantly

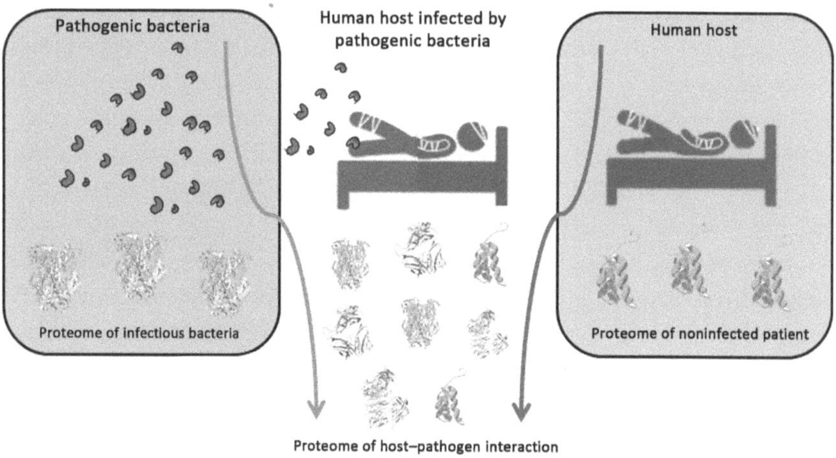

Figure 1.2 Differential protein expression observed in pathogenic bacteria (left side), human host (right side), and during the bacterial infection over human host (middle).

upregulated by bacterial infection. It was interesting to observe that there was uninterrupted activation of expression of immune-associated proteins during the infection process. Nevertheless, there was also lessening in the expression of proteins not involved in immune response (Cha et al., 2012). The modification of immune response was also observed in other host–pathogen interaction models like mice and *Francisella tularensis*, a virulent bacterial pathogen that causes the zoonosis tularemia in humans (Varnum et al., 2012). For that, a global proteomic style to characterize protein modifications in bronchoalveolar lavage fluid from mice exposed to bacteria was applied. As a result, the composition of bronchoalveolar lavage fluid proteins was altered due to the infection, including proteins related to oxidative stress, neutrophil activation, and inflammatory responses. Moreover, innate immune response components were also induced, including the complement system and the acute-phase response. This study identifies two candidate biomarkers, which were chitinase 3-like-1 and peroxiredoxin 1; this identification was also a complex main goal in proteomic studies (Varnum et al., 2012).

Swine is another model often employed for an *in vivo* approach to study pathogen–mammal interactions (Collado-Romero et al., 2012; Lu et al., 2012; Martins et al., 2012). DIGE-based proteomics was used to monitor the response of porcine mesenteric lymph nodes to *Salmonella typhimurium* infection (Martins et al., 2012). The proteome response of porcine lymph nodes to infection was associated with induction of different processes such as cytoskeleton remodeling, phagocyte infiltration, and pyroptosis. Moreover, data reported suggested that *S. typhimurium* antigens are cross-presented via MHC-I in a proteasome-dependent manner in porcine lymph nodes, suggesting that host innate and adaptive immunity act together in mesenteric lymph nodes to control bacterial dissemination in swine infections (Martins et al., 2012). *S. typhimurium* and porcine ileum mucosa interaction was also studied by 2-DE, MALDI-ToF/ToF-based approach in order to better understand the mammalian host response (Collado-Romero et al., 2012). In this study, 51 proteins involved in apoptosis, pathogen-mediated cell invasion, and immune response were identified as being differentially expressed after pathogen trial. Furthermore, anti-inflammatory signals and dendritic cell maturation downregulation were also observed. Transcriptional analysis by using RT-qPCR confirmed several proteins observed. In both cases, results derived from both studies are extremely valuable to better characterize the pathogen modulation of *in vivo* host responses (Collado-Romero et al., 2012).

Another target of immunoproteomics in host organisms consists of the better understanding of vaccine response and the discovery of new antigens for vaccine production. As an example, a model using the antigen prepared from the Gram-negative intracellular bacterium *Brucella abortus*, which causes chronic infection in humans and infectious abortion in food-producing animals, was evaluated in vaccinated cattle by proteomic techniques (Pajuaba et al., 2012). Gel analyses of the hydrophilic phase antigen revealed a wide spectrum of polypeptides. Furthermore, immunoblot evaluation showed widespread seroreactivity profile. By using this approach, potential infection markers and excluding vaccinal response were obtained. The proteomic characterization revealed 56 protein spots, with 27 of them being antigenic spots displaying differential seroreactivity profiles (Pajuaba et al., 2012). In conclusion, immunoproteomics of novel antigen preparations could enable protein characterization as a tool to develop sensitive and specific immunoassays for serodiagnosis of bacterial infection.

Additionally, the effects of exosomes have been studied in different ways (Wang et al., 2014). One example consists of the case of *Mycobacterium avium* sp. paratuberculosis-infecting macrophages in intercellular communication processes (Wang et al., 2014). In this specific case, the reactions of resting macrophages infected with this pathogen with resting macrophages treated with exosomes previously released from *M. avium*-infected macrophages were evaluated by 2-DE MS-MS. Both *M. avium* and exosomes from infected cells improved the expression of CD80 and CD86 and also the secretion of cytokines TNF-α and IFN-γ by macrophages, suggesting that exosomes from infected macrophages may be compound carriers, including bacterial antigens and/or infected macrophage components, which can elicit resting cell responses (Wang et al., 2014).

Due to the high complexity of relations between pathogen and host, the definition of the antigenic repertoire or "immunoproteome" of multiple bacterial pathogens is also a significant step toward understanding how to effectively prepare vaccines against such infections or how to reduce the lethal immune response as observed in sepsis (Silva et al., 2011). Until now, most strategies for antigen discovery were built on traditional methodologies for separation and identification of antigens from complex bacterial protein mixtures (Andersen & Doherty, 2005). Otherwise, unconventional methods have employed screening of bacterial antigens expression libraries in heterologous systems such as *Escherichia coli* with T-cell clones derived from *Mycobacterium tuberculosis*-infected individuals (Coler et al., 2009). Although these methods have been fruitful in identifying immunodominant antigens, they provide limited pathogen coverage (Kunnath-Velayudhan & Porcelli, 2013). Progress in proteome-wide selection approaches nowadays permits an extensive and unbiased investigation of antigenic targets on complex pathogens. With the advance of technologies that allow high-throughput protein synthesis, it is now possible to look for entire proteomes for antigens and, after that, to link them to the human immune response, also further evaluated by proteomics and other "omics" techniques. So far, three proteome-wide and moderately unbiased methods to recognize candidate antigens have been described. The first approach was based on the use of a peptide library designed to select potential targets of T-cell responses in infected individuals (Lindestam Arlehamn et al., 2013). The second strategy is related to protein microarrays printed with products of all pathogens expressing ORFs that could be used to screen sera from infectious disease patients for antibody reactivity (Kunnath-Velayudhan et al., 2010). Finally and no less importantly, a similar screen for serum antibody responses against bacterial proteome in human patients was also performed using traditional methods of recombinant protein expression, ELISA and MS, with a very interesting combination of immune and proteomic techniques (Li et al., 2010).

In addition to the use of animal models, only a few proteomic studies have been performed with human tissues or fluid in bacterial infectious diseases (Fu, Yi, Guan, Zhang, & Li, 2012; Lee et al., 2012). One example is a pioneer study that evaluated the qualitative and quantitative differences in sputum pulmonary protein expression submitted to tuberculosis (Fu et al., 2012). For that, 2-DE MALDI-ToF MS and an enzyme-linked immunosorbent assay were used to identify and further confirm the proteomic results. Sixty-two differentially expressed proteins were identified in the tuberculosis sputum in comparison to the controls. Furthermore, bioinformatics analysis proposed that multiple host cell pathways were involved in tuberculosis infection processes, including signal transduction, immune and acute-phase responses, and others revealing the particular human response during active pulmonary tuberculosis infection (Fu et al., 2012).

It is also important to remember that a bacterium is able to modify mammal response. Moreover, during the infection process, the bacterial pathogen is also modified in order to adapt to the host organism (Ansong et al., 2013; Melo, Schrama, Andrew, & Faleiro, 2013; Melo, Schrama, Hussey, Andrew, & Faleiro, 2013). In this case, the proteome could also be helpful in improving understanding of the interaction. In this area, Ansong et al. (2013) provided a very interesting contribution, since the description of the mature protein complement in cells seems to be crucial for a greater understanding of cell processes on a systems-wide scale. In this case, single-dimension ultra-high-pressure liquid chromatography MS was used to investigate the comprehensive "intact" proteome of *S. typhimurium*. Top-down proteomics analysis revealed hundreds of unique proteins including thousands of proteoforms yielded by PTMs, demonstrating a vast microbial top-down dataset. These data revealed several additional biological insights, such as showing that the differential use of protein S-thiolation forms S-glutathionylation and S-cysteinylation during infection-like conditions in comparison to basal circumstances (Ansong et al., 2013). This clearly described bacteria modifying several biosynthetic pathways under infection-like conditions and throughout real infection of host cells, showing that the presence of certain mammalian compounds is able to modify bacterial response, as observed here by top-down proteomics methods (Ansong et al., 2013).

6. PROSPECTS

Although several different types of research have been presented here, the use of proteomics in bacterial pathogens and host mammalian interactions in order to prevent, cure, or simply understand some infectious diseases is just beginning. At the moment, only a little information has been obtained in various animals and cell models and less work has been done in humans. For that, we must fill several gaps in knowledge and methodologies. It is obvious that classical techniques have been extremely useful, including 2-DE MS, MS/MS, and label-MS methods, as previously described. Nevertheless, such methods must be combined with others, including immunoassays, electron microscopy, histological pathology, and infectology, transforming the task of understanding the host–pathogen interaction into a multidisciplinary activity. Moreover, novel techniques and approaches are extremely welcome in improving this information, and these may include laser ablation ESI, for example (Kiss, Smith, Reschke, Powel, & Heeren, 2013), which is a novel technique for MS imaging. This technique allows lipids and small metabolites to be imaged in different samples such as tissue sections and bacterial colonies without pretreatment. Moreover, laser ablation ESI seems to be valuable also for the identification of proteins directly from sample surfaces, and this could be extremely desirable for experiments between bacteria and humans. This technique and others could be interesting to map such interactions in real time.

Therefore, the multifactor aspect of infectious diseases still needs to be better explored by a multiple strategy, which unquestionably involves a systems biology approach. To this end, next-generation sequencing combined with all proteomics methodologies demonstrated here could open a new prospect for dealing with infectious diseases. Therefore, the authors believe that synergistic use of different techniques, including genomic, transcriptomic, and proteomic technologies, will meaningfully improve the capability of bacterial detection, host immune response information, finding new biomarkers, and designing new drugs. In summary, all these techniques could clearly help in reducing the effects of infectious diseases worldwide.

REFERENCES

Aas, F. E., Egge-Jacobsen, W., Winther-Larsen, H. C., Lovold, C., Hitchen, P. G., Dell, A., et al. (2006). Neisseria gonorrhoeae type IV pili undergo multisite, hierarchical modifications with phosphoethanolamine and phosphocholine requiring an enzyme structurally related to lipopolysaccharide phosphoethanolamine transferases. *The Journal of Biological Chemistry*, *281*(38), 27712–27723.

Aebersold, R., & Mann, M. (2003). Mass spectrometry-based proteomics. *Nature*, *422*(6928), 198–207.

Andersen, P., & Doherty, T. (2005). TB subunit vaccines—Putting the pieces together. *Microbes and Infection*, 7, 911–921.

Anderson, N. L., & Anderson, N. G. (2002). The human plasma proteome: History, character, and diagnostic prospects. *Molecular & Cellular Proteomics*, *1*(11), 845–867 (Review).

Anonsen, J. H., Egge-Jacobsen, W., Aas, F. E., Borud, B., Koomey, M., & Vik, A. (2012). Novel protein substrates of the phospho-form modification system in Neisseria gonorrhoeae and their connection to O-linked protein glycosylation. *Infection and Immunity*, *80*(1), 22–30.

Ansong, C., Wu, S., Meng, D., Liu, X., Brewer, H. M., Deatherage Kaiser, B. L., et al. (2013). Top-down proteomics reveals a unique protein S-thiolation switch in Salmonella Typhimurium in response to infection-like conditions. *Proceedings of the National Academy of Sciences of the United States of America*, *110*(25), 10153–10158.

Austin, R. J., Chang, D. K., Holstein, C. A., Lee, L. W., Risler, J., Wang, J. H., et al. (2012). IQcat: Multiplexed protein quantification by isoelectric QconCAT. *Proteomics*, *12*(13), 2078–2083.

Banks, R. E., Dunn, M. J., Hochstrasser, D. F., Sanchez, J. C., Blackstock, W., Pappin, D. J., et al. (2000). Proteomics: New perspectives, new biomedical opportunities. *Lancet*, *356*(9243), 1749–1756.

Berna, M., Ott, L., Engle, S., Watson, D., Solter, P., & Ackermann, B. (2008). Quantification of NTproBNP in rat serum using immunoprecipitation and LC/MS/MS: A biomarker of drug-induced cardiac hypertrophy. *Analytical Chemistry*, *80*(3), 561–566.

Bodnar, W. M., Blackburn, R. K., Krise, J. M., & Moseley, M. A. (2003). Exploiting the complementary nature of LC/MALDI/MS/MS and LC/ESI/MS/MS for increased proteome coverage. *Journal of the American Society for Mass Spectrometry*, *14*(9), 971–979.

Bougnoux, A. C., & Solassol, J. (2013). The contribution of proteomics to the identification of biomarkers for cutaneous malignant melanoma. *Clinical Biochemistry*, *46*(6), 518–523.

Castagna, A., Polati, R., Bossi, A. M., & Girelli, D. (2012). Monocyte/macrophage proteomics: Recent findings and biomedical applications. *Expert Review of Proteomics*, *9*(2), 201–215.

Cha, I. S., Kwon, J., Park, S. H., Nho, S. W., Jang, H. B., Park, S. B., et al. (2012). Kidney proteome responses in the teleost fish Paralichthys olivaceus indicate a putative immune response against Streptococcus parauberis. *Journal of Proteomics*, *75*(17), 5166–5175.

Chang, C. J., Lin, J. H., Chang, K. C., Lai, M. J., Rohini, R., & Hu, A. (2013). Diagnosis of beta-lactam resistance in Acinetobacter baumannii using shotgun proteomics and LC-nano-electrospray ionization ion trap mass spectrometry. *Analytical Chemistry*, *85*(5), 2802–2808.

Charro, N., Hood, B. L., Faria, D., Pacheco, P., Azevedo, P., Lopes, C., et al. (2011). Serum proteomics signature of cystic fibrosis patients: A complementary 2-DE and LC-MS/MS approach. *Journal of Proteomics*, *74*(1), 110–126.

Chopra, S., Ramkissoon, K., & Anderson, D. C. (2013). A systematic quantitative proteomic examination of multidrug resistance in Acinetobacter baumannii. *Journal of Proteomics*, *84*, 17–39.

Coler, R. N., Dillon, D. C., Skeiky, Y. A., Kahn, M., Orme, I. M., Lobet, Y., et al. (2009). Identification of Mycobacterium tuberculosis vaccine candidates using human CD4+ T-cells expression cloning. *Vaccine*, *27*(2), 223–233.

Collado-Romero, M., Martins, R. P., Arce, C., Moreno, A., Lucena, C., Carvajal, A., et al. (2012). An in vivo proteomic study of the interaction between Salmonella Typhimurium and porcine ileum mucosa. *Journal of Proteomics*, *75*(7), 2015–2026.

Cox, G., Thompson, G. S., Jenkins, H. T., Peske, F., Savelsbergh, A., Rodnina, M. V., et al. (2012). Ribosome clearance by FusB-type proteins mediates resistance to the antibiotic fusidic acid. *Proceedings of the National Academy of Sciences of the United States of America*, *109*(6), 2102–2107.

Cravatt, B. F., Simon, G. M., & Yates, J. R., III (2007). The biological impact of mass-spectrometry-based proteomics. *Nature*, *450*(7172), 991–1000.

De Masi, R., Pasca, S., Scarpello, R., Idolo, A., & De Donno, A. (2013). The clinical potential of blood-proteomics in multiple sclerosis. *BMC Neurology*, *13*, 45.

Depke, M., Surmann, K., Hildebrandt, P., Jehmlich, N., Michalik, S., Stanca, S. E., et al. (2014). Labeling of the pathogenic bacterium *Staphylococcus aureus* with gold or ferric oxide-core nanoparticles highlights new capabilities for investigation of host-pathogen interactions. *Cytometry. Part A*, *85*, 140–150.

DeSouza, L. V., & Siu, K. W. (2013). Mass spectrometry-based quantification. *Clinical Biochemistry*, *46*(6), 421–431.

Domon, B., & Aebersold, R. (2006). Mass spectrometry and protein analysis. *Science*, *312*(5771), 212–217.

Elschenbroich, S., Ignatchenko, V., Sharma, P., Schmitt-Ulms, G., Gramolini, A. O., & Kislinger, T. (2009). Peptide separations by on-line MudPIT compared to isoelectric focusing in an off-gel format: Application to a membrane-enriched fraction from C2C12 mouse skeletal muscle cells. *Journal of Proteome Research*, *8*(10), 4860–4869.

Eng, J. K., Jahan, T. A., & Hoopmann, M. R. (2013). Comet: An open-source MS/MS sequence database search tool. *Proteomics*, *13*(1), 22–24.

Fonslow, B. R., Carvalho, P. C., Academia, K., Freeby, S., Xu, T., Nakorchevsky, A., et al. (2011). Improvements in proteomic metrics of low abundance proteins through proteome equalization using ProteoMiner prior to MudPIT. *Journal of Proteome Research*, *10*(8), 3690–3700.

Fonslow, B. R., Niessen, S. M., Singh, M., Wong, C. C., Xu, T., Carvalho, P. C., et al. (2012). Single-step inline hydroxyapatite enrichment facilitates identification and quantitation of phosphopeptides from mass-limited proteomes with MudPIT. *Journal of Proteome Research*, *11*(5), 2697–2709.

Franzel, B., & Wolters, D. A. (2011). Advanced MudPIT as a next step toward high proteome coverage. *Proteomics*, *11*(18), 3651–3656.

Frohlich, T., Helmstetter, D., Zobawa, M., Crecelius, A. C., Arzberger, T., Kretzschmar, H. A., et al. (2006). Analysis of the HUPO Brain Proteome reference samples using 2-D DIGE and 2-D LC-MS/MS. *Proteomics*, *6*(18), 4950–4966.

Fu, Y. R., Yi, Z. J., Guan, S. Z., Zhang, S. Y., & Li, M. (2012). Proteomic analysis of sputum in patients with active pulmonary tuberculosis. *Clinical Microbiology and Infection*, *18*(12), 1241–1247.

Gault, J., Malosse, C., Dumenil, G., & Chamot-Rooke, J. (2013). A combined mass spectrometry strategy for complete posttranslational modification mapping of Neisseria meningitidis major pilin. *Journal of Mass Spectrometry*, *48*(11), 1199–1206.

Gerber, S. A., Rush, J., Stemman, O., Kirschner, M. W., & Gygi, S. P. (2003). Absolute quantification of proteins and phosphoproteins from cell lysates by tandem MS. *Proceedings of the National Academy of Sciences of the United States of America*, *100*(12), 6940–6945.

Ghafourian, S., Sekawi, Z., Raftari, M., & Ali, M. S. (2013). Application of proteomics in lab diagnosis. *Clinical Laboratory*, *59*(5–6), 465–474.

Giltner, C. L., Habash, M., & Burrows, L. L. (2010). Pseudomonas aeruginosa minor pilins are incorporated into type IV pili. *Journal of Molecular Biology*, *398*(3), 444–461.

Gonzalez-Begne, M., Lu, B., Han, X., Hagen, F. K., Hand, A. R., Melvin, J. E., et al. (2009). Proteomic analysis of human parotid gland exosomes by multidimensional protein identification technology (MudPIT). *Journal of Proteome Research*, *8*(3), 1304–1314.

Gonzalez-Begne, M., Lu, B., Liao, L., Xu, T., Bedi, G., Melvin, J. E., et al. (2011). Characterization of the human submandibular/sublingual saliva glycoproteome using lectin affinity chromatography coupled to multidimensional protein identification technology. *Journal of Proteome Research*, *10*(11), 5031–5046.

Gorg, A., Weiss, W., & Dunn, M. J. (2004). Current two-dimensional electrophoresis technology for proteomics. *Proteomics*, *4*(12), 3665–3685.

Griffin, T. J., Gygi, S. P., Rist, B., Aebersold, R., Loboda, A., Jilkine, A., et al. (2001). Quantitative proteomic analysis using a MALDI quadrupole time-of-flight mass spectrometer. *Analytical Chemistry*, *73*(5), 978–986.

Grundmann, H., Klugman, K. P., Walsh, T., Ramon-Pardo, P., Sigauque, B., Khan, W., et al. (2011). A framework for global surveillance of antibiotic resistance. *Drug Resistance Updates*, *14*(2), 79–87.

Gygi, S. P., Rist, B., Gerber, S. A., Turecek, F., Gelb, M. H., & Aebersold, R. (1999). Quantitative analysis of complex protein mixtures using isotope-coded affinity tags. *Nature Biotechnology*, *17*(10), 994–999.

Harrison, P. M., Kumar, A., Lang, N., Snyder, M., & Gerstein, M. (2002). A question of size: The eukaryotic proteome and the problems in defining it. *Nucleic Acids Research*, *30*(5), 1083–1090.

Hoyos-Mallecot, Y., Cabrera-Alvargonzalez, J. J., Miranda-Casas, C., Rojo-Martin, M. D., Liebana-Martos, C., & Navarro-Mari, J. M. (2014). MALDI-TOF MS, a useful instrument for differentiating metallo-beta-lactamases in Enterobacteriaceae and Pseudomonas spp. *Letters in Applied Microbiology*, *58*, 325–329.

Julka, S., & Regnier, F. E. (2005). Recent advancements in differential proteomics based on stable isotope coding. *Briefings in Functional Genomics & Proteomics*, *4*(2), 158–177.

Jung, J. S., Eberl, T., Sparbier, K., Lange, C., Kostrzewa, M., Schubert, S., et al. (2013). Rapid detection of antibiotic resistance based on mass spectrometry and stable isotopes. *European Journal of Clinical Microbiology & Infectious Diseases*. In press.

Jungblut, P. R., Schiele, F., Zimny-Arndt, U., Ackermann, R., Schmid, M., Lange, S., et al. (2010). Helicobacter pylori proteomics by 2-DE/MS, 1-DE-LC/MS and functional data mining. *Proteomics*, *10*(2), 182–193.

Kiss, A., Smith, D. F., Reschke, B. R., Powel, M. J., & Heeren, R. M. (2013). Top-down mass spectrometry imaging of intact proteins by LAESI FT-ICR MS. *Proteomics*. In press.

Kolarova, M., Garcia-Sierra, F., Bartos, A., Ricny, J., & Ripova, D. (2012). Structure and pathology of tau protein in Alzheimer disease. *International Journal of Alzheimer's Disease, 2012*, 731526.

Konvalinka, A., Scholey, J. W., & Diamandis, E. P. (2012). Searching for new biomarkers of renal diseases through proteomics. *Clinical Chemistry, 58*(2), 353–365.

Kunnath-Velayudhan, S., & Porcelli, S. A. (2013). Recent advances in defining the immunoproteome of Mycobacterium tuberculosis. *Frontiers in Immunology, 4*, 335.

Kunnath-Velayudhan, S., Salamon, H., Wang, H. Y., Davidow, A. L., Molina, D. M., Huynh, V. T., et al. (2010). Dynamic antibody responses to the Mycobacterium tuberculosis proteome. *Proceedings of the National Academy of Sciences of the United States of America, 107*(33), 14703–14708.

Lee, S. W., Kim, I. J., Jeong, B. Y., Choi, M. H., Kim, J. Y., Kwon, K. H., et al. (2012). Use of MDLC-DIGE and LC-MS/MS to identify serum biomarkers for complete remission in patients with acute myeloid leukemia. *Electrophoresis, 33*(12), 1863–1872.

Lee, J. C., Lee, E. J., Lee, J. H., Jun, S. H., Choi, C. W., Kim, S. I., et al. (2012). Klebsiella pneumoniae secretes outer membrane vesicles that induce the innate immune response. *FEMS Microbiology Letters, 331*(1), 17–24.

Li, Y., Zeng, J., Shi, J., Wang, M., Rao, M., Xue, C., et al. (2010). A proteome-scale identification of novel antigenic proteins in Mycobacterium tuberculosis toward diagnostic and vaccine development. *Journal of Proteome Research, 9*(9), 4812–4822.

Lima, T. B., Pinto, M. F., Ribeiro, S. M., de Lima, L. A., Viana, J. C., Gomes Junior, N., et al. (2013). Bacterial resistance mechanism: What proteomics can elucidate. *The FASEB Journal, 27*(4), 1291–1303.

Lindestam Arlehamn, C. S., Gerasimova, A., Mele, F., Henderson, R., Swann, J., Greenbaum, J. A., et al. (2013). Memory T cells in latent Mycobacterium tuberculosis infection are directed against three antigenic islands and largely contained in a CXCR3 +CCR6+ Th1 subset. *PLoS Pathogens, 9*(1), e1003130.

Lu, Q., Bai, J., Zhang, L., Liu, J., Jiang, Z., Michal, J. J., et al. (2012). Two-dimensional liquid chromatography-tandem mass spectrometry coupled with isobaric tags for relative and absolute quantification (iTRAQ) labeling approach revealed first proteome profiles of pulmonary alveolar macrophages infected with porcine reproductive and respiratory syndrome virus. *Journal of Proteome Research, 11*(5), 2890–2903.

Mahdavi, A., Szychowski, J., Ngo, J. T., Sweredoski, M. J., Graham, R. L., Hess, S., et al. (2014). Identification of secreted bacterial proteins by noncanonical amino acid tagging. *Proceedings of the National Academy of Sciences of the United States of America, 111*, 433–438.

Maria-Neto, S., Candido Ede, S., Rodrigues, D. R., de Sousa, D. A., da Silva, E. M., de Moraes, L. M., et al. (2012). Deciphering the magainin resistance process of Escherichia coli strains in light of the cytosolic proteome. *Antimicrobial Agents and Chemotherapy, 56*(4), 1714–1724.

Martins, R. P., Collado-Romero, M., Martinez-Gomariz, M., Carvajal, A., Gil, C., Lucena, C., et al. (2012). Proteomic analysis of porcine mesenteric lymph-nodes after Salmonella typhimurium infection. *Journal of Proteomics, 75*(14), 4457–4470.

May, C., Brosseron, F., Chartowski, P., Schumbrutzki, C., Schoenebeck, B., & Marcus, K. (2011). Instruments and methods in proteomics. *Methods in Molecular Biology, 696*, 3–26.

Melo, J., Schrama, D., Andrew, P. W., & Faleiro, M. L. (2013). Proteomic analysis shows that individual Listeria monocytogenes strains use different strategies in response to gastric stress. *Foodborne Pathogens and Disease, 10*(2), 107–119.

Melo, J., Schrama, D., Hussey, S., Andrew, P. W., & Faleiro, M. L. (2013). Listeria monocytogenes dairy isolates show a different proteome response to sequential exposure to gastric and intestinal fluids. *International Journal of Food Microbiology, 163*(2–3), 51–63.

Minden, J. S. (2012). DIGE: Past and future. *Methods in Molecular Biology, 854*, 3–8 (Review).

Mitulovic, G., & Mechtler, K. (2006). HPLC techniques for proteomics analysis—A short overview of latest developments. *Briefings in Functional Genomics & Proteomics*, *5*(4), 249–260.

Nagele, E., Vollmer, M., Horth, P., & Vad, C. (2004). 2D-LC/MS techniques for the identification of proteins in highly complex mixtures. *Expert Review of Proteomics*, *1*(1), 37–46 (Review).

Oeljeklaus, S., Meyer, H. E., & Warscheid, B. (2009). Advancements in plant proteomics using quantitative mass spectrometry. *Journal of Proteomics*, *72*(3), 545–554.

Old, W. M., Meyer-Arendt, K., Aveline-Wolf, L., Pierce, K. G., Mendoza, A., Sevinsky, J. R., et al. (2005). Comparison of label-free methods for quantifying human proteins by shotgun proteomics. *Molecular & Cellular Proteomics*, *4*(10), 1487–1502 (Comparative Study).

Ong, S. E., Blagoev, B., Kratchmarova, I., Kristensen, D. B., Steen, H., Pandey, A., et al. (2002). Stable isotope labeling by amino acids in cell culture, SILAC, as a simple and accurate approach to expression proteomics. *Molecular & Cellular Proteomics*, *1*(5), 376–386.

Oswald, S., Groer, C., Drozdzik, M., & Siegmund, W. (2013). Mass spectrometry-based targeted proteomics as a tool to elucidate the expression and function of intestinal drug transporters. *The AAPS Journal*, *15*(4), 1128–1140.

Pajuaba, A. C., Silva, D. A., Almeida, K. C., Cunha-Junior, J. P., Pirovani, C. P., Camillo, L. R., et al. (2012). Immunoproteomics of Brucella abortus reveals differential antibody profiles between S19-vaccinated and naturally infected cattle. *Proteomics*, *12*(6), 820–831.

Papasergi, S., Galbo, R., Lanza-Cariccio, V., Domina, M., Signorino, G., Biondo, C., et al. (2013). Analysis of the Streptococcus agalactiae exoproteome. *Journal of Proteomics*, *89*, 154–164.

Parguina, A. F., Rosa, I., & Garcia, A. (2012). Proteomics applied to the study of platelet-related diseases: Aiding the discovery of novel platelet biomarkers and drug targets. *Journal of Proteomics*, *76*(Spec No), 275–286.

Perkins, D. N., Pappin, D. J., Creasy, D. M., & Cottrell, J. S. (1999). Probability-based protein identification by searching sequence databases using mass spectrometry data. *Electrophoresis*, *20*(18), 3551–3567.

Petriz, B. A., Gomes, C. P., Rocha, L. A., Rezende, T. M., & Franco, O. L. (2012). Proteomics applied to exercise physiology: A cutting-edge technology. *Journal of Cellular Physiology*, *227*, 885–898.

Pfortner, H., Wagner, J., Surmann, K., Hildebrandt, P., Ernst, S., Bernhardt, J., et al. (2013). A proteomics workflow for quantitative and time-resolved analysis of adaptation reactions of internalized bacteria. *Methods*, *61*(3), 244–250.

Raggiaschi, R., Lorenzetto, C., Diodato, E., Caricasole, A., Gotta, S., & Terstappen, G. C. (2006). Detection of phosphorylation patterns in rat cortical neurons by combining phosphatase treatment and DIGE technology. *Proteomics*, *6*(3), 748–756.

Ross, P. L., Huang, Y. N., Marchese, J. N., Williamson, B., Parker, K., Hattan, S., et al. (2004). Multiplexed protein quantitation in Saccharomyces cerevisiae using amine-reactive isobaric tagging reagents. *Molecular & Cellular Proteomics*, *3*(12), 1154–1169.

Russell, M. R., Achour, B., McKenzie, E. A., Lopez, R., Harwood, M. D., Rostami-Hodjegan, A., et al. (2013). Alternative fusion protein strategies to express recalcitrant QconCAT proteins for quantitative proteomics of human drug metabolizing enzymes and transporters. *Journal of Proteome Research*, *12*(12), 5934–5942.

Scherp, P., Ku, G., Coleman, L., & Kheterpal, I. (2011). Gel-based and gel-free proteomic technologies. *Methods in Molecular Biology*, *702*, 163–190.

Schmidt, A., Kellermann, J., & Lottspeich, F. (2005). A novel strategy for quantitative proteomics using isotope-coded protein labels. *Proteomics*, *5*(1), 4–15 (Comparative Study).

Seibel, J., Konig, S., Gohler, A., Doose, S., Memmel, E., Bertleff, N., et al. (2013). Investigating infection processes with a workflow from organic chemistry to biophysics: The combination of metabolic glycoengineering, super-resolution fluorescence imaging and proteomics. *Expert Review of Proteomics*, *10*(1), 25–31.

Selvaraju, S., & El Rassi, Z. (2011). Reduction of protein concentration range difference followed by multicolumn fractionation prior to 2-DE and LC-MS/MS profiling of serum proteins. *Electrophoresis*, *32*(6–7), 674–685.

Sharma, V., Eng, J. K., Maccoss, M. J., & Riffle, M. (2012). A mass spectrometry proteomics data management platform. *Molecular & Cellular Proteomics*, *11*(9), 824–831.

Silva, J. C., Denny, R., Dorschel, C. A., Gorenstein, M., Kass, I. J., Li, G. Z., et al. (2005). Quantitative proteomic analysis by accurate mass retention time pairs. *Analytical Chemistry*, *77*(7), 2187–2200.

Silva, O. N., Mulder, K. C., Barbosa, A. E., Otero-Gonzalez, A. J., Lopez-Abarrategui, C., Rezende, T. M., et al. (2011). Exploring the pharmacological potential of promiscuous host-defense peptides: From natural screenings to biotechnological applications. *Frontiers in Microbiology*, *2*, 232.

Tang, H. Y., Beer, L. A., & Speicher, D. W. (2011). In-depth analysis of a plasma or serum proteome using a 4D protein profiling method. *Methods in Molecular Biology*, *728*, 47–67.

Thakur, A., Siedlak, S. L., James, S. L., Bonda, D. J., Rao, A., Webber, K. M., et al. (2008). Retinoblastoma protein phosphorylation at multiple sites is associated with neurofibrillary pathology in Alzheimer disease. *International Journal of Clinical and Experimental Pathology*, *1*(2), 134–146.

Thierolf, M., Hagmann, M. L., Pfeffer, M., Berntenis, N., Wild, N., Roessler, M., et al. (2008). Towards a comprehensive proteome of normal and malignant human colon tissue by 2-D-LC-ESI-MS and 2-DE proteomics and identification of S100A12 as potential cancer biomarker. *Proteomics Clinical Applications*, *2*(1), 11–22.

Thompson, A., Schafer, J., Kuhn, K., Kienle, S., Schwarz, J., Schmidt, G., et al. (2003). Tandem mass tags: A novel quantification strategy for comparative analysis of complex protein mixtures by MS/MS. *Analytical Chemistry*, *75*(8), 1895–1904.

Toepfer, C., Caorsi, V., Kampourakis, T., Sikkel, M. B., West, T. G., Leung, M. C., et al. (2013). Myosin regulatory light chain (RLC) phosphorylation change as a modulator of cardiac muscle contraction in disease. *The Journal of Biological Chemistry*, *288*(19), 13446–13454.

Trombino, S., Bisio, A., Catassi, A., Cesario, A., Falugi, C., & Russo, P. (2004). Role of the non-neuronal human cholinergic system in lung cancer and mesothelioma: Possibility of new therapeutic strategies. *Current Medicinal Chemistry Anti-Cancer Agents*, *4*(6), 535–542.

Unlu, M., Morgan, M. E., & Minden, J. S. (1997). Difference gel electrophoresis: A single gel method for detecting changes in protein extracts. *Electrophoresis*, *18*(11), 2071–2077.

Varnum, S. M., Webb-Robertson, B. J., Pounds, J. G., Moore, R. J., Smith, R. D., Frevert, C. W., et al. (2012). Proteomic analysis of bronchoalveolar lavage fluid proteins from mice infected with Francisella tularensis ssp. novicida. *Journal of Proteome Research*, *11*(7), 3690–3703.

Vasudev, N. S., Ferguson, R. E., Cairns, D. A., Stanley, A. J., Selby, P. J., & Banks, R. E. (2008). Serum biomarker discovery in renal cancer using 2-DE and prefractionation by immunodepletion and isoelectric focusing; increasing coverage or more of the same? *Proteomics*, *8*(23–24), 5074–5085.

Walker, L. A., Fullerton, D. A., & Buttrick, P. M. (2013). Contractile protein phosphorylation predicts human heart disease phenotypes. *American Journal of Physiology. Heart and Circulatory Physiology*, *304*(12), H1644–H1650.

Wang, J., Chen, C., Xie, P., Pan, Y., Tan, Y., & Tang, L. (2014). Proteomic analysis and immune properties of exosomes released by macrophages infected with Mycobacterium avium. *Microbes and Infection*, *16*, 283–291.

Weeks, M. E. (2010). Urinary proteome profiling using 2D-DIGE and LC-MS/MS. *Methods in Molecular Biology*, *658*, 293–309.

Winnik, W. M., Dekroon, R. M., Jeong, J. S., Mocanu, M., Robinette, J. B., Osorio, C., et al. (2012). Analysis of proteins using DIGE and MALDI mass spectrometry. *Methods in Molecular Biology*, *854*, 47–66.

Wittmann-Liebold, B., Graack, H. R., & Pohl, T. (2006). Two-dimensional gel electrophoresis as tool for proteomics studies in combination with protein identification by mass spectrometry. *Proteomics*, *6*(17), 4688–4703.

Wu, J., Shakey, Q., Liu, W., Schuller, A., & Follettie, M. T. (2007). Global profiling of phosphopeptides by titania affinity enrichment. *Journal of Proteome Research*, *6*(12), 4684–4689 (Validation Studies).

Wu, W. W., Wang, G., Baek, S. J., & Shen, R. F. (2006). Comparative study of three proteomic quantitative methods, DIGE, cICAT, and iTRAQ, using 2D gel- or LC-MALDI TOF/TOF. *Journal of Proteome Research*, *5*(3), 651–658.

Yang, N., Feng, S., Shedden, K., Xie, X., Liu, Y., Rosser, C. J., et al. (2011). Urinary glycoprotein biomarker discovery for bladder cancer detection using LC/MS-MS and label-free quantification. *Clinical Cancer Research*, *17*(10), 3349–3359.

Zhao, Y., & Jensen, O. N. (2009). Modification-specific proteomics: Strategies for characterization of post-translational modifications using enrichment techniques. *Proteomics*, *9*(20), 4632–4641.

CHAPTER TWO

Phosphoproteomic Techniques and Applications

Ed Dudley[1], A. Elizabeth Bond
Institute of Mass Spectrometry, College of Medicine, Swansea University, Swansea, United Kingdom
[1]Corresponding author: e-mail address: e.dudley@swansea.ac.uk

Contents

Abstract

Phosphoproteomic analysis seeks to determine the overall level of protein phosphorylation, as a result of kinase and phosphatase activity, and determine the identity of proteins which are phosphorylated and the amino acid residues which hold the phosphate group. The methodologies available have improved with increased research efforts; however, the most commonly followed procedure is to enrich for phosphoproteins or peptides and undertake tandem mass spectrometric analysis focusing on specific signature losses which represent phosphopeptides. There have been many advances in this area and these are detailed both in relation to available protocols for phosphoproteomic analysis and to the widening range of biomedical fields in which such approaches are being commonly applied.

1. INTRODUCTION

High-throughput analysis has allowed for the study of changes in the biological functions of cells at a variety of levels—ranging from the study of differential gene expression, transcriptomic variations, alterations in the

Advances in Protein Chemistry and Structural Biology, Volume 95
ISSN 1876-1623
http://dx.doi.org/10.1016/B978-0-12-800453-1.00002-6

protein complement at the cellular level, and further study of the biochemical effect of these changes in the metabolite complement presented. The technologies required to undertake these individual analyses and combinations of them have continually developed over the last decade and are constantly improving in relation to their throughput, reduction in cost of analysis, the robustness of the data produced by the analyses, and the bioinformatics software solutions available to interpret the increasingly large data sets produced. The combination of the different analyses aims to fully interrogate the underlying cause behind a change in a cell's behavior, be it transformation into a tumor, differentiation into different cell types, or the production of a diagnostically or prognostically useful biomarker of a specific disorder or disease. Generally, the genome represents the overall complement of what may be produced by a cell while the transcriptome defines which of these elements is being actively utilized at any given time point. The proteome therefore is an illustration of which transcripts are being utilized at the protein level, and the metabolome can be studied in order to demonstrate the biochemical consequences of the previous analyses. In this manner, combining such data sets allows for a fully developed image of how different cells, tissues, or biological fluids may differ under different circumstances and this has a number of benefits in biomedical research and pharmacology.

Phosphoproteomics represents a subdivision of proteomic analysis concerned with the study of the phosphorylation status of proteins within a biological sample (or comparisons between biological samples). Before discussing how protein phosphorylation can be monitored in a global manner, it is worth considering the process of protein phosphorylation and its implications for the cell. The process of phosphorylation is performed by the kinase enzymes having different specificities for the target proteins which become phosphorylated and also the site of the phosphorylation. Commonly, phosphorylation occurs as a downstream effect of an external cell receptor protein interacting with its associated ligand (commonly a hormone or other circulating signaling molecule). Receptor activation causes a signal transduction cascade to be initiated which can often either activate a tyrosine kinase activity within the cytoplasmic section of the receptor protein or activate cytoplasmic protein kinases via the biosynthesis of secondary messengers such as the cyclic nucleotides. Once activated, the kinase enzymes utilize adenosine triphosphate (ATP) to add a phosphate residue to a specific amino acid within the target protein structure. In mammalian systems, only three amino acid residues are available for phosphorylation, these being threonine, tyrosine, and serine. The act of adding a phosphate group to the amino acid residue has a number of consequences for the protein phosphorylated. The addition of a phosphate residue to the amino acid sequence provides for two additional potential negative charges (arising from dissociated oxygen atoms attached covalently to the phosphate itself) and these are therefore available to either disrupt existing electrostatic interactions or provide new interactions between sections of the protein. These oxygens and hydroxyl components can also create new hydrogen bonds within the structure, and therefore conformational change of the protein's overall structure can occur due to the minor act of phosphorylation. Finally, the phosphate residue itself is derived from the hydrolysis of ATP and this hydrolysis reaction is highly exergonic and therefore provides a significant release of energy. While approximately half of this energy is required to add the phosphate residue to the protein, the other half can be conserved by the protein and/or the reactions which the protein then undertakes. As a result of these factors, the simple act of phosphorylation can lead to a significant change in the proteins activity—usually by bringing about conformational change within the proteins overall structure which can lead to increased or decreased activity of the protein and different kinetic parameters, such as K_m and V_{max}, are usually exhibited by proteins in their different conformational forms. The common signal transduction systems that result in differential protein phosphorylation vary in their net result in different cells allowing different mammalian organs to respond differently to the same circulating "signal." These systems react rapidly due to a number of factors; first, the rate-limiting step in the phosphorylation reaction itself is the concentration of the phosphate-donating substrate, ATP. However, as ATP is utilized as the cell's major energy providing metabolite, the intercellular levels of ATP are consistently maintained at a high concentration and this ensures that the amount of ATP within a cell never becomes a rate-limiting step in the phosphorylation response to a stimuli. Furthermore, the signal transduction system by which a cell responds via differential protein phosphorylation to a change in hormonal levels or similar stimulus exhibits an amplification cascade regarding the steps involved in eventual protein phosphorylation. For example, a single binding event of a single hormone at a single extracellular receptor will activate that one receptor, the activated receptor can then activate a single adenylate cyclase enzyme (as an example). This single adenylate cyclase enzyme can then produce a large number of cyclic adenosine monophosphate (cAMP) secondary messenger metabolites inside the cell. Each secondary messenger may activate a single kinase (such as protein kinase A) and this kinase in turn may activate multiple further kinases. At each step where multiple end products are produced, the single binding event is in effect amplified and therefore a single binding can cause the eventual altered phosphorylation of a significant number of individual proteins. Therefore, the response of the cell is rapid and substantial for any particular binding event. The phosphorylation of proteins in this manner is a reversible process and the removal of the phosphate from any particular target protein is undertaken by a separate series of enzymes referred to as phosphatases. Therefore, the overall protein phosphorylation status of a cell is controlled by the activity of the kinase and phosphatase enzymes which respond to stimuli which is external to the cell allowing for coordinated regulation of cellular activity within organisms. The ability of different cells to respond differently to similar stimuli is essential for different organs within mammalian systems to act in a relevant manner to any given change (usually represented at the molecular level by an altered level of a specific hormone released as a response by the given change). This specificity relies on a number of factors including whether or not the particular cell produces an extracellular receptor for the particular hormone in question and the isoform of the secondary messenger producing protein that is produced by the cell in question. Different cells produce different isoforms of the cyclase enzymes that produce the secondary messenger and also the phosphodiesterases which catabolize the cyclic nucleotide secondary messengers. These different isoforms affect how the cell responds and how long the signal that is being transduced is maintained. This variation in the enzymes in different cells also allows for different therapeutic interventions to be produced which effect selectively one specific isoform and therefore only bring about an intervention in a specific target organ or cell without effecting other tissues. A key example of this is the drug Viagra (Sildenafil), which acts as an inhibitor of the type 5 class of cyclic guanosine monophosphate (cGMP) phosphodiesterases. The inhibition of this enzyme results in elevated cGMP levels being maintained most prominently in the corpus cavernosum and the retina providing for the drugs therapeutic benefits via the maintenance of the signal transduction signal that leads to a specific alteration in the phosphorylation status of the cells proteome.

Proteins are modified posttranslationally for a number of different purposes, and protein phosphorylation represents one such modification process. Significant other posttranslational modification processes include glycosylation (the addition of a sugar residue—usually a complex polysaccharide chain), farnesylation and geranylation (the addition of hydrophobic residues to the protein), and altered redox properties of the target protein and ubiquitination (usually targeting proteins to the proteasome system for degradation). The major posttranslation modifications of interest are phosphorylation, glycosylation, and farnesylation. Ubiquitination represents a significantly wider and more diverse field of research interest. Of the stated three posttranslational modifications, a survey of research literature between 2007 and 2012 clearly demonstrates that phosphorylation has been more significantly researched compared to the other two modification processes. Protein phosphorylation articles represent over 70% of the published literature with glycosylation represented in 16% of the literature and farnesylation, ubiquitination, and acetylation in 14% (Fig. 2.1). This focus upon protein phosphorylation analysis can be attributed to a number of distinct reasons. First, as mentioned previously, the impact of protein phosphorylation is prominent in all cells and the mechanism behind the phosphorylation events is reasonably well understood and appreciated; this makes the protein phosphorylation event an attractive target therefore for the study of the development of clinical disorders and also as a therapeutic target for the treatment of any such disorder. Second, as a posttranslation modification, protein phosphorylation is a comparatively simple and consistent modification. In mammalian systems, only three amino acids represent targets of protein phosphorylation (as discussed previously) and the addition of a single phosphate to any given amino acid is the sole modification.

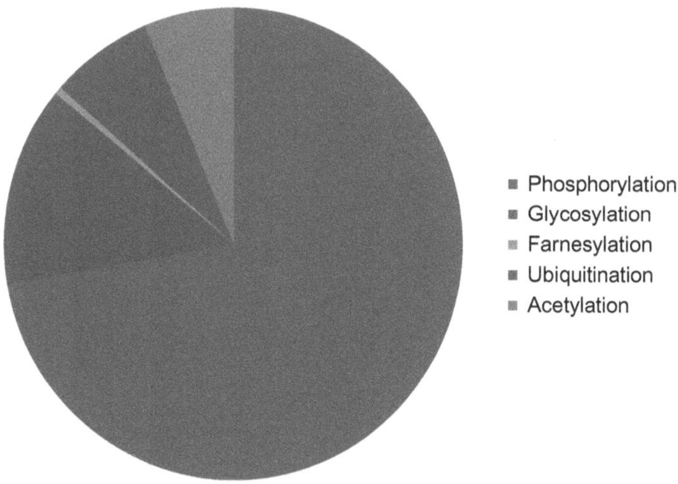

Figure 2.1 Proportions of publications covering protein and phosphorylation, glycosylation, and other posttranslational modifications as their topic between 2007 and 2012.

In comparison, glycosylation can occur at a wider number of amino acid residues and the chemical nature of the modification itself is more heterogeneous. Glycosylation status can include a small number of carbohydrate residues or a more elaborate modification with a large number of such residues attached as a long glycosyl linear chain or as a multiply branched polysaccharide chain. As the nature of the modification is so variable and comparatively complex, methods allowing for the accurate analysis of this modification on a global proteomic scale have lagged behind the development of methods that may be applied to the analysis of protein phosphorylation on the same scale. Similarly, the nature of the hydrophobic residue added when proteins are modified by processes such as farnesylation and geranylation is more varied and complex and therefore less easily applicable to high-throughput global analysis. The analysis of protein phosphorylation is therefore prominently featured in the scientific literature compared to these other posttranslational protein modification types. The application of modern analysis techniques to protein phosphorylation analysis has rapidly developed over the past decade or so as can be seen by considering the increase in the number of scientific publications related to phosphoprotein analysis published per year. As can be seen in Fig. 2.2, there has been a dramatic increase in publications in the field between 2007 and 2012, and consideration of the number of citations of scientific manuscripts detailing phosphoprotein studies gives rise to a significant number of citations per

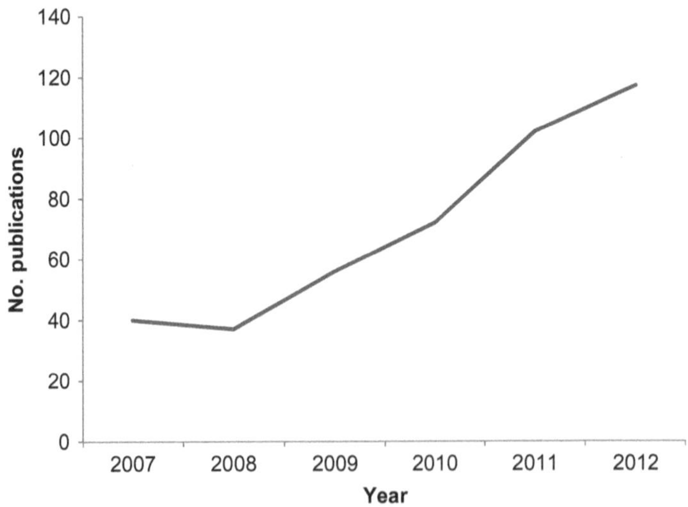

Figure 2.2 The number of publications related to phosphoproteomics between 2007 and 2012.

year. This increasing and dramatic increase in interest in protein phosphorylation analysis is not only a result of the interest in the phosphorylation events themselves (and how these play a role in disorder development and may act as therapeutic targets) but also as a result of an increased interest in the development and validation of novel methods allowing for accurate protein phosphorylation determination encompassing as much of the proteome as is possible. While a significant development in global phosphoprotein analysis has utilized mass spectrometry (MS) as a method of detecting phosphorylation status and detailing protein phosphorylation sites within proteins, other methods of analysis are also still utilized and methods to enrich phosphoproteins and phosphorylated peptides resulting from proteolytic digestion by enzymes such as trypsin have also been a major focus of research. The complexity of the analysis required in order to study phosphoproteins depends to an extent on the sample to be analyzed. Phosphoproteins and phosphopeptides have been studied from cell lines and biopsy material commonly; however, levels and differences in phosphoprotein/peptide levels present in archived tissue samples have also been attempted allowing past samples from previous patients to be analyzed and phosphoprotein levels to be correlated with patient records. As well as sampling from cellular material, the phosphoprotein complement of biological fluids has also become a major area of research interest in the biomedical field. While the majority of studies have focused upon serum as the biological fluid of interest, other studies have utilized urine, saliva, cerebrospinal fluid, and bronchoalveolar fluid as the biological source of the phosphoproteome. Beyond the challenge represented by the dynamic phosphoproteome and its accurate analysis, a further hurdle to overcome in proteomic data sets (including phosphoprotein analysis) is the analysis of the large data sets produced as a result of the high-throughput analyses available. This challenge relates not only to the amount of data that can be produced in a comparatively short experimental time period but also to the extraction of biologically relevant information from the data and the successful appreciation of the implications of the change in phosphorylation status. For many proteins, the effect of phosphorylation upon the enzymes activity is well established and therefore this can be determined easily. The impact of this posttranslational modification on the overall metabolism pathways of the cell in question is a more complex challenge, however, and therefore bioinformatics can be utilized in order to illustrate the overall impact of a phosphorylation event on multiple intersecting pathways. Protein phosphorylation has been studied in many areas including plant research (Bond, Row, & Dudley, 2011; Newton,

Brenton, Smith, & Dudley, 2004) and has been utilized to study the function of newly identified secondary messengers (Bond et al., 2007).

The aim of this review is to first discuss the methodologies available in order to detect phosphorylated proteins and investigate the site of protein phosphorylation. Methods that have been developed for the purification or enrichment of phosphorylated peptides/proteins from proteomes will also be reviewed as these can be coupled to "traditional" MS proteomic analysis platforms to provide phosphoproteomic data sets. Following the discussion of techniques available for phosphoproteomic analyses, examples of applications in a number of fields will be reviewed in order to provide the reader with an understanding of the potential of the field to inform biomedical and pharmacological studies in the future.

2. PHOSPHOPROTEOMIC METHODOLOGIES

Traditional methods for studying protein phosphorylations were laborious and time consuming and included techniques such as radiolabeling of phosphorus atoms, phosphospecific antibodies, and *in vitro* kinase assays. These techniques generally involved specific knowledge of the phosphorylation sites and phosphate groups. With the advent of MS and supporting technology, a more rapid and unbiased analysis was developed, allowing large-scale identification of phosphorylation sites from different model systems. MS phosphoprotein analysis provides both qualitative and quantitative analyses to identify and profile the abundance of thousands of phosphoproteins in a single experiment. With improvements in separation affinity media and MS selectivity and sensitivity, there has been an increase in the number of phosphorylation sites identified (Kanshin, Michnick, & Thibault, 2012) using specific phosphoproteome databases (Gruhler et al., 2005; Ham et al., 2008; Mann et al., 2002; Nagaraj, D'Souza, Cox, Olsen, & Mann, 2010; Swaney, McAlister, & Coon, 2008; Syka, Coon, Schroeder, Shabanowitz, & Hunt, 2004). Protein phosphorylation is a highly dynamic modification regulated by enzymic control of kinases and phosphatases. Both intrinsic and extrinsic processes can stimulate proteins to be phosphorylated or dephosphorylated instantly; therefore, sample preparation for phosphoprotein analyses is an important consideration when planning method strategies. Inhibitors of both phosphatases and proteases must be used, and appropriate protein isolation and extraction protocols must be designed to maintain the integrity, concentration, and phosphorylation status of the phosphoproteins from samples. Over the past decade, protocols for phosphoprotein analysis following cell fractionation with protein extraction have consisted of three main strategies:

1. phosphopeptide enrichment
2. peptide separation by HPLC
3. MS analysis.

2.1. Phosphopeptide enrichment

This section encompasses techniques such as immunoaffinity chromatography, metal oxide affinity chromatography (MOAC), and immobilized metal affinity chromatography (IMAC).

The use of selective enrichment techniques is necessary for the detection of phosphopeptides in complex biological samples. These techniques are applied following protein digestion either by trypsin or by chemical means, as they aid in protein solubilization and remove any nonphosphorylated peptides. Outlined below are the most common affinity techniques used in phosphopeptide enrichment.

2.1.1 Immunoaffinity chromatography

This technique involves the use of specific antibodies to purify phosphopeptides from the sample, with the major drawback that there are only a limited number of these antibodies commercially available. These can be classified according to recognition signal or residue-specific binding. Antibodies against phosphotyrosine are the most commonly used antibody for the enrichment of phosphotyrosine peptides in cell cultures and tissue extracts. The disadvantage with this technique is the requirement of large amounts of starting sample (10 mg) due to the low proportion of proteins phosphorylated on tyrosine (Frackelton, Ross, & Eisen, 1983). However, this technique is able to isolate and identify a number of phosphorylation sites from different proteins (Jedrychowski et al., 2011). Although this number is significantly lower than that of phosphorylated serine and threonine residues, there are commercially available antibodies for tyrosine residues, whereas there is a lack of suitable antibodies for their selective enrichment (Pandey et al., 2000). Rush et al. (2005) conducted a proteomic study of tyrosine phosphorylation in Jurkat cells. Phosphotyrosine-containing peptides from a cell digest of a tyrosine phosphatase inhibitor-treated Jurkat cells were immunoprecipitated with P-Tyr-100, a phosphotyrosine-specific antibody noncovalently coupled to protein G agarose. The enriched peptides were analyzed by LC–MS, and 688 pTyr-containing peptides and 628 pTyr-sites were identified.

2.1.2 *Metal oxide affinity chromatography*

The application of MOAC in phosphoproteomics is based on the ability of metal oxides to form complexes with a phosphate group. The most common MOAC affinity medium is TiO_2, which was first utilized for the selective retention of inorganic phosphate (Connor & McQuillan, 1999). The use of TiO_2 was then used for the purification of phosphorylated amino acids (Ikeguchi & Nakamua, 1997) and then for phosphopeptides with on-line enrichment in LC–MS analysis (Pinkse, Uitto, Hilhorst, Ooms, & Heck, 2004), and to date, the use of TiO_2 is still the most popular MOAC techniques used for the enrichment of phosphopeptides in LC–MS analysis. The method details a two-dimensional chromatographic setup with titanium dioxide-based solid-phase material (Titansphere) as the first dimension and reversed-phase material as the second dimension. Phosphorylated peptides are separated from nonphosphorylated peptides in the first dimension by trapping them under acidic conditions (0.1 M acetic acid) on the TiO_2 precolumn. Nonphosphopeptides are not retained in the first dimension but trapped in the second dimension precolumn before analysis by nanoflow LC–ESI–MS/MS. The phosphopeptides are eluted from the column under alkaline conditions (ammonium bicarbonate, pH 9.0), 125 fmol of a phosphopeptide in a 1:1 mixture of the phosphorylated and unphosphorylated form can be successfully identified with a recovery rate of above 90%. MOAC methods are characterized by high affinity for phosphopeptides and by high enrichment efficiency using loading buffers. These methods are tolerant to salts, detergents, and denaturing agents. The advantages of these metal oxides include large adsorption capacities, chemical stability when used under extreme pH ranges, mechanical stability, and unique amphoteric ion-exchange properties (Ikeguchi & Nakamura, 2000; Kawahara, Nakamura, & Nakajima, 1989; Matsuda, Nakamura, & Nakajima, 1990; Mazanek et al., 2007; Tani & Suzuki, 1997). The disadvantage of all MOAC methods is the amount of nonspecific binding, in particular of acidic peptides, which in turn decreases the enrichment efficiency and reduces the detection and identification of phosphopeptides by MS.

Larsen, Thingholm, Jensen, Roepstorff, and Jorgensen (2005) proposed the use of 2,5-dihydroxybenzoic acid (DHB) to compete with the binding of acidic peptides on TiO_2 beads while maintaining the specificity for phosphopeptides. This was demonstrated for the enrichment of phosphopeptides from a tryptic digest of casein with different concentrations of DHB in 80% acetonitrile and 0.1% TFA. The selective elution of phosphopeptides was performed by using ammonium hydroxide (pH 10.5). The presence of

DHB in the loading buffer showed an increased enrichment of phosphopeptides and suggested that the binding of acidic peptides and phosphopeptides is facilitated by different active sites on the TiO_2 surface. Mazanek et al. (2010) suggested using a mix of DHB and octanesulfonic acid (OSA), an ion pairing agent used for improved peptide separation in reversed-phase chromatography, to reduce unspecific binding. The additives were used in lower concentration and should therefore be less problematic for the following analysis. This method has since been optimized further in terms of selectivity using slightly increased concentrations of DHB and OSA, and with the addition of heptafluorobutyric acid (Sugiyama et al., 2007). However, these aromatic acids were too hydrophobic to be removed by the desalting step before LC-MS/MS analysis.

Current methods use hydrophilic hydroxylated modifiers such as lactic acid instead of DHB to improve selectivity and capacity of TiO_2 toward phosphorylated peptides. Sugiyama et al. (2007) tested different hydroxy acids in MOAC—an aliphatic hydroxy acid-modified metal oxide chromatography—and determined that lactic acid provided enhanced selectivity for the isolation of phosphopeptides from tryptic digests of HeLa cells. In addition, aliphatic hydroxy acids can be easily removed by desalting with reversed-phase cartridges, which is necessary for subsequent LC–MS/MS analyses. Large-scale phosphoproteome studies utilizing MOAC (TiO_2) have reported the variety of phosphorylation sites from different cell model systems (Hilger, Bonaldi, Gnad, & Mann, 2009; Simon, Young, Chan, Bao, & Andrews, 2008). Both multiply and singly phosphorylated peptides bind to TiO_2. Singly phosphorylated peptides are eluted with a typical eluent (pH 10–11.5), but there are several reports indicating that multiply phosphorylated peptides are also eluted under different pH conditions (Kyono, Sugiyama, Imami, Tomita, & Ishihama, 2008; Leitner, Sturm, Smått, Järn, & Lindén, 2009; Thingholm, Jorgensen, & Jensen, 2006).

Other MOAC sorbents including SnO_2 (Rivera, Choi, Vujcic, Wood, & Colón, 2009), HfO_2 (Qi, Lu, Deng, & Zhang, 2009), Ta_2O_2 (Kweon & Håkansson, 2006), ZrO_2 (Ficarro, Parikh, Blank, & Marto, 2008), Nb_2O_5 (Rivera et al., 2009), and Al_2O_3 (Wang & Bruening, 2009) have also been described for the enrichment of phosphopeptides from tryptic digests. When compared to TiO_2, these sorbents showed different populations of phosphopeptides, thus suggesting complementary selectivity for these metal oxide resins. Although these results are promising, there is no clear pattern of selectivity has been obtained for the chemical properties of phosphopeptides retained by each resin. Kweon and Håkansson (2006) enriched 100 pmol tryptic α-casein and β-casein digests on both TiO_2 and ZrO_2 microtips prior to analysis by negative-ion ESI–FT–ICR (Fourier transform ion cyclotron resonance) MS. They found more phosphorylated peptides using ZrO_2 and concluded that TiO_2 microtips were more selective for the enrichment of multiply phosphorylated peptides, whereas the ZrO_2 tips enriched primarily monophosphorylated peptides. Recently, on-plate enrichment for MALDI–MS has been developed. Wang and Bruening (2009) described a method for modification of silicon wafers, which serve as MALDI plates with 250-μm diameter microspots of phosphopeptide-binding polymer brushes enclosed by a hydrophobic poly(dimethylsiloxane) layer. Enrichment resulted in a fivefold decrease in MALDI–MS detection limits and femtomole-level sensitivity. Zhou, Xu, and Ye (2006) who used zirconium phosphonate monolayers immobilized on porous silicon observed excellent selectivity of this approach demonstrated by analyzing phosphopeptides in the digested mixture of β-casein and BSA with molar ratio of 1:100. Pipette tip-based off-line TiO_2 minicolumns have been widely used for phosphopeptide purification (Bodenmiller et al., 2008; Mohammed et al., 2008; Ovelleiro, Carrascal, Casas, & Abian, 2009; Wolschin, Wienkoop, & Weckwerth, 2005). In this regard, Agilent Technologies has introduced a chip-based device with integrated TiO_2 enrichment and RP-LC separation that is now commercially available (Raijmakers, Kraiczek, de Jong, Mohammed, & Heck, 2010).

2.1.3 Immobilized metal affinity chromatography

In 1986, Andersson and Porath reported the use of IMAC for the enrichment of phosphorylated proteins using Fe(III) immobilized via iminodiacetic acid onto a sepharose matrix (Andersson & Porath, 1986). Phosphorylated amino acids like phosphoserine, phosphothreonine, or phosphotyrosine were retained by the chromatographic material, whereas nonphosphorylated amino acids were not, or in some cases, like aspartic acid or glutamic acid, only weakly bound. The ion pair formation between Fe(III) and the negatively charged phosphate group enabled the selective retention of ovalbumin phosphoisoforms. An advantage of the method was that all steps could be carried out in water or buffer and no protein denaturing components were needed, and consequently, IMAC has been applied to a wide range of applications in phosphoproteomics (Nuhse, Yu, & Salomon, 2007). Although similar to MOAC in terms of the selective retention of phosphopeptides and binding conditions, the electrostatic interactions between phosphorylated residues and the immobilized cations favor the selective retention of phosphopeptides from a complex mixture. Over the years, a variety of different supports have been introduced for IMAC-based enrichment, and the technique has long been the most frequently used method for enrichment of phosphopeptides, also due to the fact that commercial kits are available from different suppliers. Phosphopeptides are loaded on IMAC columns using acidic buffers and eluted with high pH, EDTA, or inorganic phosphate buffers. IMAC methods are utilized for the enrichment of peptides phosphorylated on serine, threonine, and tyrosine residues. Although Fe^{3+} is predominantly used with IMAC, other coordinating metal ions such as Ga(III), Zr(IV), and Al(III) have also been described for the selective enrichment of phosphopeptides (Kokubu, Ishihama, Sato, Nagasu, & Oda, 2005; Posewitz & Tempst, 1999).

To reduce the extent to which acidic peptides can bind nonspecifically to IMAC resins, solutions containing 0.1% TFA in 50% acetonitrile are used as loading buffers (Huttlin et al., 2010). This affinity medium is entirely suited for large-scale phosphoproteomics experiments (Schreiber et al., 2012). Just 96 nonphosphopeptides and 1654 phosphopeptides were assigned by Mascot from a mouse brain sample. After manual validation, 166 phosphosites on 135 different proteins were identified using this approach. In regard to metal affinity enrichment, a novel Fe^{3+} chelate matrix based on chelate ligand called PHOS-Select Iron Affinity Gel (Sigma, St. Louis, MO, USA) should be mentioned, since it has overcome some of the problems arising with IMAC (as mentioned above). A new phosphoprotein enrichment Fe–NTA kit was introduced (Thermo Scientific, Pierce, Rockford, IL, USA), and according to manufacturer, it outperformed PHOS-Select in both number of total and unique phosphopeptides (862 vs. 430 and 178 vs. 90, respectively). The application of IMAC can also be used in quantitation experiments. More than 8000 phosphosites were shown in wild-type and PPt1-deficient yeast strains and the identity of Ser/Thr sites which are regulated by this phosphatase (Collins et al., 2005). Other notable IMAC-based studies on complex biological samples were able to identify several hundred to thousands of phosphorylation sites (Ficarro et al., 2002; Gruhler et al., 2005; Li et al., 2007). Methyl esterification of acidic residues has also been proposed to enhance the selectivity of IMAC with no apparent loss in sensitivity (Posewitz & Tempst, 1999). A disadvantage of the esterification procedure is the occurrence of side reaction products (partial hydrolysis of peptides, deamidation of asparagine and glutamine residues) that can increase sample complexity.

In 1994, Reynolds et al. used an excess of Ca^{2+} in 50% ethanol for precipitating phosphopeptides from a tryptic casein hydrolysate (Reynolds, Riley, & Adamson, 1994). At lower pH, only peptides containing multiple phosphoserines were enriched. At a pH of 8, all phosphopeptides except two monophosphorylated ones could be found in the precipitate. Although iron is used as a central ion in most IMAC methods, other metal ions have been evaluated for selective phosphate affinity. Posewitz tested different metal ions including Ga, Sn, Ge, Fe, and others for their applicability in IMAC phosphopeptide enrichment (Posewitz & Tempst, 1999). With Ga^{3+}, a better selectivity compared to conventional Fe^{3+}–IMAC was reported when analyzing a tryptic digest of phosphoproteins. An interesting approach was reported from the group of Zhou (Feng et al., 2007; Zhou et al., 2008). They used a phosphate polymer to coordinatively bind Ti^{4+} or Zr^{4+} ions, and the resulting IMAC resin was used for phosphopeptide enrichment and compared to Fe^{3+}–IMAC, TiO_2, and ZrO_2 enrichment methods. In contrast to MOAC which favors the isolation of monophosphorylated peptides, IMAC was reported to yield a higher proportion of multiply phosphorylated peptides (Jensen & Larsen, 2007). The complementary distribution of phosphopeptides obtainable by TiO_2 and Fe(III)–IMAC can be used advantageously for the combined separation of monophosphorylated and multiply phosphorylated peptides from cell digests. The sequential use of IMAC and TiO_2 also termed SIMAC (sequential elution from IMAC) gave a twofold increase in phosphopeptide identification from lysates of human mesenchymal stem cells compared to TiO_2 alone (Thingholm, Jensen, Robinson, & Larsen, 2008). Liang et al. used iTRAQ (Liang et al., 2007) to compare commercial and prototypal immobilized metal affinity chelate and metal oxide resins. They tested IMAC magnetic beads from Invitrogen (Captivate beads), Applied Biosystems (Poros 20 MC beads), and Calbiochem (ProteoExtract) against Nexus tetradentate metal chelator (Valen Biotech Inc., Atlanta, GA, USA) coupled to Dynabeads-MyOne (Invitrogen, Carlsbad, CA, USA) tosylactivated beads (U.S. patent application 20020019496).

2.2. Peptide separation by HPLC

Numerous groups have exploited the negatively charged phosphate moiety of phosphopeptides to enrich them using ion-exchange or mixed mode chromatography separation. The analytical merits of these approaches are briefly outlined here.

2.2.1 Ion-exchange chromatography

Different forms of ion-exchange chromatography are commonly used in a two-dimensional chromatography setup in proteomics. Because of the strong negative charge of the phosphate group, most phosphopeptides show particular retention behavior in ion-exchange chromatography in comparison to the majority of unmodified peptides. Ion chromatography separation of phosphopeptides has been reported for both strong anion (SAX) and strong cation (SCX) exchange resins. This type of chromatography uses the strong electrostatic interactions taking place between the ionized groups of the stationary phase and the peptide counter ions present in the sample at a given pH. For example, the interaction of peptides with the SCX resin is proposed to be (mainly) of electrostatic and (partially) of hydrophobic character, due to some residual hydrophobicity of the polymeric stationary phase, so that structurally similar peptides with the same net charge may be separated to some degree. The elution of target analytes is obtained by modulating the strength of the interactions using salts, pH, and/or organic buffers. The application of SAX for the fractionation of phosphopeptides was first demonstrated for the tryptic digest of casein (Zhang, 2006) and enabled the separation of phosphopeptides from their nonphosphorylated peptide counterparts. The application of SAX was soon utilised by different groups for the analyses of complex cell extracts including human liver tissue (Han et al., 2008) and HeLa cells (Dai et al., 2009). SAX fractionation has been mostly described with on-line reverse-phase LC–MS analysis, although a recent report described the use of an on-line RP-SAX-RP configuration to enhance the peak separation and the number of unique phosphopeptides from cell lysates (Ficarro et al., 2011). Multiple phosphorylated peptides show a higher affinity to the SAX resin than singly phosphorylated ones.

Since SAX requires that alkaline solutions are used for sample and elution buffers, precautions must be taken to avoid the formation of elimination products from phosphorylated serine and threonine residues under these conditions. Another disadvantage arising from this method is that the solvents used in SAX chromatography are not optimal for on-line LC–MS/MS coupling because the weakly acidic to neutral pH and the aqueous buffer lower the ionization efficiency when LC–ESI–MS is used.

A larger number of reports have described the application of SCX for on-line and off-line separation of phosphopeptides. The separation of peptides on SCX resins is typically performed at low pH, and the large majority of tryptic peptides containing at least one basic amino acid have an overall charge higher than two. The presence of a phosphate group reduces their effective charge and their interactions with the SCX resin, resulting in a relative enrichment of phosphopeptides in early fractions. Beausoleil et al. (2004) were the first to take advantage of this feature in a phosphoproteomic study on tryptic digests of HeLa cells where they identified more than 2000 phosphorylation sites using off-line SCX fractionation.

SCX fractionation alone is not sufficient to enrich phosphopeptides from complex cell extracts, and this technique is typically used to decrease sample complexity (Imamura, Wakabayashi, & Ishihama, 2012). But, peptides that have net zero or even negative charge, such as phosphopeptides with basic residues or multiply phosphorylated peptides, are not well retained on SCX columns. In order to capture these peptides, ultra acidic SCX exchange was recently introduced (Hennrich, van den Toorn, Groenewold, Heck, & Mohammed, 2012), in which tandem SCX is performed under two different pH conditions (usual and more acidic conditions). Further enrichment of phosphopeptides from SCX fractions is achieved using IMAC (Dephoure & Gygi, 2011; Villen, Beausoleil, Gerber, & Gygi, 2007; Zhai, Villen, Beausoleil, Mintseris, & Gygi, 2008) or MOAC (Olsen et al., 2006; Sui et al., 2008). The combination of the SCX–IMAC enrichment strategy provided up to 30-fold increase in the proportion of phosphopeptides observed from *Saccharomyces cerevisiae* compared to SCX alone (Villen & Gygi, 2008).

2.2.2 Hydrophilic interaction chromatography and electrostatic repulsion–hydrophilic interaction chromatography

Hydrophilic interaction chromatography (HILIC) can also provide an orthogonal separation to RP chromatography, and phosphopeptides can be selectively enriched due to their increased polarity (Garbis et al., 2011). In HILIC, analytes are separated according to their polarity. The sample is typically loaded on a polar stationary phase with a high concentration of organic solvent (typically 95% ACN) favoring the retention of polar phosphopeptides. Their subsequent elution is achieved by increasing the proportion of aqueous buffer leading to desorption of phosphopeptides with increasing polarity. IMAC enrichment of phosphopeptides from HILIC fractions provided 99% selectivity, as demonstrated by McNulty and Annan (2008) for HeLa cell lysate where more than 1000 unique phosphorylation sites were identified. Phosphopeptides with the highly polar phosphate group should therefore be strongly retained on the HILIC stationary phase. IMAC phosphopeptide enrichment was added after the

HILIC step to increase selectivity of the setup. Nearly 100% selectivity for phosphopeptides was achieved when IMAC enrichment was performed after HILIC chromatography, and over 1000 phosphorylation sites on 914 peptides were identified, demonstrating the use of HILIC as a powerful prefractionation tool before selective phosphopeptide enrichment is performed. Additionally, phosphopeptides elute at a buffer composition of 70% to 50% ACN containing 0.1% TFA which is asserted to be the optimal loading buffer composition for IMAC phosphopeptide loading.

The main disadvantage of HILIC is the high organic content of the fractions, which precludes its direct coupling to on-line RP-LC and HILIC is generally preferred as a prefractionation technique prior to LC–MS analysis of phosphopeptides. More recent applications of HILIC have been demonstrated in combination with size-exclusion chromatography to identify low-abundance phosphoproteins from immunodepleted plasma samples from prostate cancer patients (Garbis et al., 2011) or with IMAC and stable isotope labeling to profile the abundance of 2857 unique phosphorylation sites in 1338 phosphoproteins from 1 mg of cell lysates (Wu, Chen, Tai, & Chen, 2011). In contrast, electrostatic repulsion–hydrophilic interaction liquid chromatography (ERLIC), introduced by Alpert (2008), uses electrostatic repulsion as an additional chromatographic stationary-phase property to adjust selectivity in HILIC chromatography. ERLIC makes use of the properties of HILIC and ion-exchange chromatography whereby the selectivity is modulated by changing the pH, organic content of mobile phase, or by applying a salt gradient (Gan, Guo, Zhang, Lim, & Sze, 2008). Anionic phosphopeptides are preferentially retained on weak anion-exchange column at pH $\sim$2 while neutral and protonated peptides are eluted. At low pH, carboxyl groups of Glu and Asp and the C-terminus are largely protonated and peptides with positively charged N-termini are electrostatically repelled from the column. However, negatively charged phosphate groups of phosphopeptides interact electrostatically with WAX and their retention times are increased compared with nonphosphopeptides (Chien, Liu, & Goshe, 2011). In 2011, Chein et al. developed a method utilizing ERLIC, IMAC, and LC–MS/MS to study Marek's disease virus (MDV) infection (Chien et al., 2011). They were able to study the changes occurring in the phosphoproteome by fractionating peptides from chicken embryo fibroblast (CEF) digests using ERLIC, then IMAC enrichment to selectively target phosphorylated peptides prior to LC–MS/MS analysis. Five hundred and eighty-one unique phosphopeptides were identified from the MDV-infected CEF samples.

In addition to chromatography-based methods, tagging phosphate species with certain compounds using a specific chemical derivatization reaction is another strategy for phosphopeptide enrichment (Leitner & Leitner, 2009). Site-specific modification of phosphoseryl and phosphothreonyl residues (Jaffe, Veeranna, & Pant, 1998) using a combination of elimination and Michael addition is a way to introduce phosphosite-specific tagging. The benefits of this method are that one can selectively enrich via different types of tags available, and the tags can be isotope labeled for quantification purposes (Oda, Nagasu, & Chait, 2001) or carry functional groups to increase the ionization efficiency or to facilitate phosphorylation site determination (Arrigoni et al., 2006; Knight et al., 2003). Another labeling technique recently documented by Wijeratne, Manning, Schultz Jel, and Greis (2013) utilizes acetone-based peptide labeling or reductive alkylation by acetone. The research group investigated the regulation of FGF2 and LMW FGF2 in cardiac tissue phosphoproteome in mouse hearts. They found significant phosphorylation changes at 14 different sites on 10 distinct proteins. This study can be used in both exploratory and targeted quantification phosphoproteome studies. Multiple enzymic methods have been developed to enable consecutive digestion of samples with two or more enzymes (Hunt, Buko, Ballard, Shabanowitz, & Giordani, 1981). Filter-aided sample preparation liberates peptides after each digestion step and the remaining sample is then cleaved by the next proteinase. This method identified 40% more proteins and phosphorylation sites when compared to the one-step trypsin digest at low microgram concentration level.

2.3. MS analysis

After a digested peptide is injected into the MS, a precursor ion is fragmented into product ions. The abundance and richness of fragmentation ions are important factors for the effective identification of phosphorylated sites in shotgun proteomics. Technological development in this area has recently been very rapid, and very powerful MS instruments have become available.

2.3.1 Collision-induced dissociation

Collision-induced dissociation (CID) is a standard fragmentation technique in proteomics and phosphoproteomics. In CID, protonated peptides are accelerated by an electrical potential in the vacuum chamber of the mass spectrometer. Then a neutral gas is introduced and bond disruption occurs to generate a series of b and y ions (Schroeder, Shabanowitz, Schwartz, Hunt, & Coon, 2004). Even with low-energy CID (less than 100 eV), the O-phosphate bonds in serine- and threonine-phosphorylated peptides are labile during this process, and neutral loss (elimination of phosphate) of phosphopeptides tends to dominate over dissociation of the main peptide backbone. To prevent or minimize neutral loss, pseudo-MS^3 (Gruhler et al., 2005) or neutral-loss-directed MS^3 has been developed. In these strategies, product ions generated by neutral loss are again fragmented to cleave the peptide backbone. A possible issue in CID is intermolecular phosphate transfer reaction in the ion trap. Aguiar, Haas, Beausoleil, Rush, and Gygi (2010) used synthetic peptides to examine this issue and found that phosphate transfer does occur, but only doubly charged precursors form measurable amounts of transferred fragments. Since only a part of the ion undergoes the reaction, there is no critical effect on the precision of site determination (Dunn, Watson, & Bruening, 2006).

2.3.2 MALDI–TOF MS

After enrichment, phosphopeptides are subsequently (in the case of MALDI–TOF MS approach) spotted onto a plate. Selection of matrix crucially affects phosphopeptide signals. Typically, α-cyano-4-hydroxycinnamic acid (CHCA) and DHB are used in phosphoproteomics, but the use of other matrices has also been reported. 2,4,6-Trihydroxyacetophenone (THAP) with diammonium hydrogen citrate (DAHC) was found to overcome suppression of phosphopeptides by the nonphosphorylated peptides during positive-ion MALDI–TOF MS analysis compared to CHCA (Yang, Wu, & Kobayashi, 2004). The abundances of phosphopeptides in tryptic digests of protein kinase C-treated mouse cardiac troponin I were enhanced more than 10-fold and using THAP/DAHC leading to the identification of a unique phosphorylation site. Kjellström and Jensen (2004) have tested several organic and inorganic acids as matrix additives to enhance signal of phosphopeptides in both positive- and negative-ion modes. After examining phosphoric acid, formic acid, acetic acid, TFA, and heptafluorobutyric acid, they concluded that 1% phosphoric acid added to DHB significantly improved the resolution of MALDI mass spectra of intact proteins. According to Dunn et al. (2006), DHB/phosphoric acid typically results in stronger signal than CHCA. In our hands, DHB/phosphoric acid also yields stronger phosphopeptide signals in MS; however, CHCA is more suitable for MS/MS measurements (A.E. Bond, unpublished data).

Both MALDI and ESI–MS techniques enable the transfer of intact proteins into the gas phase without fragmentation, but that is all these two methods have in common. MALDI produces mostly singly charged ions and is preferably used with a high mass range analyzer such as the TOF mass analyzer, while ESI produces multiply charged ions (making larger proteins more accessible to analysis than MALDI does) and can be used with quadrupoles and ion traps (Fenn, Mann, & Meng, 1989). MALDI is a rapid, solid-phase technique that can be utilized, for example, in high-throughput microarrays or imaging of tissue or detection of individual cells or microorganisms. ESI, in contrast, is a liquid technique compatible with on-line chromatographic techniques and capillary electrophoresis. When coupled with FT mass spectrometers, it is more sensitive and reaches high performance indeed, although the sensitivity of ESI is reduced by the presence of salts, impurities, and organic buffers, which are more easily tolerated by MALDI.

2.3.3 Phosphoproteome sequencing by MS/MS

In general proteomics, sequencing is often performed with triple quadrupole. The first quadrupole selects the ion, which will be fragmented (precursor or parent ion). The second one is filled with an inert gas (usually argon) and the interaction of peptide ions and molecules of collision gas leads to the breakage of a peptide bond. Masses of charged fragments are subsequently measured by the third quadrupole. TOF–TOF mass spectrometer functions similarly; the first TOF analyzer selects precursor ions (which enter the collision cell located between the two) and the spectrum of fragments is measured by the second TOF analyzer.

A combination of a quadrupole and TOF analyzer (so-called QTOF) has become very popular; however, in phosphoproteomics nowadays, LTQ orbitrap and ion trap instruments play a major role due to higher sensitivity in full-scan MS/MS. It has been widely demonstrated that during CID in the positive-ion mode the labile phospho-Ser and phospho-Thr containing phosphopeptides will typically undergo β-elimination of phosphoester bond resulting in the loss of phosphoric acid (H_3PO_4; neutral loss of 98 Da), unlike Tyr residues which are more stable and preferentially lose 80 Da (HPO_3). On the other hand, in the negative-ion mode Ser, Thr, and Tyr-phosphorylated peptides form phosphopeptide-specific marker ions at m/z 79 (PO^{3-}) and m/z 63 (PO^{2-}) (Salih, 2005). Additionally, scanning for characteristic immonium ion by triple quadrupole instruments was suggested when searching for Tyr-phosphorylated peptides (m/z 216.043). Steen, Kuster, and Fernandez (2001) recommended the use of high-resolution MS such as QSTAR Pulsar Q-TOF mass spectrometer (Applied Biosystems, Foster City, CA, USA) to distinguish between a diagnostic immonium ion and those generated by other a, b, and y ions. According to the energy used for fragmentation, we can distinguish two types of dissociation; low-energy CID (<100 eV) and high-energy CID (>1000 eV). Olsen and Mann (2004) reported that efficient ion capture in a linear ion trap leads to MS^3 informative and low-background spectra allowing resolution of ambiguities in identification even at subfemtomole levels of peptide. A method commonly known as data-dependent neutral loss MS^3 analysis is a scan mode that improves acquisition of MS^3 scans only of those compounds that show the desired neutral loss; however, the production of neutral loss ions in MS/MS is almost always associated with partial fragmentation of the precursor ions. These sequence-informative fragment ions produced in MS/MS are not included when neutral loss ions are isolated for MS^3 (Boersema, Mohammed, & Heck, 2009). A new strategy, termed multistage activation, avoids the loss of sequence-informative ions and provides more fragments from the ion produced by the neutral loss. In this approach, the product ions from both the precursor and the neutral loss product activation are simultaneously stored and a composite spectrum that contains fragments from multiple precursors is generated (Schroeder et al., 2004). Savitski et al. reported that different fragmentation techniques differ strongly in their ability to localize phosphorylation sites (Savitski, Lemeer, & Boesche, 2011). At 1% false localization rate, the highest number of correctly assigned phosphopeptides was achieved by higher energy CID in combination with an Orbitrap mass analyzer followed very closely by low-resolution ion trap spectra obtained after ETD. Another option for detection of phosphopeptides is so-called postsource decay (PSD) which also takes advantage of the phosphorylation-specific losses. The molecular ion of interest is selected by ion gate and undergoes "PSD" in the first field-free region of the instrument.

3. APPLICATIONS OF PHOSPHOPROTEOMICS IN BIOMEDICINE

3.1. Applications in cancer research

Phosphoproteomic analyses have been often utilized in the study of dysregulation of proliferative pathways which lead to the onset and progression of various cancers due to the major role that protein phosphorylation usually plays in the overall control of cellular processes. Phosphoprotein analysis has commonly been used in order to attempt to obtain a global phosphoprotein data set with a view to then determining key proteins rather than selectively targeting the analysis of specific proteins in their phosphorylated or non-phosphorylated state. Winck et al. (2014) utilized MS for phosphoprotein analysis to compare the proteins phosphorylated in two strains of epithelia cell, one tumorigenic and the other a "normal," wild-type epithelial cell line as representative cell lines of tissues in which oral cancers commonly develop. Their aim was to determine key protein regulators of the process of cancer development, and to this end, proteins associated with structural and regulatory functions of the nucleus were identified as being differentially phosphorylated between the two cell lines. The analysis therefore allowed for a mechanism of tumorigenesis in these call types to be tentatively identified. Similarly, Xie et al. (2010) studied global phosphoprotein changes in two isogenic cell lines one with metastatic properties and the other without any metastatic characteristics. Using spectra counting as a semi-quantitative methodology, 27 proteins were identified as being differentially phosphorylated between the two cell lines with confirmation of the findings being undertaken by Western blot analysis. Other groups have utilized statistical clustering bioinformatics approaches in order to identify changes in complex phosphoproteomic data from cancer cell analysis such as lung cancer (Grimes, Lee, van der Maaten, & Shannon, 2013). A separate study by Frederick et al. (2011) utilized reverse-phase protein microarrays to study a large number of biopsies from head and neck cancer patients and comparative normal tissue biopsies from the same patients. The microarrays detailed the status of the end points of 60 different kinase pathways and the comparison implicated a number of these as being differentially regulated. Approximately, the same number of kinase pathways were up- and downregulated in the comparison and a specific protein kinase C isoform was identified as having a potential role in these cancers for the first time. A separate study combined data from both model cell lines and benign and cancerous biopsies from gastric cancer and analyzed these at both the phosphoproteome level and at the transcriptomic level with a view of collating the data to obtain a wider view of cellular changes (Guo et al., 2011). Almost 200 phosphoproteins were shown to be overexpressed in the cancer samples and a wide range of kinases and phosphatases could also be studied at the level of protein and transcript. Phosphorylation of p53 was shown to have a pivotal role in the cancer alongside pathways responsible for DNA damage repair. A further study utilized an animal model to provide protein samples that represented the same animal during different stages of skin cancer development (Zanivan et al., 2013). The phosphoprotein data obtained in the study were matched to known metabolic pathway networks in order to identify that the

PAK4–PKC/SRC network exhibited a role in carcinogenesis. The role of protein phosphorylation on the metastatic development of tumors has also been explored and elaborated via phosphoproteomic analysis. Wu et al. (2010) studied a subphosphoprotein profile (encompassing just tyrosine phosphorylated proteins) in lung cancer cell line models which represented differing levels of invasiveness and compared their phosphoproteomes. In doing so, they identified a known network of pathways involved in lung cancer metastasis and also identified seven novel tyrosine kinases which interact to activate these pathways which had not been previously identified and therefore act as new initiators of metastases. Within the area of brain tumors, a novel mutation within the epidermal growth factor receptor was characterized within such disorders and phosphoproteomic analysis was utilized in order to characterize the downstream effects of the mutation and thereby reveal the network involved in the eventual over proliferation caused by the novel mutations within the receptor (Pines, Huang, Zwang, White, & Yarden, 2010). Other groups, when undertaking the study of mechanisms behind tumor development and uncontrolled proliferation, have sought to combine data sets from phosphoprotein analysis with data sets from techniques studying the other aspects discussed in the introduction (transcriptomics, metabolomics, etc.). A combination of epigenetic, transcriptomic, and phosphoproteomic data was collected in respect to a cell line model of glioblastoma multiform which exhibited a mutation in the epidermal growth factor receptor in order to identify cellular networks of interactions and the effect of different compounds addition to the cell line on these network systems (Huang et al., 2013). While this study sought to combine proteomic data with data sets representing biological events earlier in the protein expression pathway (the transcription of DNA and its regulation by means of epigenetic changes to cellular structures), a separate study combined phosphoprotein expression data with metabolomic data sets (representing the downstream net effects of the differential protein phosphorylation itself). McDonnell et al. (2013) investigated large cell lymphomas which were known to produce the protein, active tyrosine kinase nucleophosmin–anaplastic lymphoma kinase. The phosphoprotein data obtained allowed for the confirmation of the phosphorylation of the anaplastic lymphoma kinase (ALK) protein as an important event in the development of the disorder, while the metabolomic data (and data obtained via the chemical induction of the process or similar analysis utilizing a cell line with a mutation in the target protein) identified increased lactate production via an elevation in the rate of aerobic respiration as a net consequence of the altered phosphoproteomic, resulting in increased cell biomass and therefore proliferation.

As well as using cancer cell lines to investigate and propose the role of proteins in cancer development, phosphoprotein analysis can also be utilized to further validate such hypotheses once proposed. In the case of lung cancer development, the protein ephrin and associated signaling pathway has been implicated in tumor development. Ståhl et al. (2011) demonstrated that downregulation of the protein in a lung cancer cell line model resulted in a reduction in the rate of cellular proliferation and therefore sought to utilize phosphoproteomic analysis to assist in determining the proteins and pathways responsible. A number of proteins phosphorylation status were shown to be dependent upon whether ephrin was expressed or depressed and erythropoietin-producing hepatocellular receptor tyrosine kinase class A2 phosphorylation was shown to required for the tumor survival and therefore further proliferation. A separate study by Iwai et al. (2013) studied collagen and downstream pathways mediated via collagen signaling as being involved in lung cancer progression via specific collagen receptors (DDRs). The study mapped theoretical pathways and protein phosphorylation targets of the pathway using *in vitro* kinase data and cell lines with DDRs mutated were further studied and one of these suggested targets shown to be correlated to tumor development dependent upon its phosphorylation status (Src homology 2 domain-containing protein tyrosine phosphatase 2).

The protein transforming growth factor-β (TGF-β) has been suggested as having a role in cancer progression in colon cancer and is usually thought to act via the protein Smad4; however, it has been hypothesized that alternative pathways controlled by TGF-β may also be important in cancer development. Ali and Molloy (2011) therefore utilized phosphoproteomic analysis in order to identify novel pathways and proteins affected by the TGF-β pathway in colonic cancer cell lines. A number of proteins including hepatoma-derived growth factor and cell division kinases were identified as targets for phosphorylation via the pathway, thereby suggesting further roles for the pathway.

As well as using phosphoproteomic analysis to study possible mechanisms of disorder development, cell lines and biopsies which represent specific cancers have also been analyzed in order to collate and provide a record of the proteins modified in these cancers. Yu et al. (2011) undertook such an analysis for the type II human lung cancer, A549, cataloging 373 phosphorylation sites on a total of 181 proteins within the cells, many of which were reported as being phosphorylated in the cancer for the first time in the study.

A separate study undertook a similar analysis in LNCaP human prostate cancer cells, and of the 116 phosphoproteins identified in the multiple analyses, 56 were newly identified as phosphorylation targets in such cancers (Myung & Sadar, 2012). A separate study sought to characterize the phosphoproteome of gastric cancer secretions (the secretome) (Yan et al., 2011), whilst other similar studies have sought to use archived clinical tissues for the study of phosphoproteins present, with tissues from a tissue repository being utilized for the study of prostate cancer protein phosphorylation (Chen, Fang, Giorgianni, Gingrich, & Beranova-Giorgianni, 2011). The study demonstrated that phosphoprotein data could still be obtained for known cancer-relevant proteins.

A further impact of phosphoproteomic analysis in cancer research is in the study of mechanisms that cause tumors to become resistant to chemotherapeutic drug regimes with a view to preventing or reducing the capacity of the tumor to survive the treatments given. A study concerning the potential role of the protein tissue inhibitor of metalloproteinase 1 (TIMP-1) in the development of resistance of breast cancers to drug treatment utilized cell lines in which the protein was up- and downregulated (Hekmat et al., 2013). A group of enzymes which are common targets of chemotherapeutic drugs, the topoisomerases, were shown to be overexpressed and/or overphosphorylated in the cells with expressed TIMP-1, and therefore drug resistance and phosphorylation of this drug target were hypothesized as conferring resistance. Another drug target for breast cancer treatment is the tyrosine kinases which can be inhibited by drugs such as lapatinib where the cells are shown to overexpress the gene HER-2. Rexer et al. (2011) produced cell lines representing breast cancer cells whose proliferation was not inhibited by the addition of lapatinib to the growth media as a model of cancer resistance to the drug and studied protein phosphorylation between cancer cells with and without this resistance. In the analysis, overphosphorylation of a group of enzymes, the Src family kinases, was exhibited in the resistant cell line and addition of Src family kinases alongside lapatinib caused the cells to lose their resistance. Therefore, such analyses can also provide possible mechanisms of overcoming resistance as well as detailing pathways involved. A separate study of cancer cell resistance to lapatinib utilized breast cell cancers exhibiting different levels of susceptibility and resistance—this was controlled by inducing overexpression of HER2 in some cells using a retrovirus vector (Vazquez-Martin, Oliveras-Ferraros, Colomer, Brunet, & Menendez, 2008). A human phospho-MAPK array proteome profiler was utilized to study cellular changes in the different cells, and a specific serine/threonine kinase was highlighted as an important mediator within the cellular processes involved. A further study focusing upon tamoxifen resistance in breast cancer combined phosphoprotein and transcriptomic data in cancer cell lines with and without resistance and then compared the data obtained to that determined from clinical biopsy data (Oyama et al., 2011). Against expectations, resistant cell lines presented a reduced protein phosphorylation status overall compared to wild-type cell lines. One particular protein, glycogen synthase kinase 3β was more often phosphorylated in the wild-type cell, and network analysis and transcriptomic data suggested that normal phosphorylation of the protein has an inhibitory effect on GSK3β at serine 9 and removal of this inhibition in drug-resistant cell lines allows for the increased activity of cAMP-responsive element-binding protein and AP-1 transcription factors causing resistance. Docetaxel resistance has also been investigated utilizing phosphoproteomic analysis as resistance can occur in 50% of prostate cancers and resistance is a major concern during treatment (Lee et al., 2014). The phosphoproteomic investigation of docetaxel-resistant prostate cancer cell lines identified specific phosphorylation sites as being modified on the protein focal adhesion kinase, and bioinformatic analysis was utilized in order to identify metabolic and cellular pathways that would be affected as a result. A similar global phosphoproteomic analysis was also applied to acute myeloid leukemia cell lines with differing degrees of resistance to kinase inhibitors in order to understand the mechanism employed by drug-resistant cancers of this type (Alcolea, Casado, Rodríguez-Prados, Vanhaesebroeck, & Cutillas, 2012).

As well as drug resistance, some tumors demonstrate radiotherapy resistance; a similar phosphoproteomic analysis was utilized to study the phosphorylated proteins affected in mammalian epithelial cells when a candidate protein required for resistance, TGF-β, is activated. The network analysis identified 14-3-3σ as a target of the growth factor pathway and phosphorylation analysis identified two novel phosphorylation sites at serines in positions 69 and 74 within the protein that were dependent on TGF-β for their phosphorylation. The network analysis suggested that such an activation would have multiple downstream effects, a complex formation required for interaction with p53 and that this may assist in the resistance to radiotreatment (Zakharchenko, Cojoc, Dubrovska, & Souchelnytskyi, 2013).

A further area of exploitation of the global analysis of protein phosphorylation events and selected protein phosphorylation analysis is in the identification of novel potential therapeutic targets in order to treat tumors or reduce proliferation rates within cancer cells (Yu, Issaq, & Veenstra, 2007). Bhola et al. (2011) studied the phosphoproteome in head and neck cancer models which were undergoing proliferation. The proliferative rate was activated by using gene silencing technology to repress the expression of the epidermal growth factor receptor (whose express is associated with reduced tumor growth) while G-protein coupled receptor activation was ensured by adding an agonist for this receptor type (as these are overexpressed in such tumors). The study of the phosphoproteomic changes that were identified as a result of such intervention allowed for the selection of the protein p70S6K as being more phosphorylated in tumors in which proliferation was activated in this manner (increased sixfold). Therefore, it was concluded in the study that this protein represents a potential downstream target of the reduced growth factor receptor/elevated G-protein coupled receptor activation of cellular proliferation processes. A recent study considered castration-resistant metastatic prostate cancers taken via biopsy compared to primary prostate tumors which were biopsied prior to any therapy (Drake et al., 2013). The study profiled phosphotyrosine peptides from the different samples and highlighted a number of phosphoproteins which correlated well with the resistance within tumors which were more aggressive, including ALK, and MAPK1/3 with a view to developing inhibitors selective for these particular kinases in such cancers. Further studies have considered myeloma cancers as the therapeutic target and studied 25 different cell strains via phosphoproteomic approaches in order to identify tyrosine kinase receptors specifically phosphorylated (and therefore activated) in such proliferative cell lines as novel targets (Tworkoski et al., 2011). Ovarian cancers have been studied via similar approaches, with 69 primary cancer cell cultures being utilized and compared (Ren et al., 2012). The study identified overphosphorylation of the protein ALK in 2–4% of cases and therefore suggested this protein phosphorylation as a novel target in a number of ovarian cancer cases.

As well as new therapeutic targets, phosphoproteomic analysis has been applied to investigate resistance mechanisms (as discussed previously) and identify targets that would reduce resistance in tumors, thereby enhancing the efficacy of the existing treatment options available to clinicians. An example of this comes from the field of gastrointestinal cancers and their treatment with the drug, imatinib. Takahashi et al. (2013) analyzed the phosphoproteome of cancers after treatment in order to identify focal adhesion kinase and associated protein phosphorylation as a key event in the reduction of the efficacy of the drug on the tumor development (confirmed using

Western blotting approaches). Addition of an inhibitor of the kinase enzyme identified in the study had a dramatic effect on the IC_{50} of imatinib and overcame the identified resistance, and therefore, such analyses can have the effect of increasing the potency of current therapeutic regimes as well as reducing the required dose to be administered.

One further area of increasing interest in the treatment of various tumors is the application of phosphoproteomic analysis to allow for the study of the treatment efficacy in different patients, allowing for a stratified approach to the treatment of patients. More specifically, phosphoproteomic analysis has been applied to study key phosphorylation changes that can be used as biomarkers or biological indicators of the predicted response to a particular anticancer agent. A bank of cancer cell lines (NCI-60) was utilized alongside a therapeutic whose mechanism involves the inhibition of phosphoinositide 3 kinase in order to determine any key phosphorylation events that could predict drug efficacy in such patients (Kwei, Baker, & Pelham, 2012). The degree of phosphorylation of two key proteins within the samples was thereby shown to be directly correlated to the *in vitro* response of the cancer cell lines to treatment with the drug and therefore acts as biomarkers that could be studied in biopsies in order to better manage patient treatment. A similar approach has also been applied to nonsmall cell cancer cell lines treated with dasatinib (a further protein kinase inhibitor), identifying 58 protein phosphorylation events that could be used to predict cancer cell vulnerability to the treatment (Klammer et al., 2012). Of the 58 signatures identified, a panel of 12 of these was sufficient in order to accurately identify cell lines which would respond well to the kinase inhibitor, and interestingly, 4 of these particular events were presented on the same protein, integrin β4. The same approach applied to hematological cancer cell lines including acute myeloid leukemia, lymphoma, and multiple myeloma allowed for the quantitation of more than 2000 protein phosphorylation events within the cell lines (Casado et al., 2013). The profiling was analyzed and regression analysis based modeling of the changes allowed for the optimal combination of signatures which differentiate cells based upon their sensitivity to kinase inhibitors without necessarily identifying the proteins involved. As well as finding a role in stratified medicine, the study of phosphoprotein and phosphopeptide signatures has also found a role in diagnosis and prognosis of patients. One example of this approach is represented by the work of Takano et al. (2010) who studied the phosphoprotein profile of serum derived from pancreatic cancer patients, control subjects, and patients suffering from nonmalignant pancreatitis. The serum phosphoprotein biomarker elucidated during the analysis provided a positive identification rate of 82% which was far superior to the 53% identification rate exhibited by the existing biomarker utilized clinically, and therefore, the potential of global phosphoproteomic analysis in identifying novel cancer serum biomarkers is of continued interest in many research groups. As discussed in this section of the review, cancer has been a major focus of phosphoproteomic analysis to date due to the particular role of kinase enzymes in the control of cellular proliferation rates and the mechanisms behind dysregulation of these in tumor cells. The phosphoproteomic data are particularly powerful when combined with other data sets both those upstream of the phosphorylation event (transcriptomics, etc.) and downstream effects of differential protein phosphorylation events in the cell (metabolomic analysis).

3.2. Applications in stem cell research

A recent application of phosphoproteomic analysis is in the field of regenerative medicine, in particular the analysis has been utilized in order to study and further understand changes at the cellular level that are involved in stem cell differentiation processes (Tobe et al., 2012).

The application of phosphoproteomics to the differentiation of embryonic stem cells was undertaken by Brill et al. (2009). Pluripotent cells were utilized in the study and analyzed before and after differentiation and a clear distinction in the phosphoproteome determined between the two cell types with hundreds of differentially phosphorylated proteins identified. The undifferentiated stem cells exhibited a larger number of phosphorylated tyrosine kinase enzymes, and these findings were further validated by undertaking a biochemical assay of the specific enzyme activities implicated in the two cell lines to show an reduction in activity during differentiation. A separate study of the phosphoproteomic changes taking place examined embryonic stem cell differentiation after initiation of the process using a diacylglycerol analog (Rigbolt et al., 2011). Of the over 6000 proteins identified, half exhibited a difference during the first 24 h after initiation of the differentiation process, suggesting a significant biochemical change in the cell's metabolism in response to the initiation of the process. As well as the expected kinase targets that would be expected to undergo differential phosphorylation during differentiation, DNA methyltransferases were also detected which were implicated in interacting with polymerase-associated factor 1 which then further interacts with specific transcription factors to control the differentiation process. This methyltransferase was therefore defined as a target protein that may be inhibited or activated as a target for interventions designed to control the differentiation process *in vitro*. The differentiation of embryonic stem cells into neuronal cell lines has been investigated at the phosphoprotein level via 2D SDS PAGE analysis of the phosphoproteome of the cells before and after differentiation (Kim et al., 2011). During differentiation, a number of specific proteins involved in the functioning of the differentiated neuronal cell line were phosphorylated including voltage-gated ion channels, vimentin, and a number of heterogeneous nuclear ribonucleoproteins.

Lo et al. (2012) utilized LC/MS analysis for the purpose of phosphoproteomic analysis of mesenchymal stem cells which display the ability to differentiate into osteoblast cells via osteogenic differentiation. The therapeutic potential of such stem cells in the future treatment of conditions such as osteoporosis could be further developed with increased understanding of differentiation into such cell lines. Three time points during differentiation were studied, and an initial hypomodification status of the proteome with respect to phosphorylation at the early time point monitored was reported and related to proteins involved in proliferative processes. During further differentiation, a number of ion channel proteins and transcription factors were shown to become differentially phosphorylated and it was therefore suggested that these proteins may be key factors in the conversion to osteoblast cells. The study of the phosphoproteome of human CD34(+) hematopoietic progenitor cells has also been undertaken in isolation with a view to understanding the underlying proteomic pathways activated in such cells. Of the proteome studied, more than 3000 proteins were shown to be phosphorylated at one amino acid residue or more, and bioinformatic analysis was then applied to the data set to assist in identifying the pathways found to be active in the specific stem cell type (Guo et al., 2013).

The role of phosphorylation of specific known proteins with a suggested function within the stem cell development process as well as the effect of a suspected protein involved in differentiation initiation has also utilized phosphoproteomics as a mechanism to further add to researcher's understanding of the biochemical mechanisms behind the process. Nestin is a filament protein which has been implicated as a neural stem cell/progenitor biomarker in past studies and a recent study sought to study the phosphorylation of this protein in the central nervous system and in the vascular system in neural cells and in bone-marrow-derived progenitor cells, respectively. The nestin determined from the neuronal cell line was shown to be multiply phosphorylated while the vascular isoform did not present any phosphorylation sites being actively modified, suggesting a different function of this protein within the two different cell types based upon their eventual function (Namiki, Suzuki, Masuda, Ishihama, & Okano, 2012). Fibroblast growth factor 2 (FGF-2) acts as a growth factor in human embryonic stem cells in order to allow for their efficient and effective expansion and production. The addition of FGF-2 therefore allows for the stimulation of the stem cells, and this process was monitored by studying phosphoprotein changes in the cells, after which 40% of the detected phosphoproteome demonstrated a significant change in their modification status. The identified proteins included proteins involved in the self-renewal processes underlying cell proliferation processes and proteins whose expression is regulated by transcription factors previously implicated in pluripotency (Zoumaro-Djayoon et al., 2011). Finally, the role of the phosphoproteome in the clinical disorder, hemoglobin E/β thalassemia, in which increased apoptosis of erythrocytes leads to the symptoms of the disorder was undertaken by comparing stem cells that differentiate to provide these cells on a continual basis in patients with the disorder compared to control hematopoietic stem cells (Ponnikorn et al., 2011). The research found that 229 phosphoproteins were found to be differently presented in the normal and ineffective stem cell lines of which many were of importance in such cells including cytochrome *c* and caspase 6—suggesting its direct role in the observed apoptosis symptoms. Within the discipline of regenerative medicine and further understanding of stem cell differentiation and production, phosphoproteomics is therefore beginning to show its value as a targeted method which can be utilized to obtain information regarding the function of this posttranslational modification on a global scale. Further work in this field is envisaged to allow for stem cells to be more efficiently produced in the laboratory setting and also to provide a more refined methodology for the control of the differentiation process (in relation to controlling when this process is initiated and also the nature of the resulting differentiated cell).

3.3. Applications in cardiac research

Phosphoproteomics has also found a role in the investigation of cardiovascular illnesses, mainly in the study of the phosphoproteome of cardiac muscle in normal conditions and during or after stress or myocardial infarction. A recent review highlighted the application of phosphoproteomics in the global and high-throughput identification of proteins phosphorylated in the different subcellular compartments of cardiac tissues, outlining the potential biological insight that may be gained by such analysis of the heart and its proper functioning (Edwards, Cordwell, & White, 2011). At around the same time, Deng et al. (2011) reported on the analysis of the phosphoproteome of murine cardiac mitochondria. Given the changing energy requirements of myocardial tissue, the mitochondria and the mechanisms that underlie its ability to adapt to energy requirements brought about by differential muscle contraction needs are a major factor in the health of the heart. LC/MS with a combined fragmentation analysis using both CID and ETD allowed for a more comprehensive study of the phosphopeptides present after enzymatic proteolysis with some phosphopeptides being solely identified by one or other of the fragmentation approaches. As would be expected, key components of the electron transport chain and the tricarboxylic acid cycle were shown to be important regulatory sites within the mitochondria, and kinases associated with these steps were also shown to be regulated by phosphorylation status. Other kinases, such as Scr, which had previously not been associated with mitochondrial function, however, were also should be present and represent phosphorylation targets within the subcellular compartment, suggesting further novel mechanisms of regulation that are as yet not fully understood. A further study sought to study the phosphorylation status of a single protein present in the heart, cardiac myosin-binding protein C, as it has been suggested that dephosphorylation of this particular protein is linked directly with contractile dysfunction in patients (Kooij, Holewinski, Murphy, & Van Eyk, 2013). The study identified the N terminal section of the protein as being the most phosphorylated section of the protein, containing the majority of the 17 phosphorylation sites identified; however, a specific dephosphorylation of a serine amino acid residue at position 284 was shown to be the most commonly dephosphorylated site in patients with symptoms of heart failure. In relation to the study of heart defects or pressure-derived damage via phosphoproteomic analysis, the effect of pressure overload has been studied in murine heart tissue using transverse aortic banding in order to bring about increased aortic pressure and sampling myocardial tissue after 10, 30, and 60 min and also at a 2-week time point (Chang et al., 2013). In total, 360 phosphorylation sites were shown to be differentially modified as a result of the banding with some being exhibited during acute pressure overload while others were linked to chronic increases in pressure. Dynamin-related protein 1 (DRP-1) was demonstrated to be of interest with the banding bringing about phosphorylation-dependent translocation of the DRP-1 to the mitochondria where it plays a role in mitochondrial fission. Furthermore, inhibition of this protein was shown to reduce the hypertrophic response undergone by the myocardial tissue when presented with increased pressure, thereby implicating the protein as having a role in the disruptive consequences of pressure overload on the heart. As well as focusing on the myocardial tissue, phosphoproteomic analyses have also widened their target in cardiovascular studies to incorporate the prolonged impact of subarachnoid hemorrhage on the longer-term health and functioning of cerebral artery systems. Parker, Larsen, Edvinsson, and Povlsen (2013) demonstrated longer-term changes in phosphorylation status in such arteries after hemorrhage and identified key proteins whose phosphorylation appeared to be linked to the longer-term damage and ischemia. Inhibition of one of the kinases identified was shown to lead to improved cerebrovascular outcomes and suggest a novel target in treating longer-term deficiencies brought about by increased pressure within the cerebral artery system. A further study utilized phosphoproteomic analysis to investigate the role of oxidized phospholipids in the development of atherosclerotic plaques leading to hypertension and eventual myocardial infarction. The study identified proteins which were specifically phosphorylated in response to the oxidized phospholipids, suggesting a role in cellular signaling as well as plaque formation due to deposition of lipid during atherosclerosis (Zimman et al., 2010). As well as cardiac damage brought about by tissue damage and hypertension, the effect of pharmaceutical interventions for other disorders on the heart has also been investigated from a phosphoproteomic perspective. The chemotherapeutic family of drugs, the anthracyclines—such as doxorubicin, have limited use due to their effect on the heart as a side effect of the therapy, and recently, phosphoproteomics has been applied in order to further determine the mechanism behind these detrimental cardiac specific side effects. Rat heart tissue utilized as a model for the heart was infused with doxorubicin at two concentrations—the usual clinically relevant concentration and an elevated concentration, five times greater than the clinical dose. The phosphoproteome was then studied using 2D SDS PAGE with phosphoprotein specific staining for protein visualization and MS protein identification (Gratia et al., 2012). Differences in the phosphoproteome, identified by MS, were verified by Western blotting and proteins associated with the functions of the mitochondria constituted 40% of the differences suggesting this as the subcellular site of action of the drug, and the implicated functions impaired by the therapy were suggested as being energy balance and myofibrillar organization processes. Phosphorylation of specific proteins has also been suggested as representing potential biomarkers of myocardial infarction and dysfunction. Dubois et al. (2011) and Dubois-Deruy et al. (2013) investigated the efficacy of the phosphorylation status of the circulating protein, troponin, at a serine at position 208 in the protein's amino acid sequence compared to existing biomarkers of cardiac dysfunction in rat models of the disorder. The level of phosphorylation of troponin was shown to be reduced in rats with induced myocardial infarction compared to rats with no dysfunction and this reduced phosphorylation status was reversed in treated rats upon treatment with a heart rate reducing drug, alleviating the symptoms. The phosphorylation of this circulating protein therefore has the ability to assist in the diagnosis of poor cardiac function and also allows for the study of the efficacy of treatment regimes. As well as studying the role of phosphoproteins in disorder progression, the same type of analysis has also been applied to the study of mechanisms behind protective measures that reduce the risk or impact of myocardial damage. Isoflurane was shown to provide a protective function to heart tissue in a rat model and protein extraction, 2D SDS PAGE, and immunoblotting with a phosphor—serine/threonine/tyrosine stain allowed for the comparative phosphoproteomes of the respective mitochondria to be studied. Eleven phosphoproteins in total were shown to provide differences in their phosphorylation status after treatment with isoflurane. Further mass spectrometric analysis identified a novel phosphorylation site in an adenosine nucleotide cotranslocator and mutation of this amino acid site in yeast resulted in impaired growth (Feng et al., 2008). A further mechanism thought to improve the protection of myocardial tissue is the inhibition of protein phosphatase 1 (Nicolaou et al., 2009). The group produced a transgenic mouse model with an inducible inhibitor of the phosphatase enzyme and demonstrated that the increased expression of the inhibitor allowed for improved contractile function in the mice. Phosphoproteomic analysis was also performed and demonstrated an increased phosphorylation of phospholamban which allowed for improved calcium transport, suggesting a mechanism by which the change in phosphorylation might bring about the observed protective effects.

3.4. Applications in immunity research

Protein phosphorylation analysis has also been applied to the study of immune responses in a few studies. Such an analysis was utilized in order to study the signal transduction cascade activated by the binding of thymic stromal lymphopoietin to its cytokine receptor, as the ligand's over-expression has been linked to the development of asthma in patients. A quantitative phosphoproteomic approach was used in order to study the proteins involved in bringing about the cytokines cellular effects (Zhong et al., 2012). The study identified 226 proteins whose phosphorylation status was altered in response to the addition of the cytokine including specific kinases involved in the amplification of the ligand-binding event. An earlier study utilized an antiserum to probe the phosphoproteome of activated T cells and demonstrated phosphorylation of pro-interleukin 16 on a specific serine residue (position 144) allowing the determination of Erk 1/2 kinase as the modifying enzyme involved, thereby identifying the kinase as a novel target for therapeutic regimes that seek to limit the interleukin production and release (Laurence, Astoul, Hanrahan, Totty, & Cantrell, 2004). While this study focused on the specific protein identified as being phosphorylated, immunoprecipitation and phosphoprotein analysis in a more recent study identified 2814 phosphopeptides after T cell activation with roles in cytoskeletal restructuring and GTPase activation, suggested as having a role in the formation of the immune synapse (Ruperez, Gago-Martinez, Burlingame, & Oses-Prieto, 2012). The approach of studying protein phosphorylation has also been of benefit when investigating antibody-mediated rejection of organs during organ transplant treatments. Jindra et al. (2008) developed a mouse model for heart transplant rejection and studied protein phosphorylation during the rejection process, identifying the class-I signaling pathway in the *in vivo* experiment and the relationship of this pathway with the pathway which is the therapeutic target of the drug, rapamycin. The improved knowledge as to the role of protein phosphorylation events in mediating immune response has also led to the development of phosphopeptides as drugs which act by blocking the normal phosphorylation process. For example, one such drug was developed which acts by binding the SH2 domain of STAT3 to prevent its role in the normal response (McMurray, Mandal, Liao, Klostergaard, & Robertson, 2012) while a separate study developed a cell-permeable phosphopeptide drug which bound the SH2 domain within an inducible T cell kinase whose overactivation can lead to severe lung inflammation (Guimond et al., 2013).

4. DISCUSSION

Following on from advances in the application of modern proteomic techniques to categorize proteomes and the various changes in expression levels of proteins, the ability to accurately characterize the protein posttranslational modification status, especially in terms of protein phosphorylation, has also improved both in relation to the robustness of the data obtained and the throughput of the analytical protocols undertaken. MS remains a key analytical device in such global phosphoproteomic analysis, and advances in the development of mass spectrometers have led to improvements in the rate of phosphoprotein discovery. The applications of the ability to monitor protein phosphorylation status on a global proteomic scale have been applied to many distinct and varied areas of biomedicine. While cancer phosphoproteomics has been very well researched (due to the role of signal transduction cascades and kinase activity in the unregulated cell proliferation process itself), further areas of study within the biomedical field have utilized the same technologies (as detailed in the select few highlighted areas covered within this review). Furthermore, the global importance of protein phosphorylation as a mediator and regulator of cellular metabolism and adaptation means that the areas within biomedicine (and beyond) which will benefit from such analytical techniques are ever expanding.

REFERENCES

Aguiar, M., Haas, W., Beausoleil, S. A., Rush, J., & Gygi, S. P. (2010). Gas-phase rearrangements do not affect site localization reliability in phosphoproteomics data sets. *Journal of Proteome Research*, *9*, 3103–3107.

Alcolea, M. P., Casado, P., Rodríguez-Prados, J. C., Vanhaesebroeck, B., & Cutillas, P. R. (2012). Phosphoproteomic analysis of leukemia cells under basal and drug-treated conditions identifies markers of kinase pathway activation and mechanisms of resistance. *Molecular & Cellular Proteomics*, *11*(8), 453–466.

Ali, N. A., & Molloy, M. P. (2011). Quantitative phosphoproteomics of transforming growth factor-β signaling in colon cancer cells. *Proteomics*, *11*(16), 3390–3401.

Alpert, A. J. (2008). Electrostatic repulsion hydrophilic interaction chromatography for isocratic separation of charged solutes and selective isolation of phosphopeptides. *Analytical Chemistry*, *80*, 62–76.

Andersson, L., & Porath, J. (1986). Isolation of phosphoproteins by immobilized metal ($Fe3+$) affinity chromatography. *Analytical Biochemistry*, *154*, 250–254.

Arrigoni, G., Resjo, S., Levander, F., Nilsson, R., Degerman, E., Quadroni, M., et al. (2006). Chemical derivatization of phosphoserine and phosphothreonine containing peptides to increase sensitivity for MALDI-based analysis and for selectivity of MS/MS analysis. *Proteomics*, *6*, 757–766.

Beausoleil, S. A., Jedrychowski, M., Schwartz, D., Elias, J. E., Villén, J., Li, J., et al. (2004). Large-scale characterization of HeLa cell nuclear *phosphoproteins. Proceedings of the National Academy of Sciences of the United States of America*, *101*, 12130–12135.

Bhola, N. E., Thomas, S. M., Freilino, M., Joyce, S., Sahu, A., Maxwell, J., et al. (2011). Targeting GPCR-mediated p70S6K activity may improve head and neck cancer response to cetuximab. *Clinical Cancer Research*, *17*(15), 4996–5004.

Bodenmiller, B., Campbell, D., Gerrits, B., Lam, H., Jovanovic, M., & Picotti, P. (2008). PhosphoPep—A database of protein phosphorylation sites in model organisms. *Nature Biotechnology*, *26*, 1339–1340.

Boersema, P. J., Mohammed, S., & Heck, A. J. (2009). Phosphopeptide fragmentation and analysis by mass spectrometry. *Journal of Mass Spectrometry*, *44*, 861–878.

Bond, A. E., Dudley, E., Tuytten, R., Lemière, F., Smith, C. J., Esmans, E. L., et al. (2007). Mass spectrometric identification of Rab23 phosphorylation as a response to challenge by cytidine 3',5'-cyclic monophosphate in mouse brain. *Rapid Communications in Mass Spectrometry*, *21*(16), 2685–2692.

Bond, A. E., Row, P. E., & Dudley, E. (2011). Post-translation modification of proteins; methodologies and applications in plant sciences. *Phytochemistry*, *72*(10), 975–996.

Brill, L. M., Xiong, W., Lee, K. B., Ficarro, S. B., Crain, A., Xu, Y., et al. (2009). Phosphoproteomic analysis of human embryonic stem cells. *Cell Stem Cell*, *5*(2), 204–213.

Casado, P., Alcolea, M. P., Iorio, F., Rodríguez-Prados, J. C., Vanhaesebroeck, B., Saez-Rodriguez, J., et al. (2013). Phosphoproteomics data classify hematological cancer cell lines according to tumor type and sensitivity to kinase inhibitors. *Genome Biology*, *14*(4), R37.

Chang, Y. W., Chang, Y. T., Wang, Q., Lin, J. J., Chen, Y. J., & Chen, C. C. (2013). Quantitative phosphoproteomic study of pressure-overloaded mouse heart reveals dynamin-related protein 1 as a modulator of cardiac hypertrophy. *Molecular & Cellular Proteomics*, *12*(11), 3094–3107.

Chen, L., Fang, B., Giorgianni, F., Gingrich, J. R., & Beranova-Giorgianni, S. (2011). Investigation of phosphoprotein signatures of archived prostate cancer tissue specimens via proteomic analysis. *Electrophoresis*, *32*(15), 1984–1991.

Chien, K. Y., Liu, H. C., & Goshe, M. B. (2011). Development and application of a phosphoproteomic method using electrostatic repulsion-hydrophilic interaction chromatography (ERLIC), IMAC, and LC-MS/MS analysis to study Marek's disease virus infection. *Journal of Proteome Research*, *10*, 4041–4053.

Collins, M. O., Yu, L., Coba, M. P., Husi, H., Campuzano, L., Blackstock, W. P., et al. (2005). Robust enrichment of phosphorylated species in complex mixtures by sequential protein and peptide metal-affinity chromatography and analysis by tandem mass spectrometry. *The Journal of Biological Chemistry*, *280*, 5972–5982.

Connor, P. A., & McQuillan, A. J. (1999). Phosphate adsorption onto TiO2 from aqueous solutions: An in situ internal reflection infrared spectroscopic study. *Langmuir*, *15*, 2916–2921.

Dai, J., Wang, L. S., Wu, Y. B., Sheng, Q. H., Wu, J. R., & Shieh, C. H. (2009). Fully automatic separation and identification of phosphopeptides by continuous pH-gradient anion exchange online coupled with reversed-phase liquid chromatography mass spectrometry. *Journal of Proteome Research*, *8*, 133–141.

Deng, N., Zhang, J., Zong, C., Wang, Y., Lu, H., Yang, P., et al. (2011). Phosphoproteome analysis reveals regulatory sites in major pathways of cardiac mitochondria. *Molecular & Cellular Proteomics*, *10*(2), M110.000117.

Dephoure, N., & Gygi, S. P. (2011). A solid phase extraction-based platform for rapid phosphoproteomic analysis. *Methods*, *54*, 379–386.

Drake, J. M., Graham, N. A., Lee, J. K., Stoyanova, T., Faltermeier, C. M., Sud, S., et al. (2013). Metastatic castration-resistant prostate cancer reveals intrapatient similarity and interpatient heterogeneity of therapeutic kinase targets. *Proceedings of the National Academy of Sciences of the United States of America*, *110*, E4762–E4769.

Dubois, E., Richard, V., Mulder, P., Lamblin, N., Drobecq, H., Henry, J. P., et al. (2011). Decreased serine207 phosphorylation of troponin T as a biomarker for left ventricular remodelling after myocardial infarction. *European Heart Journal*, *32*(1), 115–123.

Dubois-Deruy, E., Belliard, A., Mulder, P., Chwastyniak, M., Beseme, O., Henry, J. P., et al. (2013). Circulating plasma serine208-phosphorylated troponin T levels are indicator of cardiac dysfunction. *Journal of Cellular and Molecular Medicine*, *17*, 1335–1344. http://dx.doi.org/10.1111/jcmm.12112.

Dunn, J. D., Watson, J. T., & Bruening, M. L. (2006). Techniques for phosphopeptide enrichment prior to analysis by mass spectrometry. *Analytical Chemistry*, *78*, 1574–1580.

Edwards, A. V., Cordwell, S. J., & White, M. Y. (2011). Phosphoproteomic profiling of the myocyte. *Circulation. Cardiovascular Genetics*, *4*(5), 575.

Feng, H., Ye, M., Zhou, H., Jiang, X., Zou, H., & Gong, B. (2007). Immobilized zirconium ion affinity chromatography for specific enrichment of phosphopeptides in phosphoproteome analysis. *Molecular Cellular Proteomics*, *6*, 1656–1665.

Feng, J., Zhu, M., Schaub, M. C., Gehrig, P., Roschitzki, B., Lucchinetti, E., et al. (2008). Phosphoproteome analysis of isoflurane-protected heart mitochondria: Phosphorylation of adenine nucleotide translocator-1 on Tyr194 regulates mitochondrial function. *Cardiovascular Research*, *80*(1), 20–29.

Fenn, J. B., Mann, M., & Meng, C. K. (1989). Electrospray ionization for mass spectrometry of large biomolecules. *Science*, *246*, 64–71.

Ficarro, S. B., McCleland, M. L., Stukenberg, P. T., Burke, D. J., Ross, M. M., & Shabanowitz, J. (2002). Phosphoproteome analysis by mass spectrometry and its application to Saccharomyces cerevisiae. *Nature Biotechnology*, *20*, 301–305.

Ficarro, S. B., Parikh, J. R., Blank, N. C., & Marto, J. A. (2008). Niobium(V) oxide (Nb2O5): Application to phosphoproteomics. *Analytical Chemistry*, *80*, 4606–4613.

Ficarro, S. B., Zhang, Y., Carrasco-Alfonso, M. J., Garg, B., Adelmant, G., & Webber, J. T. (2011). Online nanoflow multidimensional fractionation for high efficiency phosphopeptide analysis. *Molecular and Cellular Proteomics*, *10*, O111.011064.

Frackelton, A. R., Jr., Ross, A. H., & Eisen, H. N. (1983). Characterization and use of monoclonal antibodies for isolation of phosphotyrosyl proteins from retrovirus-transformed cells and growth factor-stimulated cells. *Molecular and Cellular Biology*, *3*, 1343–1352.

Frederick, M. J., VanMeter, A. J., Gadhikar, M. A., Henderson, Y. C., Yao, H., Pickering, C. C., et al. (2011). Phosphoproteomic analysis of signaling pathways in head and neck squamous cell carcinoma patient samples. *The American Journal of Pathology*, *178*(2), 548–571.

Gan, C. S., Guo, T., Zhang, H., Lim, S. K., & Sze, S. K. (2008). A comparative study of electrostatic repulsion–hydrophilic interaction chromatography (ERLIC) versus SCX-IMAC-based methods for phosphopeptide isolation/enrichment. *Journal of Proteome Research*, 7, 4869–4877.

Garbis, S. D., Roumeliotis, T. I., Tyritzis, S. I., Zorpas, K. M., Pavlakis, K., & Constantinides, C. A. (2011). A novel multidimensional protein identification technology approach combining protein size exclusion prefractionation, peptide zwitterion-ion hydrophilic interaction chromatography, and nano-ultraperformance RP chromatography/nESI-MS2 for the in-depth analysis of the serum proteome and phosphoproteome: Application to clinical sera derived from humans with benign prostate hyperplasia. *Analytical Chemistry*, *83*, 708–718.

Gratia, S., Kay, L., Michelland, S., Sève, M., Schlattner, U., & Tokarska-Schlattner, M. (2012). Cardiac phosphoproteome reveals cell signaling events involved in doxorubicin cardiotoxicity. *Journal of Proteomics*, *75*(15), 4705–4716.

Grimes, M. L., Lee, W. J., van der Maaten, L., & Shannon, P. (2013). Wrangling phosphoproteomic data to elucidate cancer signaling pathways. *PLoS One*, *8*(1), e52884.

Gruhler, A., Olsen, J. V., Mohammed, S., Mortensen, P., Faergeman, N. J., Mann, M., et al. (2005). Quantitative phosphoproteomics applied to the yeast pheromone signalling pathway. *Molecular and Cellular Proteomics*, *4*, 310–327.

Guimond, D. M., Cam, N. R., Hirve, N., Duan, W., Lambris, J. D., Croft, M., et al. (2013). Regulation of immune responsiveness in vivo by disrupting an early T-cell signaling event using a cell-permeable peptide. *PLoS One*, *8*(5), e63645.

Guo, H., Isserlin, R., Chen, X., Wang, W., Phanse, S., Zandstra, P. W., et al. (2013). Integrative network analysis of signaling in human CD34(+) hematopoietic progenitor cells by global phosphoproteomic profiling using TiO2 enrichment combined with 2D LC-MS/MS and pathway mapping. *Proteomics*, *13*(8), 1325–1333.

Guo, T., Lee, S. S., Ng, W. H., Zhu, Y., Gan, C. S., Zhu, J., et al. (2011). Global molecular dysfunctions in gastric cancer revealed by an integrated analysis of the phosphoproteome and transcriptome. *Cellular and Molecular Life Sciences*, *68*(11), 1983–2002.

Ham, B. M., Yang, F., Jayachandran, H., Jaitly, N., Monroe, M. E., & Gritsenko, M. A. (2008). The influence of sample preparation and replicate analyses on HeLa cell phosphoproteome coverage. *Journal of Proteome Research*, *7*, 2215–2221.

Han, G., Ye, M., Zhou, H., Jiang, X., Feng, S., & Jiang, X. (2008). Large-scale phosphoproteome analysis of human liver tissue by enrichment and fractionation of phosphopeptides with strong anion exchange chromatography. *Proteomics*, *8*, 1346–1361.

Hekmat, O., Munk, S., Fogh, L., Yadav, R., Francavilla, C., Horn, H., et al. (2013). TIMP-1 increases expression and phosphorylation of proteins associated with drug resistance in breast cancer cells. *Journal of Proteome Research*, *12*(9), 4136–4151.

Hennrich, M. L., van den Toorn, H. W., Groenewold, V., Heck, A. J., & Mohammed, S. (2012). Ultra acidic strong cation exchange enabling the efficient enrichment of basic phosphopeptides. *Analytical Chemistry*, *84*, 1804–1808.

Hilger, M., Bonaldi, T., Gnad, F., & Mann, M. (2009). Systems-wide analysis of a phosphatase knock-down by quantitative proteomics and phosphoproteomics. *Molecular and Cellular Proteomics*, *8*, 1908–1920.

Huang, S. S., Clarke, D. C., Gosline, S. J., Labadorf, A., Chouinard, C. R., Gordon, W., et al. (2013). Linking proteomic and transcriptional data through the interactome and epigenome reveals a map of oncogene-induced signaling. *PLoS Computational Biology*, *9*(2), e1002887.

Hunt, D. F., Buko, A. M., Ballard, J. M., Shabanowitz, J., & Giordani, A. B. (1981). Sequence analysis of polypeptides by collision activated dissociation on a triple quadrupole mass spectrometer. *Biomedical Mass Spectrometry*, *8*, 397–408.

Huttlin, E. L., Jedrychowski, M. P., Elias, J. E., Goswami, T., Rad, R., & Beausoleil, S. A. (2010). A tissue-specific atlas of mouse protein phosphorylation and expression. *Cell*, *143*, 1174–1189.

Ikeguchi, Y., & Nakamua, H. (1997). Determination of organic phosphates by column-switching high performance anion-exchange chromatography using on-line preconcentration on titania. *Analytical Sciences*, *13*, 479–483.

Ikeguchi, Y., & Nakamura, H. (2000). Selective enrichment of phospholipids by titania. *Analytical Sciences*, *16*, 541–543.

Imamura, H., Wakabayashi, M., & Ishihama, Y. (2012). Analytical strategies for shotgun phosphoproteomics: Status and prospects. *Seminars in Cell & Developmental Biology*, *23*, 836–842.

Iwai, L. K., Payne, L. S., Luczynski, M. T., Chang, F., Xu, H., Clinton, R. W., et al. (2013). Phosphoproteomics of collagen receptor networks reveals SHP-2 phosphorylation downstream of wild-type DDR2 and its lung cancer mutants. *The Biochemical Journal*, *454*(3), 501–513.

Jaffe, H., Veeranna, & Pant, H. C. (1998). Characterization of the phosphorylation sites of human high molecular weight neurofilament protein by electrospray ionization tandem mass spectrometry and database searching. *Biochemistry*, *37*, 16211–16224.

Jedrychowski, M. P., Huttlin, E. L., Haas, W., Sowa, M. E., Rad, R., & Gygi, S. P. (2011). Evaluation of HCD- and CID-type fragmentation within their respective detection platforms for murine phosphoproteomics. *Molecular and Cellular Proteomics*, *10*, 1–19.

Jensen, S. S., & Larsen, M. R. (2007). Evaluation of the impact of some experimental procedures on different phosphopeptide enrichment techniques. *Rapid Communications in Mass Spectrometry*, *21*, 3635–3645.

Jindra, P. T., Hsueh, A., Hong, L., Gjertson, D., Shen, X. D., Gao, F., et al. (2008). Anti-MHC class I antibody activation of proliferation and survival signaling in murine cardiac allografts. *Journal of Immunology*, *180*(4), 2214–2224.

Kanshin, E., Michnick, S., & Thibault, P. (2012). Sample preparation and analytical strategies for large-scale phosphoproteomics experiments. *Seminars in Cell & Developmental Biology*, *23*, 843–853.

Kawahara, M., Nakamura, H., & Nakajima, T. (1989). Group separation of ribonucleosides and deoxyribonucleosides on a new ceramic titania column. *Analytical Sciences*, *5*, 485–486.

Kim, J., Kim, J. S., Kim, H. E., Jeon, Y. J., Kim, D. W., Soh, Y., et al. (2011). Proteomic analysis of phosphotyrosyl proteins in human embryonic stem cell-derived neural stem cells. *Neuroscience Letters*, *499*(3), 158–163.

Kjellström, S., & Jensen, O. N. (2004). Phosphoric acid as a matrix additive for MALDI MS analysis of phosphopeptides and phosphoproteins. *Analytical Chemistry*, *76*, 5109–5117.

Klammer, M., Kaminski, M., Zedler, A., Oppermann, F., Blencke, S., Marx, S., et al. (2012). Phosphosignature predicts dasatinib response in non-small cell lung cancer. *Molecular & Cellular Proteomics*, *11*(9), 651–668.

Knight, Z. A., Schilling, B., Row, R. H., Kenski, D. M., Gibson, B. W., & Shokat, K. M. (2003). Phosphospecific proteolysis for mapping sites of protein phosphorylation. *Nature Biotechnology*, *21*, 1047–1054.

Kokubu, M., Ishihama, Y., Sato, T., Nagasu, T., & Oda, Y. (2005). Specificity of immobilized metal affinity-based IMAC/C18 tip enrichment of phosphopeptides for protein phosphorylation analysis. *Analytical Chemistry*, 77, 5144–5154.

Kooij, V., Holewinski, R. J., Murphy, A. M., & Van Eyk, J. E. (2013). Characterization of the cardiac myosin binding protein-C phosphoproteome in healthy and failing human hearts. *Journal of Molecular and Cellular Cardiology*, *60*, 116–120.

Kwei, K. A., Baker, J. B., & Pelham, R. J. (2012). Modulators of sensitivity and resistance to inhibition of PI3K identified in a pharmacogenomic screen of the NCI-60 human tumor cell line collection. *PLoS One*, 7(9), e46518.

Kweon, H. K., & Håkansson, K. (2006). Selective zirconium dioxide-based enrichment of phosphorylated peptides for mass spectrometric analysis. *Analytical Chemistry*, *78*, 1743–1749.

Kyono, Y., Sugiyama, N., Imami, K., Tomita, M., & Ishihama, Y. (2008). Successive and selective release of phosphorylated peptides captured by hydroxy acid-modified metal oxide chromatography. *Journal of Proteome Research*, 7, 4585–4593.

Larsen, M. R., Thingholm, T. E., Jensen, O. N., Roepstorff, P., & Jorgensen, T. J. (2005). Highly selective enrichment of phosphorylated peptides from peptide mixtures using titanium dioxide microcolumns. *Molecular and Cellular Proteomics*, *4*, 873–886.

Laurence, A., Astoul, E., Hanrahan, S., Totty, N., & Cantrell, D. (2004). Identification of pro-interleukin 16 as a novel target of MAP kinases in activated T lymphocytes. *European Journal of Immunology*, *34*(2), 587–597.

Lee, B. Y., Hochgräfe, F., Lin, H. M., Castillo, L., Wu, J., Raftery, M. J., et al. (2014). Phosphoproteomic profiling identifies focal adhesion kinase as a mediator of docetaxel resistance in castrate resistant prostate cancer. *Molecular Cancer Therapeutics*, *13*, 190–201.

Leitner, A., & Leitner, W. (2009). Chemical tagging strategies for mass spectrometry-based phosphoproteomics. *Methods in Molecular Biology*, *527*, 229–243.

Leitner, A., Sturm, M., Smått, J. H., Järn, M., & Lindén, M. (2009). Optimizing the performance of tin dioxide microspheres for phosphopeptide enrichment. *Analytica Chimica Acta*, *6*(38), 51–57.

Li, X., Gerber, S. A., Rudner, A. D., Beausoleil, S. A., Haas, W., Villén, J., et al. (2007). Large-scale phosphorylation analysis of alpha-factor-arrested Saccharomyces cerevisiae. *Journal of Proteome Research*, *6*, 1190–1197.

Liang, X. Q., Fonnum, G., Hajivandi, M., Stene, T., Kjus, N. H., & Ragnhildstveit, E. (2007). Quantitative comparison of IMAC and TiO_2 surfaces used in the study of regulated, dynamic protein phosphorylation. *Journal of the American Society for Mass Spectrometry*, *18*, 1932–1944.

Lo, T., Tsai, C. F., Shih, Y. R., Wang, Y. T., Lu, S. C., Sung, T. Y., et al. (2012). Phosphoproteomic analysis of human mesenchymal stromal cells during osteogenic differentiation. *Journal of Proteome Research*, *11*(2), 586–598.

Mann, M., Ong, S. E., Gronborg, M., Steen, H., Jensen, O. N., & Pandey, A. (2002). Analysis of protein phosphorylation using mass spectrometry: Deciphering the phosphoproteome. *Trends in Biotechnology*, *20*, 261–268.

Matsuda, H., Nakamura, H., & Nakajima, T. (1990). New ceramic titania selective adsorbent for organic phosphates. *Analytical Sciences*, *6*, 911–912.

Mazanek, M., Mitulovic, G., Herzog, F., Stingl, C., Hutchins, J. R. A., Peters, J.-M., et al. (2007). Titanium dioxide as a chemo-affinity solid phase in offline phosphopeptide chromatography prior to HPLC-MS/MS analysis. *Nature Protocols*, *2*, 1059–1069.

Mazanek, M., Roitinger, E., Hudecz, O., Hutchins, J. R. A., Hegemann, B., Mit-ulovic, G., et al. (2010). A new acid mix enhances phosphopeptide enrichment on titanium- and zirconium dioxide for mapping of phosphorylation sites on protein complexes. *Journal of Chromatography B*, *878*, 515–524.

McDonnell, S. R., Hwang, S. R., Rolland, D., Murga-Zamalloa, C., Basrur, V., Conlon, K. P., et al. (2013). Integrated phosphoproteomic and metabolomic profiling reveals NPM-ALK-mediated phosphorylation of PKM2 and metabolic reprogramming in anaplastic large cell lymphoma. *Blood*, *122*(6), 958–968.

McMurray, J. S., Mandal, P. K., Liao, W. S., Klostergaard, J., & Robertson, F. M. (2012). The consequences of selective inhibition of signal transducer and activator of transcription 3 (STAT3) tyrosine705 phosphorylation by phosphopeptide mimetic prodrugs targeting the Src homology 2 (SH2) domain. *JAKSTAT*, *1*(4), 263–347.

McNulty, D. E., & Annan, R. S. (2008). Hydrophilic interaction chromatography reduces the complexity of the phosphoproteome and improves global phosphopeptide isolation and detection. *Molecular and Cellular Proteomics*, 7, 971–980.

Mohammed, S., Kraiczek, K., Pinkse, M. W. H., Lemeer, S., Benschop, J. J., & Heck, A. J. R. (2008). Chip-based enrichment and nanoLC-MS/MS analysis of phosphopeptides from whole lysates. *Journal of Proteome Research*, 7, 1565–1571.

Myung, J. K., & Sadar, M. D. (2012). Large scale phosphoproteome analysis of LNCaP human prostate cancer cells. *Molecular Biosystems*, *8*(8), 2174–2182.

Nagaraj, N., D'Souza, R. C., Cox, J., Olsen, J. V., & Mann, M. (2010). Feasibility of large-scale phosphoproteomics with higher energy collisional dissociation fragmentation. *Journal of Proteome Research*, *9*, 6786–6794.

Namiki, J., Suzuki, S., Masuda, T., Ishihama, Y., & Okano, H. (2012). Nestin protein is phosphorylated in adult neural stem/progenitor cells and not endothelial progenitor cells. *Stem Cells International*, *2012*, 430138.

Newton, R. P., Brenton, A. G., Smith, C. J., & Dudley, E. (2004). Plant proteome analysis by mass spectrometry: Principles, problems, pitfalls and recent developments. *Phytochemistry*, *65*(11), 1449–1485.

Nicolaou, P., Rodriguez, P., Ren, X., Zhou, X., Qian, J., Sadayappan, S., et al. (2009). Inducible expression of active protein phosphatase-1 inhibitor-1 enhances basal cardiac function and protects against ischemia/reperfusion injury. *Circulation Research*, *104*(8), 1012–1020.

Nuhse, T., Yu, K., & Salomon, A. (2007). Isolation of phosphopeptides by immobilized metal ion affinity chromatography. *Current Protocols in Molecular Biology*, (edited by F. M. Ausubel, Chapter 18: Unit 18.13).

Oda, Y., Nagasu, T., & Chait, B. T. (2001). Enrichment analysis of phosphorylated proteins as a tool for probing the phosphoproteome. *Nature Biotechnology*, *19*, 379–382.

Olsen, J. V., Blagoev, B., Gnad, F., Macek, B., Kumar, C., & Mortensen, P. (2006). Quantitative phosphoproteomics reveals widespread full phosphorylation site occupancy during mitosis. *Cell*, *127*, 635–648.

Olsen, J. V., & Mann, M. (2004). Improved peptide identification in proteomics by two consecutive stages of mass spectrometric fragmentation. *Proceedings of the National Academy of Sciences of the United States of America*, *101*, 13417–13422.

Ovelleiro, D., Carrascal, M., Casas, V., & Abian, J. (2009). LymPHOS: Design of a phosphosite database of primary human T cells. *Proteomics*, *9*, 3741–3751.

Oyama, M., Nagashima, T., Suzuki, T., Kozuka-Hata, H., Yumoto, N., Shiraishi, Y., et al. (2011). Integrated quantitative analysis of the phosphoproteome and transcriptome in tamoxifen-resistant breast cancer. *The Journal of Biological Chemistry*, *286*(1), 818–829.

Pandey, A., Podtelejnikov, A. V., Blagoev, B., Bustelo, X. R., Mann, M., & Lodish, H. F. (2000). Analysis of receptor signalling pathways by mass spectrometry: Identification of vav-2 as a substrate of the epidermal and platelet-derived growth factor receptors. *Proceedings of the National Academy of Sciences of the United States of America*, *97*, 179–184.

Parker, B. L., Larsen, M. R., Edvinsson, L. I., & Povlsen, G. K. (2013). Signal transduction in cerebral arteries after subarachnoid hemorrhage—A phosphoproteomic approach. *Journal of Cerebral Blood Flow and Metabolism*, *33*(8), 1259–1269.

Pines, G., Huang, P. H., Zwang, Y., White, F. M., & Yarden, Y. (2010). EGFRvIV: A previously uncharacterized oncogenic mutant reveals a kinase autoinhibitory mechanism. *Oncogene*, *29*(43), 5850–5860.

Pinkse, M. W., Uitto, P. M., Hilhorst, M. J., Ooms, B., & Heck, A. J. (2004). Selective isolation at the femtomole level of phosphopeptides from proteolytic digests using 2D-nano-LC-ESI-MS/MS and titanium oxide precolumns. *Analytical Chemistry*, *76*, 3935–3943.

Ponnikorn, S., Panichakul, T., Sresanga, K., Wongborisuth, C., Roytrakul, S., Hongeng, S., et al. (2011). Phosphoproteomic analysis of apoptotic hematopoietic stem cells from hemoglobin E/β-thalassemia. *Journal of Translational Medicine*, *9*, 96.

Posewitz, M. C., & Tempst, P. (1999). Immobilized gallium(III) affinity chromatography of phosphopeptides. *Analytical Chemistry*, *71*, 2883–2892.

Qi, D., Lu, J., Deng, C., & Zhang, X. (2009). Development of core-shell structure $Fe_3O_4@Ta_2O_5$ microspheres for selective enrichment of phosphopeptides for mass spectrometry analysis. *Journal of Chromatography A*, *1216*, 5533–5539.

Raijmakers, R., Kraiczek, K., de Jong, A. P., Mohammed, S., & Heck, A. J. R. (2010). Exploring the human leukocyte phosphoproteome using a microfluidic reversed-phase-TiO_2-reversed-phase high-performance liquid chromatography phosphochip coupled to a quadrupole time-of-flight mass spectrometer. *Analytical Chemistry*, *82*, 824–832.

Ren, H., Tan, Z. P., Zhu, X., Crosby, K., Haack, H., Ren, J. M., et al. (2012). Identification of anaplastic lymphoma kinase as a potential therapeutic target in ovarian cancer. *Cancer Research*, *72*(13), 3312–3323.

Rexer, B. N., Ham, A. J., Rinehart, C., Hill, S., Granja-Ingram Nde, M., González-Angulo, A. M., et al. (2011). Phosphoproteomic mass spectrometry profiling links Src family kinases to escape from HER2 tyrosine kinase inhibition. *Oncogene*, *30*(40), 4163–4174.

Reynolds, E. C., Riley, P. F., & Adamson, N. J. (1994). A selective precipitation purification procedure for multiple phosphoseryl-containing peptides and methods for their identification. *Analytical Biochemistry, 217*, 277–284.

Rigbolt, K. T., Prokhorova, T. A., Akimov, V., Henningsen, J., Johansen, P. T., Kratchmarova, I., et al. (2011). System-wide temporal characterization of the proteome and phosphoproteome of human embryonic stem cell differentiation. *Science Signaling, 4*(164), rs3.

Rivera, J. G., Choi, Y. S., Vujcic, S., Wood, T. D., & Colón, L. A. (2009). Enrichment/isolation of phosphorylated peptides on hafnium oxide prior to mass spectrometric analysis. *Analyst, 134*, 31–33.

Ruperez, P., Gago-Martinez, A., Burlingame, A. L., & Oses-Prieto, J. A. (2012). Quantitative phosphoproteomic analysis reveals a role for serine and threonine kinases in the cytoskeletal reorganization in early T cell receptor activation in human primary T cells. *Molecular & Cellular Proteomics, 11*(5), 171–186.

Rush, J., Moritz, A., Lee, K. A., Guo, A., Goss, V. L., Spek, E. J., et al. (2005). Immunoaffinity profiling of tyrosine phosphorylation in cancer cells. *Nature Biotechnology, 23*, 94–101.

Salih, E. (2005). Phosphoproteomics by mass spectrometry and classical protein chemistry approaches. *Mass Spectrometry Reviews, 24*, 828–846.

Savitski, M. M., Lemeer, S., & Boesche, M. (2011). Confident phosphorylation site localization using the Mascot Delta Score. *Molecular & Cellular Proteomics, 10*, M110.003830.

Schreiber, T. B., Mausbacher, N., Soroka, J., Wandinger, S. K., Buchner, J., & Daub, H. (2012). Global analysis of phosphoproteome regulation by the Ser/Thr phosphatase Ppt1 in Saccharomyces cerevisiae. *Journal of Proteome Research, 11*, 2397–2408.

Schroeder, M. J., Shabanowitz, J., Schwartz, J. C., Hunt, D. F., & Coon, J. J. (2004). A neutral loss activation method for improved phosphopeptide sequence analysis by quadrupole ion trap mass spectrometry. *Analytical Chemistry, 76*, 3590–3598.

Simon, E. S., Young, M., Chan, A., Bao, Z. Q., & Andrews, P. C. (2008). Improved enrichment strategies for phosphorylated peptides on titanium dioxide using methyl esterification and pH gradient elution. *Analytical Biochemistry, 377*, 234–242.

Ståhl, S., Branca, R. M., Efazat, G., Ruzzene, M., Zhivotovsky, B., Lewensohn, R., et al. (2011). Phosphoproteomic profiling of NSCLC cells reveals that ephrin B3 regulates pro-survival signaling through Akt1-mediated phosphorylation of the EphA2 receptor. *Journal of Proteome Research, 10*(5), 2566–2578.

Steen, H., Kuster, B., & Fernandez, M. (2001). Detection of tyrosine phosphorylated peptides by precursor ion scanning quadrupole TOF mass spectrometry in positive ion mode. *Analytical Chemistry, 73*, 1440–1448.

Sugiyama, N., Masuda, T., Shinoda, K., Nakamura, A., Tomita, M., & Ishihama, Y. (2007). Phosphopeptide enrichment by aliphatic hydroxy acid-modified metal oxide chromatography for nano-LC–MS/MS in proteomics applications. *Molecular and Cellular Proteomics, 6*, 1103–1109.

Sui, S., Wang, J., Yang, B., Song, L., Zhang, J., & Chen, M. (2008). Phosphoproteome analysis of the human Chang liver cells using SCX and a complementary mass spectrometric strategy. *Proteomics, 8*, 2024–2034.

Swaney, D. L., McAlister, G. C., & Coon, J. J. (2008). Decision tree-driven tandem mass spectrometry for shotgun proteomics. *Nature Methods, 5*, 959–964.

Syka, J. E., Coon, J. J., Schroeder, M. J., Shabanowitz, J., & Hunt, D. F. (2004). Peptide and protein sequence analysis by electron transfer dissociation mass spectrometry. *Proceedings of the National Academy of Sciences of the United States of America, 101*, 9528–9533.

Takahashi, T., Serada, S., Ako, M., Fujimoto, M., Miyazaki, Y., Nakatsuka, R., et al. (2013). New findings of kinase switching in gastrointestinal stromal tumor under imatinib using phosphoproteomic analysis. *International Journal of Cancer, 133*(11), 2737–2743.

Takano, S., Sogawa, K., Yoshitomi, H., Shida, T., Mogushi, K., Kimura, F., et al. (2010). Increased circulating cell signalling phosphoproteins in sera are useful for the detection of pancreatic cancer. *British Journal of Cancer, 103*(2), 223–231.

Tani, K., & Suzuki, Y. (1997). Investigation of the ion-exchange behaviour of titania-application as a packing material for ion chromatography. *Chromatographia, 46*, 623–627.

Thingholm, T. E., Jensen, O. N., Robinson, P. J., & Larsen, M. R. (2008). SIMAC (sequential elution from IMAC), a phosphoproteomics strategy for the rapid separation of monophosphorylated from multiply phosphorylated peptides. *Molecular and Cellular Proteomics, 7*, 661–671.

Thingholm, T. E., Jorgensen, T. J., & Jensen, O. N. (2006). Highly selective enrichment of phosphorylated peptides using titanium dioxide. *Nature Protocols, 1*, 1929–1935.

Tobe, B. T., Hou, J., Crain, A. M., Singec, I., Snyder, E. Y., & Brill, L. M. (2012). Phosphoproteomic analysis: An emerging role in deciphering cellular signaling in human embryonic stem cells and their differentiated derivatives. *Stem Cell Reviews, 8*(1), 16–31.

Tworkoski, K., Singhal, G., Szpakowski, S., Zito, C. I., Bacchiocchi, A., Muthusamy, V., et al. (2011). Phosphoproteomic screen identifies potential therapeutic targets in melanoma. *Molecular Cancer Research, 9*(6), 801–812.

Vazquez-Martin, A., Oliveras-Ferraros, C., Colomer, R., Brunet, J., & Menendez, J. A. (2008). Low-scale phosphoproteome analyses identify the mTOR effector p70 S6 kinase 1 as a specific biomarker of the dual-HER1/HER2 tyrosine kinase inhibitor lapatinib (Tykerb) in human breast carcinoma cells. *Annals of Oncology, 19*(6), 1097–1109.

Villen, J., Beausoleil, S. A., Gerber, S. A., & Gygi, S. P. (2007). Large-scale phosphorylation analysis of mouse liver. *Proceedings of the National Academy of Sciences of the United States of America, 104*, 1488–1493.

Villen, J., & Gygi, S. P. (2008). The SCX/IMAC enrichment approach for global phosphorylation analysis by mass spectrometry. *Nature Protocols, 3*, 1630–1638.

Wang, W. H., & Bruening, M. L. (2009). Phosphopeptide enrichment on functionalized polymer microspots for MALDI-MS analysis. *Analyst, 134*, 512–518.

Wijeratne, A. B., Manning, J. R., Schultz Jel, J., & Greis, K. D. (2013). Quantitative phosphoproteomics using acetone-based peptide labeling: Method evaluation and application to a cardiac ischemia/reperfusion mode. *Journal of Proteome Research, 12*, 4268–4279.

Winck, F. V., Belloni, M., Pauletti, B. A., de Lima Zanella, J., Domingues, R. R., Sherman, N. E., et al. (2014). Phosphoproteome analysis reveals differences in phosphosite profiles between tumorigenic and non-tumorigenic epithelial cells. *Journal of Proteomics, 96*, 67–81 S1874-3919(13)00554-X.

Wolschin, F., Wienkoop, S., & Weckwerth, W. (2005). Enrichment of phosphorylated proteins and peptides from complex mixtures using metal oxide/hydroxide affinity chromatography (MOAC). *Proteomics, 5*, 4389–4397.

Wu, C. J., Chen, Y. W., Tai, J. H., & Chen, S. H. (2011). Quantitative phosphoproteomics studies using stable A isotope dimethyl labelling coupled with IMAC-HILIC-nanoLC–MS/MS for estrogen-induced transcriptional regulation. *Journal of Proteome Research, 10*, 1088–1097.

Wu, H. Y., Tseng, V. S., Chen, L. C., Chang, H. Y., Chuang, I. C., Tsay, Y. G., et al. (2010). Identification of tyrosine-phosphorylated proteins associated with lung cancer metastasis using label-free quantitative analyses. *Journal of Proteome Research, 9*(8), 4102–4112.

Xie, X., Feng, S., Vuong, H., Liu, Y., Goodison, S., & Lubman, D. M. (2010). A comparative phosphoproteomic analysis of a human tumor metastasis model using a label-free quantitative approach. *Electrophoresis, 31*(11), 1842–1852.

Yan, G. R., Ding, W., Xu, S. H., Xu, Z., Xiao, C. L., Yin, X. F., et al. (2011). Characterization of phosphoproteins in gastric cancer secretome. *OMICS, 15*(1–2), 83–90.

Yang, X. F., Wu, X. P., & Kobayashi, T. (2004). Enhanced ionization of phosphorylated peptides during MALDI TOF mass spectrometry. *Analytical Chemistry*, *76*, 1532–1536.

Yu, L. R., Issaq, H. J., & Veenstra, T. D. (2007). Phosphoproteomics for the discovery of kinases as cancer biomarkers and drug targets. *Proteomics Clinical Applications*, *1*(9), 1042–1057.

Yu, G., Xiao, C. L., Lu, C. H., Jia, H. T., Ge, F., Wang, W., et al. (2011). Phosphoproteome profile of human lung cancer cell line A549. *Molecular Biosystems*, 7(2), 472–479.

Zakharchenko, O., Cojoc, M., Dubrovska, A., & Souchelnytskyi, S. (2013). A role of TGFß1 dependent 14-3-3σ phosphorylation at Ser69 and Ser74 in the regulation of gene transcription, stemness and radioresistance. *PLoS One*, *8*(5), e65163.

Zanivan, S., Meves, A., Behrendt, K., Schoof, E. M., Neilson, L. J., Cox, J., et al. (2013). In vivo SILAC-based proteomics reveals phosphoproteome changes during mouse skin carcinogenesis. *Cell Reports*, *3*(2), 552–566.

Zhai, B., Villen, J., Beausoleil, S. A., Mintseris, J., & Gygi, S. P. (2008). Phosphoproteome analysis of Drosophila melanogaster embryos. *Journal of Proteome Research*, 7, 1675–1682.

Zhang, K. (2006). From purification of large amounts of phospho-compounds (nucleotides) to enrichment of phosphopeptides using anion-exchanging resin. *Analytical Biochemistry*, *357*, 225–231.

Zhong, J., Kim, M. S., Chaerkady, R., Wu, X., Huang, T. C., Getnet, D., et al. (2012). TSLP signaling network revealed by SILAC-based phosphoproteomics. *Molecular & Cellular Proteomics*, *11*(6), M112.017764.

Zhou, H., Xu, S., & Ye, M. (2006). Zirconium phosphonate-modified porous silicon for highly specific capture of phosphopeptides and MALDI-TOF MS analysis. *Journal of Proteome Research*, *5*, 2431–2437.

Zhou, H., Ye, M., Dong, J., Han, G., Jiang, X., Wu, R., et al. (2008). Specific phosphopeptide enrichment with immobilized titanium ion affinity chromatography adsorbent for phosphoproteome analysis. *Journal of Proteome Research*, 7, 3957–3967.

Zimman, A., Chen, S. S., Komisopoulou, E., Titz, B., Martínez-Pinna, R., Kafi, A., et al. (2010). Activation of aortic endothelial cells by oxidized phospholipids: A phosphoproteomic analysis. *Journal of Proteome Research*, *9*(6), 2812–2824.

Zoumaro-Djayoon, A. D., Ding, V., Foong, L. Y., Choo, A., Heck, A. J., & Muñoz, J. (2011). Investigating the role of FGF-2 in stem cell maintenance by global phosphoproteomics profiling. *Proteomics*, *11*(20), 3962–3971.

CHAPTER THREE

Recent Advances in Mass Spectrometry-Based Glycoproteomics

Dustin C. Frost[*], **Lingjun Li**[*,†,1]
[*]School of Pharmacy, University of Wisconsin, Madison, Wisconsin, USA
[†]Department of Chemistry, University of Wisconsin, Madison, Wisconsin, USA
[1]Corresponding author: e-mail address: lli@pharmacy.wisc.edu

Contents

Abstract

Protein glycosylation plays fundamental roles in many biological processes as one of the most common, and the most complex, posttranslational modification. Alterations in glycosylation profile are now known to be associated with many diseases. As a result, the discovery and detailed characterization of glycoprotein disease biomarkers is a primary interest of biomedical research. Advances in mass spectrometry (MS)-based glycoproteomics and glycomics are increasingly enabling qualitative and quantitative approaches for site-specific structural analysis of protein glycosylation. While the complexity presented by glycan heterogeneity and the wide dynamic range of clinically relevant samples like plasma, serum, cerebrospinal fluid, and tissue make comprehensive analyses of the glycoproteome a challenging task, the ongoing efforts into the development of glycoprotein enrichment, enzymatic digestion, and separation strategies

Advances in Protein Chemistry and Structural Biology, Volume 95
ISSN 1876-1623
http://dx.doi.org/10.1016/B978-0-12-800453-1.00003-8

combined with novel quantitative MS methodologies have greatly improved analytical sensitivity, specificity, and throughput. This review summarizes current MS-based glycoproteomics approaches and highlights recent advances in its application to cancer biomarker and neurodegenerative disease research.

1. INTRODUCTION

Glycosylation is the most frequent posttranslational modification (PTM) of proteins, with over 50% of proteins featuring covalently attached glycans (Apweiler, Hermjakob, & Sharon, 1999), and is undoubtedly the most structurally complex in the types and linkage patterns of these glycans. This structural diversity serves to impart functional variance, placing cell surface and secreted glycosylated proteins into vital roles in a wide variety of biological processes including molecular recognition, cellular adhesion, intra- and intercellular signaling, fertilization, immunity, and host–pathogen interactions (Copeland, Han, & Hart, 2013; Helenius & Aebi, 2004; Lux & Nimmerjahn, 2011; Varki, 1993). Alterations in glycan composition can significantly modify the activity and function of a glycoprotein, and aberrant glycosylation has long been known to be involved in the progression of disease, including cancer and neurodegenerative diseases (Dube & Bertozzi, 2005; Fuster & Esko, 2005). A crucial first step in investigating the involvement of glycoproteins in disease is the unambiguous identification, detailed characterization, and accurate quantitation of glycoproteins and their glycan features using sensitive and robust methods. Thus, glycoproteomics and glycomics have become increasingly relevant areas of interest in biomedical research for the initial phase of disease biomarker discovery as the starting point for diagnosing and treating disease. Mass spectrometry (MS), in particular, is an extremely versatile and powerful tool for the investigation of complex biological problems and provides a rapid and sensitive means of structural elucidation of peptides and glycans. However, comprehensive profiling of glycoproteins in clinically relevant samples like plasma and serum by MS-based methods is still an elaborate and difficult task. The tremendous dynamic range of protein abundance in human plasma poses a technical challenge in that the top 22 most abundant proteins represent nearly 99% of the total protein mass, while glycoproteins of diagnostic or therapeutic value are likely to be low in abundance and heterogeneous in nature (Anderson & Anderson, 2002). Furthermore, the move from proteomics to comprehensive glycoproteomics comes with an exponential increase in the amount of information encoded by glycan structures.

The complexity of glycan moieties presents a significant challenge to glycoproteomics analysis. Glycans exist as polysaccharides that vary widely in composition, linkage, and branching, all of which define their structural diversity. Seven monosaccharides constitute these structures in humans: mannose (Man), glucose (Glc), galactose (Gal), *N*-acetylglucosamine (GlcNAc), *N*-acetylgalactosamine (GalNAc), fucose (Fuc), and *N*-acetylneuraminic acid (Neu5Ac), also referred to as sialic acid (SA). Due to the stereoisomeric nature of monosaccharides, the glycosidic bonds that connect them exist in two anomeric forms, denoted as α- or β-linked (Mariño, Bones, Kattla, & Rudd, 2010). Linear or branching glycans are covalently attached to amino acid residues on the protein backbone, and two classes of glycosylation, N- and O-linked, are of greatest interest for biomedical studies. N-linked glycosylation occurs at the amino group of asparagine residues within a consensus sequence of Asn-X-Ser/Thr, in which X may be any amino acid residue except proline. O-linked glycosylation occurs most commonly at the hydroxyl group of Ser or Thr residues but lacks a specific amino acid sequence. N-linked glycans begin with a conserved $GlcNAc_2Man_3$ chitobiose core structure and can be categorized into high-mannose, complex, and hybrid subgroups, while O-linked glycans do not feature a common core structure but exist in eight common formations (Mariño et al., 2010). O-linked monosaccharide β-*N*-acetylglucosamine (O-GlcNAc) is a dynamic PTM that is similar to phosphorylation and plays central roles in healthy biological processes. O-GlcNAcylation is mutually exclusive to phosphorylation at many Ser/Thr sites and can modulate phosphorylation-dependent pathways (Copeland et al., 2013). Figure 3.1 illustrates the common N- and O-linked glycan structures using the Consortium for Functional Glycomics (CFG) notation.

The complexity of profiling the glycoproteome is further compounded by the microheterogeneity at which glycans occupy specific sites along the polypeptide chain of a glycoprotein (Hua, An, et al., 2011). That is, a protein with a single site of glycosylation can display a range of different glycans and glycan isoforms. It has been suggested that a glycoprotein with just three glycosylation sites displaying 10 different glycans at each site could realize a thousand different glycoforms of the protein (An, Froehlich, & Lebrilla, 2009). Moreover, macroheterogeneity arises from the observation that a glycosite may be only partially occupied or vacant entirely (Mariño et al., 2010).

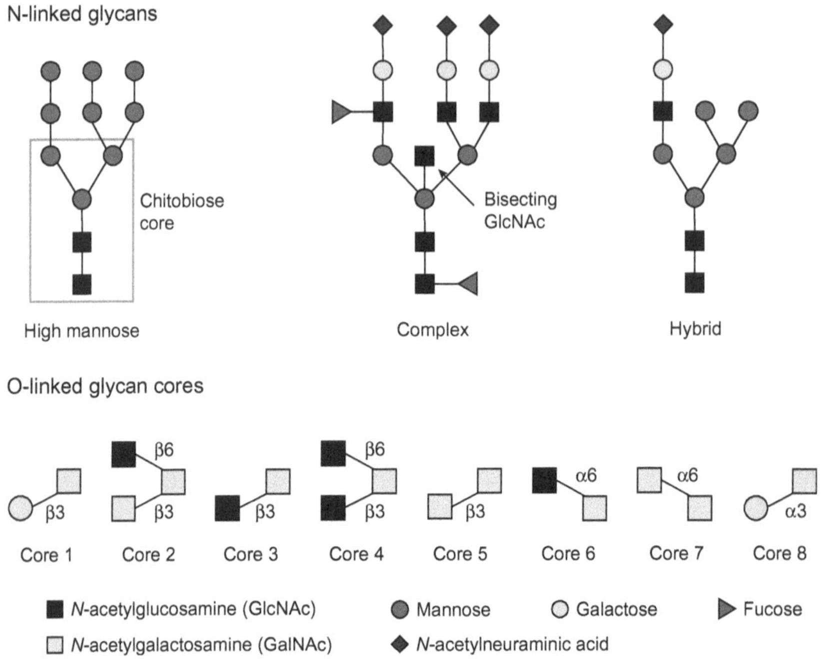

Figure 3.1 Basic structures of high-mannose, complex, and hybrid N-linked glycans and the eight common O-linked glycan cores depicted using the CFG notation. *Adapted with permission from Alley, Mann, and Novotny (2013). Copyright 2013 American Chemical Society.*

One of the principal goals of biomedical research is preclinical biomarker discovery. Glycoproteins can act as biomarkers for disease through deviations in their secreted expression levels in plasma, serum, urine, or other bodily fluids. Irregularities in glycosylation site occupancy patterns or aberrance in glycan composition or structure can also serve as indicators of disease. Thus, elucidating biomarkers in the glycoproteome requires a comprehensive approach that includes glycomics and the ability to form not only qualitative but also quantitative conclusions, aiming to provide the identification and relative abundances of glycoproteins, the locations and degree of occupancy of glycosites, and detailed characterization of glycans and their abundance. Due to recent technological advances, MS-based glycoproteomics has become an ideal platform for the discovery of disease-associated glycoproteins and glycoforms. Modern workflows using glycoprotein or glycopeptide enrichment and multidimensional chromatographic separation followed by rapid and sensitive detection via high-resolution, high mass accuracy MS have decreased limits of detection and increased analytical dynamic range of glycoproteomics analyses of complex biological samples, and MS-based quantitative profiling of the glycoproteome is increasingly being served by stable isotopic- or isobaric-labeling strategies that have been introduced in the past decade. Still, truly comprehensive proteomics methods are rare. Most strategies focus on only one or two parts of the equation. A protein-based approach may enrich for glycoproteins or glycopeptides, deglycosylate them, and proceed with standard glycoproteomics workflow to achieve protein identification and reveal basic glycosite indication at the expense of glycan structure information. On the other hand, a glycan-based approach separates glycans from their glycopeptide counterparts to achieve detailed glycan characterization at the expense of information on specific glycosite origin. The integration of the two approaches is a work in progress for the glycoproteomics field as a whole but is necessary for effective application of MS-based glycoproteomics to biomedical studies that aim to understand certain biological processes, discover biomarkers for disease, determine drug targets, and develop therapeutic agents. The aim of this review is to summarize the current state of MS-based glycoproteomics and highlights recent advances in their contribution to disease biomarker research.

2. GLYCOPROTEOMIC PROFILING BY MS

2.1. Glycoproteomics methodology

Because the success of any MS experiment relies heavily on analyte purity, the ultimate aim of sample preparation in an MS-based glycoproteomics workflow is to simplify or purify a sample to facilitate sensitive detection of peptides and glycans by the mass spectrometer. Once proteins are harvested from biological specimens, glycoproteins must be isolated from nonglycosylated proteins. Top-down MS analysis of purified glycoproteins can be performed, but bottom-up strategies, in which glycoproteins are digested into peptides and glycopeptides prior to MS analysis, are most common. Mixtures of glycopeptides and nonglycosylated peptides present a problem, however. The hydrophilic nature of attached glycans significantly impairs the ionization of glycopeptides, and the nonglycosylated peptides are preferentially ionized and detected by a great degree. The combination of enrichment and chromatography serves to sufficiently isolate the glycopeptides of interest. Glycan cleavage, followed by derivatization and separation, allows detailed glycomics characterization of composition, structure, linkages, and isomers by tandem mass spectrometry (MS/MS), though the relationship of the glycans to peptide glycosylation sites is lost. Likewise, MS/MS analysis of deglycosylated peptides provides more sensitive analysis of peptide sequence, but only limited glycosylation site information is obtained. Depending on the acquisition parameters, analysis of native, intact glycopeptides may provide only glycan composition or peptide sequence with glycosite indication, though recent technological advances in alternative digestion strategies, instrumentation, and bioinformatics allow more complete site-specific glycosylation information.

The general glycoproteomics workflow consists of glycoprotein enrichment, proteolytic digestion, multidimensional chromatographic separation, MS/MS analysis, and bioinformatic data processing. Enrichment may be performed at the glycoprotein level or the glycopeptide level. Intact glycopeptides can be analyzed directly by MS/MS, under specific acquisition parameters discussed later in this review, to obtain glycan composition and peptide sequence information for glycoprotein identification. Alternatively, the glycopeptides can be deglycosylated through enzymatic or chemical means prior to separation and MS/MS analysis to obtain protein identification and glycosylation site assignment, and the released glycans typically undergo chemical derivatization prior to separation and MS/MS analysis to determine glycan structure. The raw MS/MS spectral data then rely heavily on bioinformatics software and database searching to provide peptide sequencing and protein identification, glycosylation site assignment, glycan characterization, and quantitation. A schematic diagram of a general glycoproteomics and glycomics workflow is illustrated in Fig. 3.2.

2.2. Affinity enrichment

Generally, proteins of diagnostic or therapeutic interest exist in far lower abundance compared to the rest of the proteins in biological samples. Thus, reducing sample complexity via selective, affinity-based enrichment of proteins and peptides is an essential step in MS-based proteomics methods. Several detailed reviews discussing affinity enrichment techniques for general proteomics have been recently published (Hage et al., 2012; Medvedev, Kopylov, Buneeva, Zgoda, & Archakov, 2012; Ongay, Boichenko, Govorukhina, & Bischoff, 2012; Pernemalm, Lewensohn, & Lehtiö, 2009; Selvaraju & Rassi, 2011; Zhang, Lu, & Yang, 2009). A common approach for plasma, serum, and cerebrospinal fluid (CSF) samples is the antibody-based depletion of several highly abundant proteins prior to downstream enrichment techniques, whereby the removal of over 90% of the

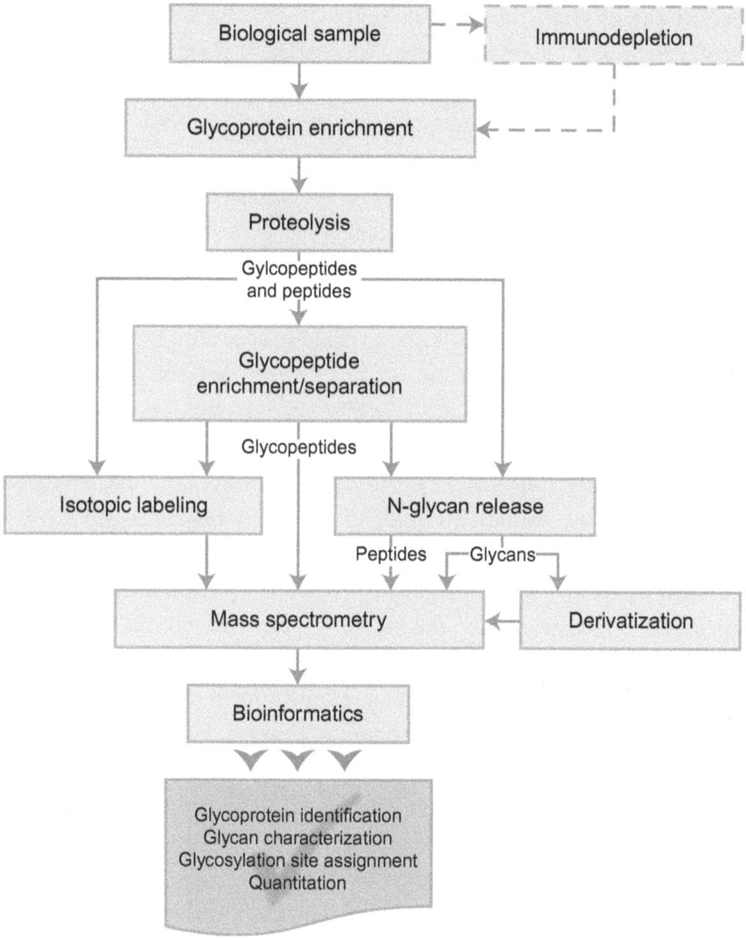

Figure 3.2 Schematic diagram of an integrated glycoproteomics and glycomics workflow.

original protein content greatly facilitates downstream analysis of low abundance, potentially interesting proteins (Plavina, Wakshull, Hancock, & Hincapie, 2007; Tep, Hincapie, & Hancock, 2012). Glycoprotein or glycopeptide enrichment is then widely performed using lectin affinity chromatography (LAC) (Kaji et al., 2003; Sparbier, Koch, Kessler, Wenzel, & Kostrzewa, 2005; Wang, Wu, & Hancock, 2006) or hydrazide capture (Liu et al., 2005; Zhang, Li, Martin, & Aebersold, 2003), and boronic acid (Xu et al., 2009) and titanium dioxide (Larsen, Jensen, Jakobsen, & Heegaard, 2007) are also used.

LAC is the primary method of glycoprotein enrichment and is often applied to glycopeptide enrichment. An in-depth review of LAC methods has recently been published (Fanayan, Hincapie, & Hancock, 2012). Lectins are a diverse group of proteins that recognize and reversibly bind specific sugar groups. More than 60 lectins with different binding affinities are commercially available, some of which have specificity that broadly covers the plasma and serum glycoproteome while others have very narrow specificities toward small glycoproteomic subsets. This flexibility allows researchers to select lectins whose affinities are either wide for exploratory biomarker discovery studies or strict for a known disease-specific glycoprotein target. The most extensively used lectin, concanavalin A (Con A), binds a vast number of N-glycoproteins at the trimannosyl core of accessible high-mannose glycans and at branched α-mannosidic groups of hybrid and complex biantennary glycans; wheat germ agglutinin (WGA) binds chitobiose *N*-acetylglucosamine and sialic acid; and jacalin binds O-linked glycans and galactosyl (β1–3) *N*-acetylgalactosamine. Release of bound glycoproteins or glycopeptides is accomplished with an elution buffer containing appropriate sugars that disrupt the lectin–glycan interaction through competitive binding, with acidic conditions, or a combination of both. The use of nonionic detergents at low concentrations in a technique called detergent-assisted lectin affinity chromatography has been reported by Wei et al. to enhance lectin binding and elution of glycoproteins, especially hydrophobic and membrane glycoproteins, facilitating their enrichment from tissue samples (Wei, Dulberger, & Li, 2010). Importantly, salts and sugars introduced during LAC must be removed and pH is adjusted prior to proteolytic digest of glycoproteins or analysis of glycopeptides via MS. Lectins are commonly immobilized on agarose, silica, or polyhydroxylate polymer (POROS™) supports for use in centrifugal filter units, pipet tips, high-performance liquid chromatography (HPLC) columns, and microarrays (Gupta, Surolia, & Sampathkumar, 2010; Kullolli, Hancock, & Hincapie, 2008; Zielinska, Gnad, Wiśniewski, & Mann, 2010).

Because the affinities of individual lectins make them unable to bind the entire glycoproteome, comprehensive enrichment strategies benefit from using several different lectins with contrasting binding specificities to achieve more complete coverage. Serial lectin affinity chromatography (SLAC) (Cummings & Kornfeld, 1982) uses single lectin enrichments of a sample in succession to simultaneously target different glycoprotein subsets, enabling the comparison of glycosylation patterns or determination of glycoform structural changes in glycoprotein biomarkers. Multilectin affinity chromatography (MLAC) (Yang & Hancock, 2004) combines several different lectins into a single enrichment format to increase glycoproteome coverage by targeting a more diverse subproteome of N- and O-glycoproteins. Elution of glycoproteins in an MLAC strategy can be performed *en masse* by using an elution solution containing a mixture of all appropriate eluting sugars (Yang, Hancock, Chew, & Bonilla, 2005; Yang, Harris, Palmer-Toy, & Hancock, 2006) or in a serial fashion by using elution solutions separately for each lectin (Yang & Hancock, 2005), though overlap in the fractions will be observed for glycoproteins bound by multiple lectins due to multiple glycosites or microheterogeneity of specific glycosites. The MLAC strategy has been extended to HPLC column format as high-performance lectin affinity chromatography (HP-MLAC) (Kullolli et al., 2008), and modern platforms combine depletion of highly abundant proteins followed by inline HP-MLAC and reversed-phase (RP) cleanup on a single HPLC system for automated, high-throughput sample enrichment (Gbormittah et al., 2013; Kullolli, Hancock, & Hincapie, 2010; Zeng et al., 2011).

The lectin array has been used for rapid, sensitive, and high-throughput profiling of glycosylation. The lectin microarray, recently reviewed in depth elsewhere (Gupta et al., 2010; Hirabayashi, Yamada, Kuno, & Tateno, 2013; Yue & Haab, 2009), consists of a glass slide containing many different immobilized lectin spots, onto which fluorescently labeled proteins are bound, detected, and the extent of binding to the different lectin spots based on fluorescent signal intensity allows glycoform characterization without the liberation of glycans. While microarrays are not a technique for enrichment, they can serve as an initial probe into the glycomic profile of a sample in order to guide an appropriate lectin enrichment approach prior to LC–MS/MS analysis, a strategy which has been used recently in several glycoproteomics studies of cancer (Kaji et al., 2013; Li, Wen, et al., 2013; Zhu, He, Liu, Simeone, & Lubman, 2012). The idea has been adapted for high-throughput glycoprotein enrichment using magnetic bead-immobilized lectins and microwell plates for parallel isolation of several sub-glycoproteomes from a sample, followed by LC–MS/MS analysis (Choi, Loo, Dennis, O'Leary, & Hill, 2011; Loo, Jones, & Hill, 2010).

Hydrazide capture is another common glycoprotein and glycopeptide enrichment method. Here, glycans are covalently coupled to a resin displaying immobilized hydrazide groups through periodate oxidation of glycan *cis*-diol groups (Zhang et al., 2003). In contrast to lectin affinity, hydrazide capture is nonspecific, allowing the enrichment of all glycoconjugates.

Glycoprotein-level hydrazide capture is followed by proteolytic digestion, washing of nonglycopeptides, and enzymatic release of glycopeptides by peptide-N4-(acetyl-β-glucosaminyl)-asparagine amidase N-glycosidase F (PNGase F), a glycosidase which specifically cleaves N-linked glycans at the asparagine-bonded GlcNAc (except those carrying α(1–3)-linked core fucose, Liu et al., 2005). Glycopeptide-level hydrazide capture shows greater specificity and yield for glycopeptide enrichment owing to better accessibility to N-glycosites compared to the glycoprotein-level approach (Zhou, Aebersold, & Zhang, 2007), though glycoprotein-level enrichment may result in greater numbers of glycopeptide and glycoprotein identifications (Berven, Ahmad, Clauser, & Carr, 2010; Wang et al., 2012). Recently, hydrazide resin has been integrated into pipet tips for rapid, automated solid-phase extraction of N-linked glycopeptides (Chen, Shah, & Zhang, 2013).

Some shortcomings to hydrazide capture have been identified. While the hydrazide capture is nondiscriminatory, the recovery and downstream analysis of captured glycopeptides is limited by the release method. Additionally, since glycans remain bound to the hydrazide resin, structural and glycosite occupancy information is lost, making comparative glycan biomarker research difficult. One method addresses this issue to an extent for sialylated N- and O-glycopeptides by replacing PNGase F cleavage with acid hydrolysis of sialic acid glycosidic bonds using formic acid, which retains the glycans with the exception of terminal sialic acid (Nilsson & Larson, 2013; Nilsson et al., 2009), but consequently does not allow downstream analysis of sialylation or of nonsialylated glycopeptides. Hydrolysis with ice-cold 1 M HCl, however, appears to retain sialic acids (Kurogochi et al., 2010). Specific release of O-GlcNAc peptides by hydroxylamine has been described (Klement, Lipinszki, Kupihár, Udvardy, & Medzihradszky, 2010), and a modified hydrazide capture by O-GlcNAc derivatization with 2-keto-galactose (GalNAz) and 3-ethynylbenzaldehyde (3EBA), rather than periodate oxidation, was recently devised to enable reversible hydrazine chemistry (Nishikaze, Kawabata, Iwamoto, & Tanaka, 2013). Still, general and routine hydrazide enrichment of O-linked glycopeptides remains difficult due in part to the lack of enzymes for cleaving O-linked glycans (Klement et al., 2010). Chemical release of O-linked glycopeptides from hydrazide resin, by β-elimination, for example, is destructive to peptides and has proved generally impractical. Thus, hydrazide capture is relegated primarily to N-linked glycopeptides, though efforts are continually being made to improve the method for O-linked glycopeptide applications.

Boronic acid chemistry has been used for glycopeptide enrichment based on its covalent, yet reversible, chemical reaction with 1-2 and 1-3 *cis*-diol containing saccharides (e.g., Man, Glc, and Gal) to form stable cyclic esters (Sparbier et al., 2005; Sparbier, Wenzel, & Kostrzewa, 2006). Binding occurs under basic or nonaqueous conditions, and elution under acidic conditions yields the glycopeptides with native glycans still attached. Boronic acid recognition of glycans is nonspecific and tolerant of the various branching and linear glycans as well as monosaccharide modifications, enabling unbiased enrichment of a wide range of N- and O-linked glycopeptides. The covalent interaction with glycosylated peptides allows stringent washing conditions at pH >8. Boronic acid can be easily functionalized to a variety of supports such as mesoporous silica (Xu et al., 2009), monoliths (Huang et al., 2013), and nanoparticles (Pan, Sun, Zheng, & Yang, 2013; Zhang, Xu, et al., 2009; Zhou et al., 2008) for use with HPLC and capillary columns (Zhang et al., 2008, 2007), pipette tips (Takátsy et al., 2009), and matrix-assisted laser desorption–ionization (MALDI) plates (Tang et al., 2009; Xu, Zhang, Lu, & Yang, 2010).

Titanium dioxide (TiO_2) is used in glycopeptide enrichment and solid-phase extraction (SPE) applications due to its affinity for sialic acid (Larsen et al., 2007; Palmisano et al., 2010; Zhang, Sheng, et al., 2011). As both phosphopeptides and glycopeptides bind to TiO_2, phosphatase pretreatment to removed phosphate modifications benefits glycopeptide enrichment efficiency (Larsen et al., 2007). Binding of sialic acid to TiO_2 occurs by way of negative charges on the carboxylic acid and hydroxyl groups of sialic acid that form a multidentate chelating ligand to Ti^{4+}. The specificity toward sialic acid is especially attractive in that increased glycan sialylation has been associated with cancer progression, hepatitis, and inflammation (Larsen et al., 2007; Mondal, Chatterjee, Chawla, & Chatterjee, 2011; Nie, Li, & Sun, 2012).

Antibody-based strategies are especially useful when a single glycoprotein target needs to be isolated. However, because glycans are poor antigens, it is difficult to obtain antiglycan antibodies with sufficient affinity and specificity to use for enrichment purposes. Still, a number of antibodies with relevant antigens in O-GlcNAc (Comer, Vosseller, Wells, Accavitti, & Hart, 2001; Wang, Pandey, & Hart, 2007), O-GalNAc (Nakada et al., 1991), sialyl Lewis^X (Cho, Jung, & Regnier, 2008), and polysialic acid (Liedtke et al., 2001) have been used effectively for glycoprotein enrichment. Recently, Teo et al. were able to procure three monoclonal antibodies against O-GlcNAc using a synthetic antigen and enrich three subsets of potentially

O-GlcNAcylated glycoproteins from human embryonic kidney HEK293T cell lysate followed by MS analysis to identify over 200 proteins (Teo et al., 2010).

Selecting a proper affinity enrichment strategy depends on the aim of the study. For most, depletion as a first step is likely to benefit analysis of glycoproteins, especially low-abundance potential biomarker candidates. On the other hand, protein–protein interactions in complex samples or nonspecific interactions with the solid phase could result in unintended losses of low-abundance proteins. Nonbiased methods like boronic acid and several separation strategies discussed later that do not rely on distinct structural characteristics of glycoproteins or glycopeptides are best for overall comprehensive enrichment. Many biomedical studies, whether discovery-based or diagnostic in nature, are interested in a subset of the glycoproteome containing a biomarker candidate displaying particular glycan elements. Such studies may be better served by a carefully selected lectin affinity strategy. Even more effective are combinations of affinity strategies in conjunction with chromatographic separation and SPE.

2.3. Glycoprotein digestion

Upon isolation of glycoproteins, digestion into peptides using proteolytic enzymes is the next step in bottom-up approaches. For most proteins, specific proteases like trypsin cleave at well-defined sites, resulting in peptides in length that are readily ionized, well fragmented by collision-induced dissociation (CID) tandem MS, and have predictable sequences for protein database searching. However, some drawbacks with using trypsin for glycoprotein digestion have been reported. While some glycoproteins contain cleavage sites in abundance, others, such as transmembrane glycoproteins that densely populate lipid bilayers of cells, may contain few cleavage sites and produce long glycopeptides upon digestion that are difficult to detect by MS due to decreased ionization efficiency or instrument limitations. Such long peptides may also contain several glycosylation sites, fatally confounding glycan assignment (Hua, Hu, et al., 2013). Additionally, glycans themselves can sterically hinder access to nearby tryptic cleavage sites and cause missed cleavages (Dodds, Seipert, Clowers, German, & Lebrilla, 2009).

Alternative specific proteases, nonspecific proteases, and multiple-protease digestion strategies can be employed to overcome these limitations and provide increased coverage of glycosylation sites upon MS analysis.

Proteins that are poorly digested by trypsin alone have been successfully analyzed following digestion with chymotrypsin (Grass, Pabst, Chang, Wozny, & Altmann, 2011; Nyalwidhe et al., 2013), pepsin (Taga, Kusubata, Ogawa-Goto, & Hattori, 2013), and Glu C–trypsin mix (Pompach, Chandler, Lan, Edwards, & Goldman, 2012). In a complex glycoproteomics experiment, Chen et al. demonstrated that pepsin and thermolysin digestion complemented trypsin digestion for human liver tissue samples, increasing the number of identified glycosites by half (Chen et al., 2009). The nonspecific proteinase K and broadly specific pronase (a protease cocktail) produce short glycopeptides three to eight amino acids in length that are perhaps more useful for site-specific glycosylation analysis (Clowers, Dodds, Seipert, & Lebrilla, 2007; Temporini et al., 2007). The resulting glycans with short amino acid sequence "tags" are then appropriate for proved glycan separation techniques like hydrophilic interaction chromatography (HILIC) or porous graphitized carbon (PGC) (Froehlich et al., 2011; Zauner, Koeleman, Deelder, & Wuhrer, 2010). Recently, Plomp et al. used trypsin, proteinase K, and chymotrypsin to digest polyclonal IgE and were able to determine site-specific assignments and structural characterization of all six N-linked glycans as a result of the complementary peptide sequences (Plomp et al., 2013). Schiel et al. employed extended pronase digestion of RNase B to achieve universal proteolysis and obtain N- and O-linked single amino acid glycans, which were then permethylated and subjected to MS^n analysis (discussed later in this review) to identify detailed isomeric structure information. This alternative glycan "release" strategy mitigates some limitations to traditional glycan cleavage strategies (see below), though peptide sequence and glycosite identification are compromised (Schiel, Smith, & Phinney, 2013). Hua et al. were able to achieve site-specific, isomeric, and quantitative glycan profiling with rapid, in-solution proteinase K, pronase, and subtilisin digestion to yield short glycopeptides in a strategy called glycoanalytical multispecific proteolysis (Glyco-AMP) (Hua, Hu, et al., 2013).

2.4. Glycan release

Once glycopeptides are obtained, glycans may be enzymatically or chemically released to facilitate separate analyses of stripped peptides by traditional shotgun proteomics and/or glycans by glycomics strategies. The enzyme PNGase F is widely used for complete cleavage of high-mannose, complex, and hybrid N-glycans (except those with α(1–3)-linked core fucose) from the asparagine side-chain amide, converting the asparagine to aspartic acid through a deamidation process and introducing a mass shift of 0.9840 Da. While these deamidation modifications can act as an indicator of a glycosylation site, spontaneous deamidation reactions can occur during sample preparation and produce false-positives. To increase confidence in site assignment, performing the deglycosylation reaction in $H_2{}^{18}O$ to impose a mass shift of 2.9890 Da through the incorporation of ^{18}O at glycosylation sites has been proposed (Küster & Mann, 1999). However, this has recently been further investigated, and it was shown in a large-scale N-glycoproteomics experiment that uncertainty remains as chemical deamidation at N-linked consensus sites can occur with incorporation of ^{18}O and is dependent on factors such as pH, temperature, reaction time, and proximity to glycine and serine (Palmisano, Melo-Braga, Engholm-Keller, Parker, & Larsen, 2012). Furthermore, partial incorporation of ^{18}O at the C-terminus of a peptide may also confound site identification (Lin, Lo, Simeone, Ruffin, & Lubman, 2012). Thus, the interpretation of a deamidation modification for N-glycan site assignment still requires discretion.

An alternative family of enzymes is endo-β-*N*-acetylglucosaminidase (ENGase) which specifically hydrolyzes the glycosidic bond between the two GlcNAc residues of the N-linked chitobiose core while retaining a terminal GlcNAc residue at the asparagine, which can be detected by a 203.0793-Da mass shift, as an unambiguous marker of glycosylation. Whereas PNGase F cleaves nearly all N-linked glycans, ENGases are not as widely specific but provide complementary site identification. For example, Endoglycosidase H (Endo H) cleaves only at high-mannose and hybrid glycans but is tolerant of the core fucosylation sometimes present on hybrid and complex glycans, so detection of core fucosylation by a 349.14-Da mass shift provides indication of a hybrid glycan (Zhang, Wang, Zhang, Yao, & Yang, 2011). Increased core fucosylation has been implicated in inflammation and cancer and can be more sensitive and specific than corresponding protein abundance (Drake et al., 2011; Miyoshi, Moriwaki, & Nakagawa, 2008), making Endo H a potentially useful tool in fucosylation biomarker studies. Endo M, on the other hand, does not cleave in the presence of core fucosylation but includes biantennary complex glycans (Segu, Hussein, Novotny, & Mechref, 2010). Endo D is limited to certain trimannosyl glycans with tolerance of fucose. Endo F1 cleaves high mannose, hybrid, and GlcNAc-bisected hybrid; Endo F2 cleaves high mannose and biantennary complex glycans; and Endo F3 cleaves bi- and triantennary complex glycans, with fucose position-dependent specificity (Gerlach, Kilcoyne, Farrell, Kane, & Joshi, 2012). Exoglycosidases β-galactosidase, neuraminidase, and *N*-acetyl-β-glucosaminidase have been used in conjunction with Endo D, Endo H, and Endo M to enable site assignment of complex glycans (Hägglund et al., 2007; Segu et al., 2010), though the exoglycosidase treatment limits glycan characterization. In a recent study, Lin et al. used both PNGase F and Endo F3 for comprehensive site-specific N-glycosylation and core fucosylation analysis of alpha-2-macroglobulin, identifying six out of eight potential N-glycosylation sites and characterizing glycoforms for three sites; Endo F3 provided five site assignments and uniquely revealed core fucosylation at three sites (Lin et al., 2012). The range of specificities and the confidence of glycosylation site assignment afforded by the preservation of GlcNAc and fucosylated GlcNAc make the ENGase family a versatile, though perhaps underexplored, alternative for N-glycan release and site-specific study.

The release of O-linked glycans is commonly performed through chemical β-elimination due to the lack of broadly specific enzymes for O-linked glycan core structures. The classic reductive β-elimination method (Carlson, 1968), though still widely used, results in loss of the glycan reducing end and suffers from low sensitivity due to excessive salt cleanup (Goetz, Novotny, & Mechref, 2009). Milder, nonreductive β-elimination methods have been developed which are better suited for sensitive glycan MS analysis and yield either permethylated or pyrazolone-derivatized O-glycans that can be separated by RP-HPLC or purified by PGC SPE (Furukawa et al., 2011; Goetz et al., 2009; Wang, Fan, Zhang, Wang, & Huang, 2011; Zauner, Koeleman, Deelder, & Wuhrer, 2012). The method described by Furukawa et al. also derivatized the deglycosylated peptides at the O-linked glycosylation sites and phosphorylation sites, allowing some site specificity to be determined. Hydrazinolysis is another method that for releasing O-glycans with free-reducing termini, undesirable and destructive "peeling" remains a problem (Kozak, Royle, Gardner, Fernandes, & Wuhrer, 2012). Nonspecific digestion of O-glycoproteins with pronase followed by PGC SPE can yield O-glycans attached to very short peptide "tags" that enable site-specific, isomer-specific, and quantitative O-glycan analysis by chip-based PGC nano-LC–MS/MS (Hua, Nwosu, et al., 2011; Nwosu et al., 2011). A recent review rigorously covering O-glycosylation analysis has been published (Zauner, Kozak, et al., 2012).

2.5. Chromatographic separation and SPE

Separation of glycopeptides from nonglycosylated peptides based on their physicochemical properties by chromatographic means serves to further simplify complex samples to allow sensitive downstream analysis by MS. Following tryptic digestion of glycoprotein samples, glycopeptides make up only 2–5% of the peptide mixture (Alvarez-Manilla et al., 2006). Established RP and strong cation exchange separation methods for general proteomics applications are less effective for separating native, intact glycopeptides due mainly to the size and hydrophilicity of the attached glycans. Glycosylated peptides are poorly retained on hydrophobic RP stationary compared to their deglycosylated counterparts, and separation of a complex glycopeptide mixture is mainly based on peptide sequence. Efficient separation of glycopeptide glycoforms displaying differences in glycan composition but similar glycan size is generally not observed due to similar hydrophobicity; rather, separation occurs based mainly on glycan size (Otvos, Urge, & Thurin, 1992). Coelution of glycoforms of similar mass can be problematic in that abundant glycoforms can suppress the signals of less-abundant glycoforms. Instead, chromatographic methods based on size-exclusion chromatography (SEC), HILIC, electrostatic repulsion hydrophilic interaction chromatography (ERLIC), or using PGC are commonly used for native glycopeptide separation. The ability of a chromatographic technique to separate isomeric glycopeptides or isomeric glycan structures is especially useful for biomarker studies in which specific glycan isomers or alterations in isomeric abundance signal a disease state.

SEC allows separation of N-linked glycopeptides in particular from nonglycosylated peptides based on the considerable amount of added bulk of N-glycans. This technique has been shown to give a threefold increase in observed glycosylation sites (Atwood et al., 2005).

HILIC is a variation of a normal-phase HPLC using a polar, hydrophilic stationary phase with a less polar mobile phase of organic solvent (typically acetonitrile) in an aqueous buffer at concentrations between 50% and 95% ACN. Most glycopeptides can be well retained on the hydrophilic stationary and well separated with an eluting gradient of increasing aqueous buffer, though highly hydrophobic glycopeptides are not retained (Alley, Mechref, & Novotny, 2009a). For example, zwitterionic HILIC (ZIC-HILIC) functionalized with sulfobetaine groups—one of many functionalized HILIC phases—was shown to separate sialylated N-glycopeptides with isomeric tri- and tetraantennary N-glycans (Takegawa et al., 2006). The retention mechanism and selectivity can vary greatly depending on solid support and functional group as well as mobile phase composition. ERLIC combines HILIC mode of separation on ion-exchange stationary. At low pH, retention acts by hydrophilic interaction for glycopeptides displaying noncharged glycans and by charge-based repulsion forces for those displaying charged glycans with sialic acid. Nonmodified peptides flow through, and an elution gradient of increasing aqueous buffer separates glycopeptides well. Phosphopeptides are also retained by ERLIC, but phosphatase treatment prior to separation eliminates copurification. Hydrophilic interaction chromatography is now a popular approach to glycopeptide and glycan separation and purification due to its efficient yet flexible modes of separation. Recent, extensive reviews of HILIC and ERLIC stationary phases and their current applications to glycoproteomics and glycomics are available elsewhere for further information (Chen, Su, Huang, Chen, & Tai, 2014; Ongay et al., 2012; Zauner, Deelder, & Wuhrer, 2011).

PGC is a highly effective material for separation and SPE of glycans and glycopeptides. Used in SPE cartridges, glycopeptides are retained and nonglycopeptides flow through. Glycopeptide retention is a function of both peptide and glycan structure in that retention of small peptides is controlled more by the glycan and retention of large peptides is less controlled by the glycan, so glycopeptide separation by PGC is most advantageous for short peptides made by non- or broadly specific proteases like proteinase K or pronase. It has been shown to be particularly useful in separating isomeric glycoforms (Mechref & Novotny, 2002). The introduction of PGC in chip-based nanoflow LC (Alley, Mechref, & Novotny, 2009b) has enabled rapid and sensitive online separation and MS analysis of pronase and proteinase K glycopeptides to provide detailed site-specific glycosylation information (Froehlich et al., 2011; Hua, Nwosu, et al., 2011). Microfluidic chip-based PGC combined with nano-LC–MS has been recently used by Hua et al. to separate and quantify native N-glycans from the serum of prostate cancer and ovarian cancer patients and allow rapid and detailed compositional and structure-specific profiling of potential glycan biomarkers (Hua, An, et al., 2011; Hua, Williams, et al., 2013).

Purification of glycans released from peptides and their chromatographic separation are important steps for sensitive glycan-centric analyses by MS. Isolation of glycans from peptides can be done with C_{18} or C_8 sorbents in the RP, where peptides are bound and glycans flow through. Both purification and chromatographic separation are commonly fulfilled by HILIC or PGC for native glycans. In a recent example, Hua et al. used PGC SPE to both purify PNGase F-released native N-glycans from mouse serum proteins and separate them online using chip-based PGC nano-LC for MS and MS/MS analysis, enabling isomer-specific structural analysis (Hua, Williams, et al., 2013). Parker et al. used both PGC separation of native

N-glycans and ZIC-HILIC purification of N-linked glycopeptides followed by orthogonal offline (pH ~8) and online (pH ~3) RP-HPLC glycopeptide separation in a multidimensional approach for site-specific glycan/glycopeptide characterization by nano-LC–MS/MS (discussed further in Section 2.6) (Parker et al., 2013). However, in contrast to the examples just described, glycans are commonly first derivatized by permethylation, sialic acid modification, or reducing end modification to increase their hydrophobicity, which can facilitate retention, improve recovery, and enhance separation (Walker, Carlisle, & Muddiman, 2012). As a detailed summary of glycan-specific separation techniques is beyond the scope of this review, the reader is referred to other recent publications (Alley et al., 2013; Harvey, 2011; Ruhaak et al., 2010; Yang & Zhang, 2012).

2.6. Mass spectrometry

The most widely used ionization methods for glycopeptide and glycan analysis by MS are MALDI and electrospray ionization (ESI). In MALDI analysis, the analyte is combined with a matrix which facilitates ionization into singly charged species, usually via a sodium ion. In ESI analysis, analytes in solution are aerosolized into multiply charged species. ESI is a gentler ionization technique and benefits from the ability to be interfaced with online liquid chromatography techniques. MALDI, on the other hand, can cause source dissociation of labile glycosidic bonds, especially those containing glycans with sialic acids or fucose, so derivatization is usually a prerequisite for MALDI MS analysis (Leymarie & Zaia, 2012). While ESI is capable of native glycan ionization, derivatization benefits both ionization methods as the inherent hydrophilicity of glycans results in poor ionization and signal suppression during ESI. Derivatization of glycans at hydroxyl groups, sialic acids, or reducing ends prior to MS analysis increases their hydrophobicity, which facilitates their ionization and detection. Permethylation is the most common derivatization, which modifies hydrogens on hydroxyl groups, carboxyl groups, and amines by replacing them with methyl groups (Ciucanu & Kerek, 1984). This not only stabilizes sialic acids for MALDI analysis but also renders acidic glycans neutral, facilitating positive-mode MS analysis (Guillard et al., 2009) while also enabling cross-ring MS/MS fragmentation mechanisms for linkage/branching structural elucidation (Prien, Ashline, Lapadula, Zhang, & Reinhold, 2009). Derivatization of the glycan reducing end by reductive amination for incorporation of hydrophobic tags, UV/fluorescent tags, or stable-isotope-labeled tags for quantitation are common, as are pyrazolone and hydrazone derivatization (Walker et al., 2012). Comprehensive reviews covering glycan derivatization, chromatographic separation, and MS analysis specifically have been published recently (Alley et al., 2013; Harvey, 2011; Kailemia, Ruhaak, Lebrilla, & Amster, 2014; Wuhrer, 2012).

Direct tandem mass analysis of intact glycopeptides to glean information on the peptide sequence, glycosite location, and glycan characteristics is a complex and challenging task. Typically, a single fragmentation mode or stage offers only one piece of information. Tandem mass fragmentation of glycopeptides by CID results predominantly in cleavage of the glycan but leaves the peptide backbone relatively intact, revealing glycan composition based on B- and Y-type fragmentation of glycosidic linkages at the expense of peptide sequence and glycosylation site information. Ion trap instruments capable of multiple-stage tandem mass (MS^n) events can provide peptide backbone fragment ion spectra by following the MS/MS scan with an MS^3 scan in which the remaining intact peptide ion is isolated and fragmented to produce B- and Y-type peptide backbone fragment ions. Partial retention of N-linked GlcNAc on some fragments allows determination of glycosylation site. Higher orders of MS^n can be used for analysis of released glycans to elucidate linkage and branching of structural isomers (Jiao, Zhang, & Reinhold, 2011; Prien et al., 2009). Quadrupole-time-of-flight (Q-TOF) instruments produce different glycopeptide fragmentation characteristics based on applied collision energy. At low energy, predominantly glycosidic bond cleavage is observed; at high energy, peptide backbone cleavage prevails with few observed glycan fragments, though retention of N-linked GlcNAc may be evident depending on peptide sequence (Wuhrer, Catalina, Deelder, & Hokke, 2007). Higher energy collision-induced dissociation (HCD) in the C-trap of Orbitrap instruments generates intense, distinct y_1 ions of the peptide + GlcNAc which can serve as a good marker for glycosylation site identification, especially when detected at high mass accuracy (<1 ppm) in the Orbitrap. Additionally, because the C-trap lacks the low mass cutoff inherent to ion traps, strong HexNAc oxonium ion signals are observed in the low mass range, which also serve as markers for glycosylation. However, CID offers stronger glycan fragmentation signal in the higher mass region (Segu & Mechref, 2010).

Electron-activated dissociation (ExD) techniques like electron capture dissociation (ECD) and electron transfer dissociation (ETD) fragment the peptide backbone while retaining the glycan intact, allowing simultaneous peptide sequencing and glycosylation site determination based on c- and z-type peptide fragment ions and observed mass shifts attributed to the attached glycan (Håkansson et al., 2001). In Fourier transform ion cyclotron resonance instruments (FT-ICR), soft fragmentation of glycopeptides by ECD is carried out at low kinetic energy (1.5 eV), resulting in peptide backbone fragments (Adamson & Håkansson, 2006). Hot-ECD at moderate kinetic energy (9 eV) has been applied to permethylated glycans, resulting in glycosidic and cross-ring fragmentation, and at high kinetic energy (14 eV), electronic excitation dissociation takes place, resulting in more extensive cross-ring fragmentation to elucidate glycan composition, branching, and linkage information in detail (Yu, Huang, Lin, & Costello, 2012; Yu et al., 2013). Other electron-aided fragmentation methods applied to glycopeptides in negative mode include electron detachment dissociation (Leach et al., 2012) and negative electron transfer dissociation (Wolff et al., 2010).

Strategies that alternate or combine multiple fragmentation modes have recently been reported for more complete intact glycopeptide analysis, as have concerted approaches that separately analyze released glycans and their deglycosylated peptide counterparts in order to obtain more detailed information on each of the parts. Scott et al. employed a strategy of parallel CID, HCD, and ETD for N-linked glycoproteomic analysis of *Campylobacter jejuni* to identify 130 glycopeptides with 75 glycosylation sites following ZIC-HILIC enrichment of glycopeptides (Scott et al., 2011). Singh et al. utilized HCD and product ion-triggered ETD fragmentation of N-linked glycopeptides, based on observed oxonium ions from the initial HCD MS/MS spectra, to identify glycopeptide sequence, glycan localization, and glycan structure from RNase B and IgG digests (Singh, Zampronio, Creese, & Cooper, 2012). Yin et al. compared alternating HCD–ETD and HCD product ion-triggered ETD MS/MS for analysis of ZIC-HILIC-enriched derivatized glycopeptides, which were either analyzed directly or treated with PNGase F in $H_2^{18}O$ and subjected to separate peptide and glycan analysis. They found that HCD product ion-triggered ETD detected more complex/hybrid glycoforms and lower abundant species compared to alternating HCD–ETD, with alternating HCD–ETD selecting larger glycopeptides at higher intensities, higher charge, and smaller m/z, resulting in little overlap of glycopeptide identification between the two acquisition methods. They also observed little overlap in identifications between direct glycopeptide analysis and separate peptide and glycan analysis (Yin et al., 2013). Halim et al. combined complementary CID-MS^2/MS^3 and ECD/ETD acquisition for O-glycan structure characterization, glycopeptide identification, and O-glycosylation site determination in a nano-LC–MS/MS analysis of hydrazide-enriched glycopeptides from human CSF (Halim, Rüetschi, Larson, & Nilsson, 2013) in an approach that the authors had previously demonstrated for human urinary glycoprotein characterization (shown in Fig. 3.3) (Halim et al., 2012). An integrated, multidimensional glycoproteomics and glycomics approach to site-specific analysis of N-linked glycosylation heterogeneity in rat brain using complementary fragmentation

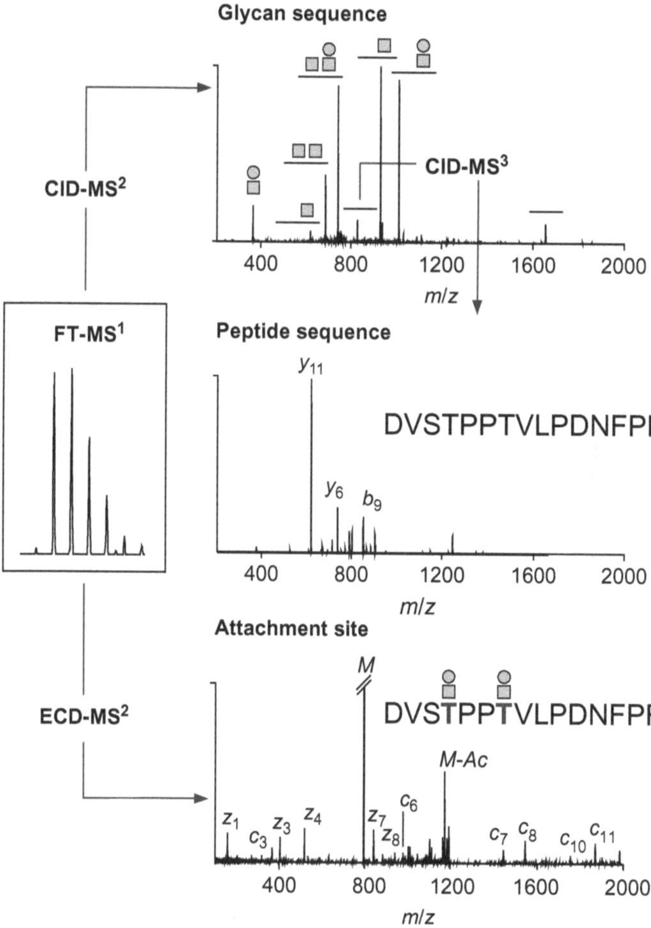

Figure 3.3 Schematic of mass spectrometric characterization of glycan sequence, peptide sequence, and glycosylation site identification by CID-MS^2, CID-MS^3, and ECD, using a desialylated O-glycosylated tryptic peptide as an example. Yellow circle (gray circle in print version), Hex; yellow square (gray square in print version), HexNAc. *Adapted with permission from Halim, Nilsson, Rüetschi, Hesse, and Larson (2012). Copyright 2012 American Society for Biochemistry and Molecular Biology.*

modes was recently reported by Parker et al. The investigators first used PGC LC–MS/MS for global characterization of PNGase F released native N-glycans to create a database consisting of 71 structurally unique N-glycans. Then, tryptic glycopeptides were enriched by ZIC-HILIC, fractionated by offline orthogonal RP-HPLC (pH 7.9), and analyzed by online nano-LC–MS/MS on a Thermo LTQ Orbitrap Velos in two separately acquired sample sets as either intact glycopeptides or PNGase F-deglycosylated peptides. Intact glycopeptides were subjected to CID, HCD, and ETD fragmentation, and deglycosylated peptides were subjected to HCD fragmentation. Glycopeptide sequences were determined by HCD spectra, and glycans were assigned from the database based on glycopeptide CID spectra linked to HCD spectra containing an oxonium ion signal or based on ETD spectra. This multifaceted, comprehensive approach enabled confident identification of 863 unique N-linked glycopeptides from 161 rat proteins (Parker et al., 2013). Further information on fragmentation strategies can be found in recent reviews covering MS/MS analysis of glycopeptides (Dodds, 2012), glycans (Zaia, 2010), and both (Leymarie & Zaia, 2012) have been published recently.

Ion mobility mass spectrometry (IMS-MS) offers a unique method of separation of gas-phase ions during MS acquisition. As ions travel through a specialized drift tube containing an inert buffer gas, ions separate based on their mobilities through the gas, which depends on the shape and conformation of the ions. Isobaric glycan isomers differing by their glycosidic linkage arrangements and branching can be distinguished based on their drift times and resolved prior to MS detection. Clemmer and coworkers used ESI-IMS-TOF/MS to separate and characterize PNGase F released ovalbumin N-linked glycans, revealing 19 different glycan structures with 42 distinct isomeric features (Plasencia, Isailovic, Merenbloom, Mechref, & Clemmer, 2008). McLean and coworkers demonstrated MALDI-IM-MS separation of singly charged peptides and glycans from a PNGase F-treated RNase B digest, as well as ESI-IM-MS separation of singly charged peptides, singly charged glycans, doubly charged peptides, and higher order charged species (Fenn & McLean, 2009). The same researchers later showed structural separation of isobaric positional and structural isomers using a library of synthetic glycan isomers and combined pre- and post-IM fragmentation to improve confidence in MS identification (Fenn & McLean, 2011). Harvey et al. demonstrated the versatility of traveling wave ion mobility mass spectrometry for negative mode native N-glycan analysis by ESI-MS/MS on a Synapt G2 Q-TOF. A complex mass spectrum of N-glycans released from bovine fetuin by hydrazinolysis could be extracted into discrete spectra of singly, doubly, or triply charged glycans, as well as doubly, triply, and quadruply charged glycopeptides. N-glycan spectra contaminated by PEG could likewise be extracted into discrete singly, doubly, and triply charged spectra and properly identified. Small isomeric glycans from ovalbumin separated well and could be characterized in detail by their CID fragmentation; large glycans, the authors note, require higher resolution ion mobility separation than currently available in commercial instruments (Harvey et al., 2013). Isomeric glycopeptides with different glycan site localization can also be characterized by IMS. Creese et al. showed that isomeric O-linked GalNAc glycopeptides from mucin 5AC that coelute during online LC can be separated by high-field asymmetric wave ion mobility spectrometry at different compensation voltage ranges and fragmented by supplemental activation ETD to reveal the peptide structure and O-linked glycosylation sites (Creese & Cooper, 2012).

2.7. Quantitation

The correlation between changes in glycosylation and disease makes relative quantitative assessment an important part of glycoproteomics studies of disease states. Relative quantitation of two or more similar samples by MS requires special considerations in that things like ionization efficiencies, differences in sample matrix, instrument to instrument variability, instrument performance variability, and sample preparation all introduce errors into the quantitative analysis. In general, MS-based quantitative approaches either incorporate stable isotope labels onto analytes or are label-free. Label-free strategies are attractive in that they require little to no modification to a glycoproteomics workflow, instead relying more heavily on data analysis and bioinformatics to interpret the semiquantitative value of acquired data through normalized ion intensities or spectral counting, but generally ought to be considered a method for screening of potential biomarkers rather than for definitive quantitative assessment (Orlando, 2013). Isotopic mass difference labeling strategies enable relative quantitation of a few chemically derivatized peptide samples within an MS run through comparison of heavy- and light-labeled peptide ion peak abundances, while isobaric-labeling strategies involve chemically labeling several peptide samples with a multiplexed set of isobaric reagents that are indistinguishable in the MS scan but form discrete reporter ion peaks during MS/MS fragmentation, enabling relative quantitation based on reporter ion peak abundances

(Iliuk, Galan, & Tao, 2008). These approaches require extra sample preparation by derivatization with costly stable isotope reagents but compensate for variations in ionization efficiencies and instrument to instrument variability for coeluting labeled analytes; still, since isotopic labels are incorporated during sample preparation, losses due to sample preparation can introduce errors (Orlando, 2013).

Both glycopeptides and glycans can be labeled and quantified using stable isotope labels. The abundance of enriched glycoproteins can be determined using established proteomics labeling strategies, like stable isotope dimethylation (Hsu, Huang, Chow, & Chen, 2003), ICAT (Gygi et al., 1999), iTRAQ (Ross et al., 2004), or TMT (Thompson et al., 2003) by labeling digested glycopeptides, combining samples to be compared, and analyzing by ESI LC–MS/MS. A study comparing protein-level and peptide-level isobaric labeling for serum glycoprotein quantification was recently published by Nie et al. under the justification that variance in sample preparation prior to peptide labeling contributes to quantitative error that can be eliminated by labeling at the protein level. Three commercial isobaric tags for tandem mass quantitation—iTRAQ 4-plex, TMT 6-plex, and iTRAQ 8-plex—were compared, optimal solvent conditions for protein labeling were tested, and several proteases—trypsin, trypsin–GluC, chymotrypsin, and Asp-N—for digestion of labeled proteins were explored. Under the optimal conditions, immunodepleted human serum enriched for glycoproteins with AAL LAC was labeled in 50 mM TEAB:DMSO, combined, digested with trypsin and Asp-N, deglycosylated with PNGase F, and analyzed via ESI LC–MS/MS on a Thermo Orbitrap Elite mass spectrometer using HCD fragmentation. Alternatively, enriched glycoproteins were digested with trypsin prior to labeling for comparison. The combination of trypsin and Asp-N resulted in a 30% increase in quantified proteins over the other proteases, and iTRAQ 4-plex gave a 20% increase in identified and quantified proteins over the other labeling strategies. Protein-level labeling performed slightly better than peptide labeling, with 169 vs. 140 identified proteins and 135 vs. 125 quantified proteins at a higher quantitative precision (RSD = 11% vs. 15%) (Nie et al., 2013). Unfortunately, quantitative approaches like this may have limited implication for disease biomarker studies without glycosylation site or glycoform analysis. In another study, using bovine fetuin as a model glycoprotein for qualitative and quantitative characterization of both glycopeptides and glycoforms, Ye et al. combined TMT labeling of tryptic glycopeptides and LC–MS/MS analysis using a Thermo LTQ Orbitrap XL instrument, acquiring tandem mass spectra via alternating CID, HCD, and ETD fragmentation, to identify 23 glycoforms from five glycopeptides. Glycopeptide sequences and glycosylation sites were determined from ETD and HCD spectra, glycopeptide quantitation was determined by TMT reporter ions from HCD spectra, and N- and O-linked glycoforms were identified at high mass accuracy (<3 ppm) from high-resolution (100,000 resolving power) CID spectra. Poor ETD glycopeptide fragment ion signal and poor HCD reporter ion signal were overcome by using an Advion BioSciences Triversa NanoMate to collect and reinfuse 15s target fractions, and 50–100 scans were averaged to obtain high-quality ETD and HCD spectra. It was observed that changing the HCD collision energy yielded different sets of ions—at high energy, b and y ions were observed, and reporter ion signal was strong; at low energy, glycosidic bond cleavage was preferred (Ye, Boyne, Buhse, & Hill, 2013).

Because changes in glycoforms can be biomarkers for disease, differential labeling techniques for glycan quantification have also been introduced. In this approach, glycans are released from two glycopeptide samples (e.g., healthy and disease state), labeled with "light" and "heavy" stable isotope reagents, combined, and then analyzed in a single LC–MS run, yielding peak pairs in the MS parent scan whose abundances can be used for relative quantitation. Permethylation of N-glycans by $^{12}CH_3I$ and $^{13}CH_3I$ allows quantitation of the resulting light and heavy differentially labeled samples upon LC–MS/MS analysis of the combined sample, though this method imparts variable different mass shifts depending on glycan structure and cannot quantify differences between isomeric glycans (Alvarez-Manilla et al., 2007, 2010). By instead labeling with $^{13}CH_3I$ and $^{13}CH_2DI$, which impart much smaller mass differences, and analyzing with high-resolution MS/MS, isomeric glycoforms with the same nominal mass are isolated and fragmented together but can be differentiated and quantitated by differences in product ions in the MS/MS spectra (Atwood et al., 2008). The drawbacks to these methods are that if permethylation efficiency is not identical between the two samples, an error of 3–10% can be introduced depending on glycan size. Also, a heavy-labeled glycan carrying multiple deuteriums can experience a retention time shift during online RPLC separation, causing it to elute at a different time from its light-labeled counterpart and introduce analytical variability (Walker, Budhathoki-Uprety, Novak, & Muddiman, 2011). However, deuterium effects have been evaluated for ZIC-HILIC separations, and under acidic mobile phase conditions (pH 3.5), the effect is eliminated (Di Palma, Raijmakers, Heck, & Mohammed, 2011). A nonreductive isotopic labeling of glycans with nondeuterated (d_0) and deuterated (d_6) Girard's reagent P for quantitative glycomics was reported for ESI direct infusion and ESI HILIC LC–MS (Wang et al., 2013). Another glycan-labeling approach incorporates light (^{12}C) and heavy ($^{13}C_6$) hydrazide biphenyl reagents (4-phenetyl-benzohydrazide; P2GPN) on the reducing end of glycans, which increases their hydrophobicity and ESI ionization efficiency while avoiding chromatographic shifts (Walker et al., 2011). Recently, the same authors demonstrated a method, individuality normalization when labeling with isotopic glycan hydrazide tags (INLIGHT), to compensate for quantitative inaccuracies due to overlapping isotopic envelopes of light and heavy P2GPN-labeled species (Walker, Taylor, & Muddiman, 2013). An isobaric stable isotope labeling reagent, glyco-TMT, was recently reported for mass difference or isobaric tandem mass quantitation of glycans (Fig. 3.4). The reagent consists of an isotopic reporter group and a mass balance group and bears a carbonyl-reactive aminooxy group for reaction with the glycan reducing end. Light (TMT^0) and heavy (TMT^6) amino oxy labels were demonstrated to have an accessible dynamic range of 1:20 by MALDI-TOF analysis of N-glycans from ovalbumin. The isobaric versions TMT^6-128 and TMT^6-131, which instead yield discrete reporter ions in the low mass region upon tandem mass fragmentation, do not add spectral complexity to the parent scan, but were demonstrated to show limited dynamic range by MALDI-TOF/TOF analysis compared to the mass difference labels, though the choice of instrumentation was likely the limiting factor (Hahne et al., 2012). Gong et al. recently demonstrated duplex labeling of N-glycans from a monoclonal antibody with commercially available amine-reactive TMT^0 and TMT^6. Under basic conditions (pH 8.3) in 50 mmol TEAB buffer, PNGase F release of N-glycans results in stable N-glycosylamines which are readily labeled by the NHS-ester of TMT reagents. The dynamic range for three coeluting TMT^6:TMT^0-labeled glycans was shown to be accurate to 1:20 by LC–MS/MS analysis on a Thermo LTQ Orbitrap XL (Gong et al., 2013). Detailed reviews of glycan-labeling strategies for quantitative glycomics have been recently published elsewhere (Mechref, Hu, Desantos-Garcia, Hussein, & Tang, 2013; Ruhaak et al., 2010).

Once candidate biomarkers have been determined and sufficiently isolated from the complex sample, targeted, label-free MS quantitation can be performed via selected or multiple reaction monitoring (SRM or MRM) using a triple quadrupole (QqQ), Q-Trap, or hybrid QqTOF mass spectrometers. In this approach, MS analysis time is focused exclusively on analytes of interest and all others are excluded in order to increase sensitivity by several orders of magnitude. This is achieved by isolating precursor ions of interest,

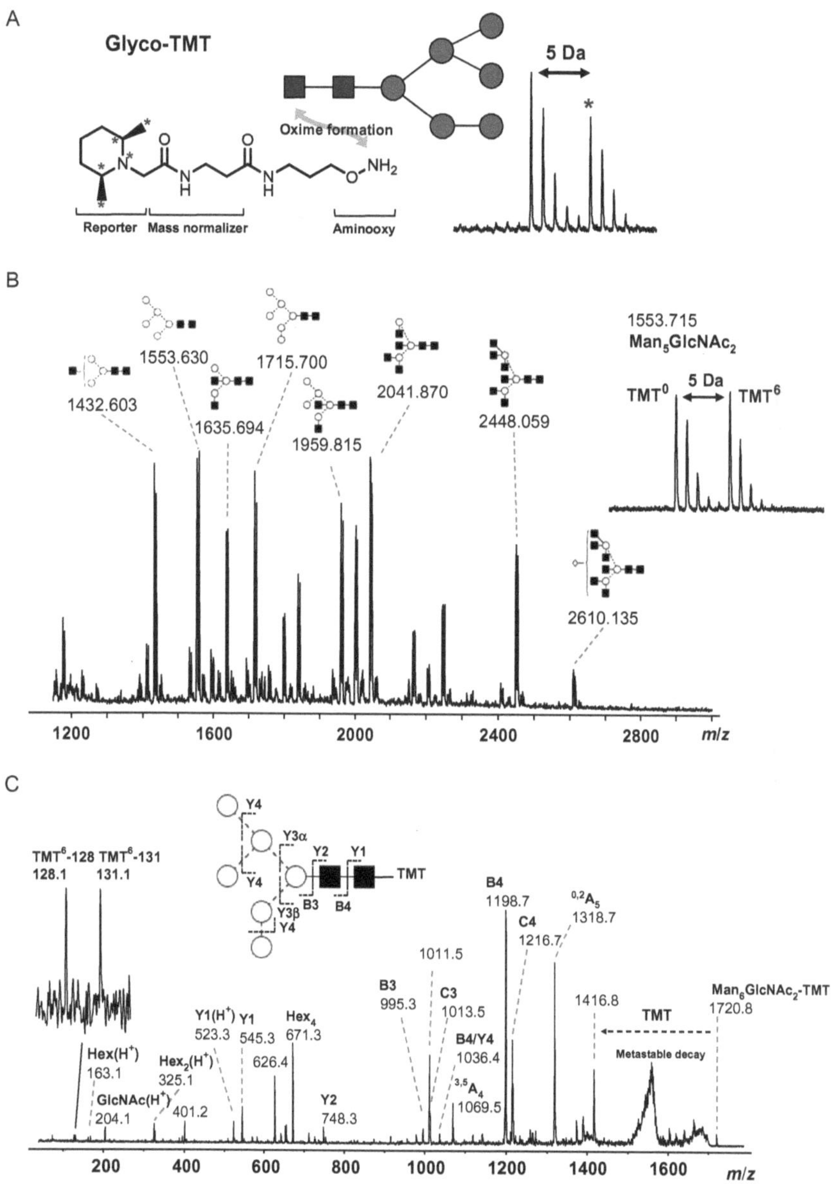

Figure 3.4 (A) Structure of Glyco-TMT reagent with reporter group, mass normalizer, and aminooxy functional group. Locations of isotopic substitution in the Glyco-TMT reporter structure are indicated (*). A mass shift of 5 Da is incorporated between heavy- and light-labeled glycans; the heavy-labeled species in the spectra is indicated (*). (B) MALDI-TOF MS^1 level quantitation of heavy- and light-labeled (TMT^0/TMT^6 1:1) N-glycans from ovalbumin. (C) MALDI-TOF/TOF MS^2 level reporter ion-based quantitation and structural determination of isobarically labeled N-linked ovalbum glycan Man_6-$GlcNAc_2$ (*m/z* 1720.708, aminooxy TMT^6-128/131, 1:1). Circle, Man; square, GlcNAc; diamond, NeuNAc. *Adapted with permission from Hahne et al. (2012). Copyright 2012 American Chemical Society.*

fragmenting them, and selecting specific product fragment ions for detection and scheduling precursor/product "transitions" based on retention times over an LC run. Absolute quantitation is accomplished by spiking synthetic, stable isotope-labeled versions of each analyte peptide in known concentrations and comparing signal intensities of the analyte and stable isotope reference to determine protein concentration (Gillette & Carr, 2013). Recent glycoproteomics studies have employed MRM approaches for quantification of core fucosylated peptides in hepatocellular carcinoma (HCC) patient serum (Zhao et al., 2011), sialylated peptides from mouse serum (Kurogochi et al., 2010), IgG glycopeptides and their site-specific glycoform abundances in serum (Hong, Lebrilla, Miyamoto, & Ruhaak, 2013), haptoglobin glycopeptides and their site-specific glycoform abundances in liver disease (Sanda et al., 2013), and serum glycopeptides based on selection of oxonium ion transitions rather than peptide ion transitions (Song, Pyreddy, & Mechref, 2012). Targeted MS-based quantitation by MRM is seeing wide clinical and bioanalytical use in biomarker studies and has been reviewed extensively (Boja & Rodriguez, 2012; Gillette & Carr, 2013; Kitteringham, Jenkins, Lane, Elliott, & Park, 2009; Lemoine et al., 2012; Liebler & Zimmerman, 2013; Meng & Veenstra, 2011; Percy, Parker, & Borchers, 2013).

A recently developed alternative method of label-free quantitation is SWATH-MS, a data-independent acquisition approach that rapidly cycles through defined 25-Da precursor isolation windows (swaths) to obtain broad fragmentation data for all analytes using Q-TOF, QqTOF, or triple TOF mass spectrometers. This technique has been shown to provide qualitative and quantitative performance comparable to SRM (Gillet et al., 2012). Aebersold and coworkers used SWATH-MS for quantitative analysis of N-linked glycoproteins in human plasma and observed similar quantitative accuracy, reproducibility, and dynamic range between SWATH-MS and SRM, though SRM was the more sensitive approach (Liu et al., 2013).

2.8. Bioinformatics

Following MS-based glycoproteomics analyses, the acquired tandem mass spectral data contains a wealth of information that must be interpreted using bioinformatics software. Such tools determine protein identifications based on sequencing of peptide fragment ions and protein database search, glycosylation site assignment based on peptide fragment ions containing markers of glycosylation (e.g., Asn->Asp conversion following PNGase F treatment, N-linked GlcNAc, retained glycans from ETD), and glycan characterization based on oxonium ions, glycosidic cleavage ions, and cross-ring cleavage ions and glycan structural database matching. Given the vast amounts of data acquired during large-scale glycoproteomics experiments, the importance of bioinformatics in the glycoproteomics workflow cannot be overstated. While a detailed summary of current bioinformatics tools is beyond the scope of this review, several recent reviews are available for further information (Dallas, Martin, Hua, & German, 2013; Li, Glinskii, & Glinsky, 2013; Woodin, Maxon, & Desaire, 2013). Recent bioinformatics tools introduced since the publication of these reviews include: SweetSEQer, a simple, open source tool for *de novo* analysis of glycoconjugate MS/MS spectra with annotations (Serang et al., 2013); GlycoFragwork, a framework for N-glycopeptide scoring and glycan sequencing that combines LC–MS/MS data set alignment, scoring of CID, HCD, and ETD spectra, and an algorithm for elucidating glycan structure based on peaks in the CID spectrum with scoring, ranking, and FDR reporting for potential glycans (Mayampurath et al., 2014); PTM MarkerFinder, a tool for screening database search output of HCD and ETD data for marker ions of glycosylation (e.g., HexNAc, Hex) (Nanni et al., 2013); GlycoPep Detector, a tool for assigning glycopeptide composition based on ETD spectra (Zhu, Hua, Clark, Go, & Desaire, 2013); and SRMAtlas, a library of SRM assays of N-glycosites for targeted quantitative analyses of N-glycoprotein biomarkers (Hüttenhain et al., 2013).

3. MS-BASED GLYCOPROTEOMICS IN DISEASE RESEARCH

Glycoproteomics and glycomics methodology development is advancing at an explosive rate. The combination of selective enrichment, efficient separation, tailored digestion strategies, quantitative labeling, and rapid, sensitive, high-resolution MS is increasingly enabling sophisticated, novel applications in a wide range of disease research fields. This review highlights just two of these fields, cancer biomarker and neurodegenerative disease research, with a focus on recent studies.

3.1. Cancer biomarker research

It has been established that aberrant glycoforms detected in plasma, serum, tissues, and bodily fluids can be associated with various types of cancers, and MS-based glycoproteomics and glycomics approaches have been used to study changes in N-and O-linked glycosylation and glycan isoforms in glycoprotein biomarkers for prostate cancer, ovarian cancer, breast cancer, colon cancer, pancreatic cancer, lung cancer, and liver cancer. Several cancer biomarkers have been identified: prostate-specific antigen for prostate cancer; CA125 and sialylated Lewis^{X} glycans for ovarian cancer; HER2, CA27-29, and O-glycan sialylation of CA15-3, and sialylated Lewis^{X} glycans for breast cancer; carcinoembryonic antigen for colon cancer; CA19-9 and fucosylated haptoglobin for pancreatic cancer; sialylated Lewis^{X} glycans for lung cancer; alpha-fetoprotein (AFP) and AFP-L3 core fucosylation for HCC (Adamczyk, Struwe, Ercan, Nigrovic, & Rudd, 2013; Kuzmanov, Kosanam, & Diamandis, 2013; Mechref et al., 2012). In a recent study by Hua et al., chip-based PGC nano-LC-TOF/MS was used to quantitatively profile N-glycans from 15 different cancer cell lines isolated from ovarian, breast, lung, cervical, and lymphatic cancer patients to uncover significant relative abundance changes in broad glycan classes (high mannose, complex/hybrid fucosylated, complex/hybrid sialylated, etc.) and individual glycan structures to differentiate between cell lines (Figs. 3.5 and 3.6) (Hua et al., 2014). Once cancer-associated glycoforms are identified,

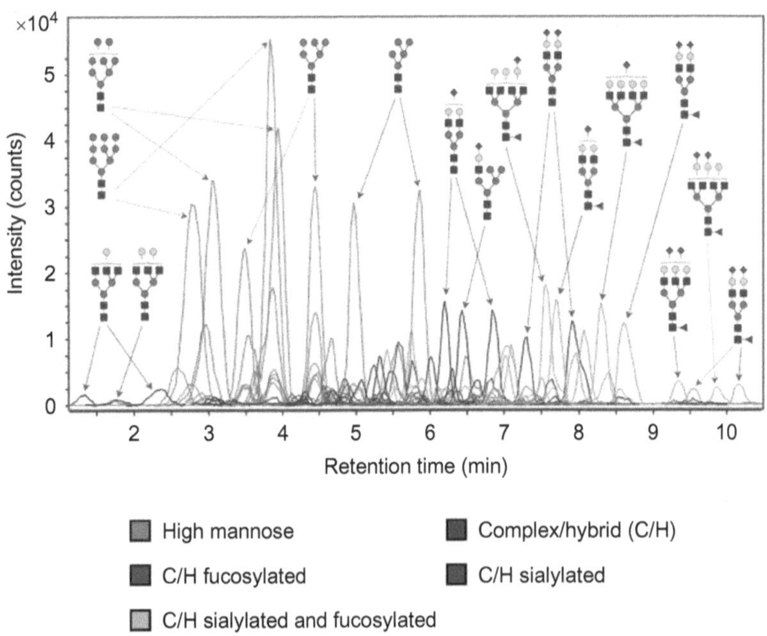

Figure 3.5 Extracted compound chromatograms of cell membrane N-glycans identified on non-CD4 T cells from human blood. Colors denote the biosynthetic class of each glycan. Glycan structures are putative, based on known biosynthetic pathways. *Adapted with permission from Hua et al. (2014). Copyright 2013 American Chemical Society.*

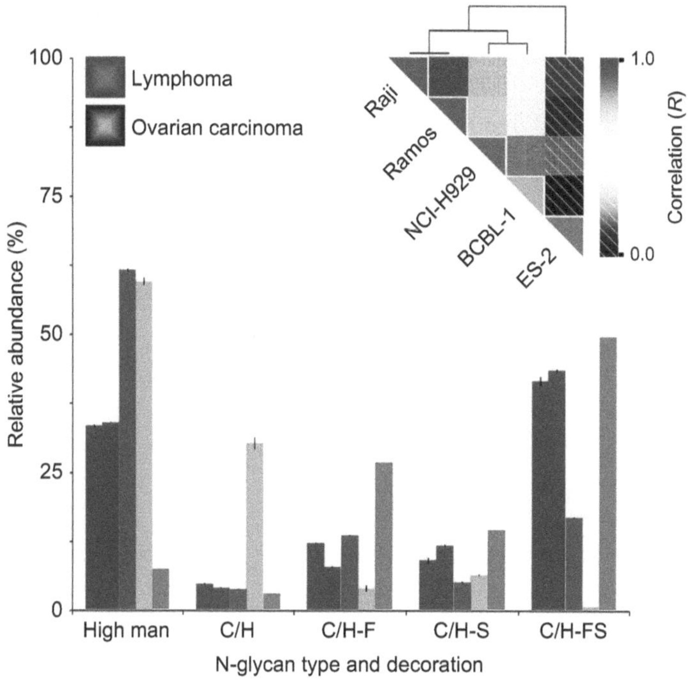

Figure 3.6 Glycan class profiles of four B-cell lymphoma cell lines (Raji, Ramos, NCI-H929, and BCBL-1) and one ovarian carcinoma cell line (ES-2). Relative abundances are shown for each glycan biosynthetic class. Inset: color-coded representation of the Pearson correlation coefficient (R) between each pair of cell lines, ranging from red (high correlation; gray top of scale in print version) to blue (low correlation; black bottom of scale with hash lines in print version), along with hierarchical clustering trees. *Adapted with permission from Hua et al. (2014). Copyright 2014 American Chemical Society.*

MS-based diagnostic protocols using specific affinity enrichment or chromatographic separations that target that glycoform may be developed.

LAC has seen wide use for MS-based cancer glycoproteomics. Many studies use LAC as an initial step for broad cancer biomarker discovery. For example, HCC cell lines have been studied using Con A-immobilized magnetic nanoparticles (Tang et al., 2010; Yang et al., 2013); breast cancer biomarker candidates have been determined using SLAC (Selvaraju & Rassi, 2012) and MLAC (Zeng et al., 2011) as initial enrichment methods; and pancreatic cancer biomarker candidates have been determined from pancreatic cyst fluids using MLAC enrichment (Gbormittah et al., 2013). Lectins have also been used to target specific glycan characteristics as biomarkers. Drake et al. reported a LAC workflow using *Aleuria aurantia* lectin (AAL) and *Sambucus nigra* agglutinin to enrich glycoproteins with fucose and sialic acid from several luminal (less aggressive) and triple negative (more aggressive) breast cancer cell lines, followed by deglycosylation with PNGase F and LC–MS/MS analysis on a Thermo LTQ Orbitrap XL. Their statistical analysis of over 1000 glycosites on 533 identified glycoproteins revealed that 100 glycosites were solely detected in several triple negative lines, and differential expression of fucosyl- and sialyltransferases between the two line subtypes suggests that changes in glycosylation may act to indicate putative biomarkers (Drake et al., 2012). Ahn et al. also used AAL to enrich fucosylated glycoproteins from small-cell lung cancer (SCLC) patient sera samples. A combination of label-free and iTRAQ-labeled quantitative MS analysis revealed decreases in abundance of serum paraoxonase (PON1) alongside increases in the degree of N-linked glycan fucosylation of PON1 (Ahn, Sung, et al., 2013), findings which support previous reports of increased fucosylation and sialylation of serum PON1 in liver cirrhosis patients (Sun et al., 2012) and suggest PON1 glycosylation patterns as a biomarker for both HCC and SCLC.

IMS-MS can uniquely facilitate MS-based cancer biomarker glycoproteomics research due to its ability to separate glycan isomers during MS analysis, prior to detection. Isaiolovic et al. demonstrated the potential of IMS-MS for cancer biomarker discovery by characterizing serum N-linked glycans from patients with cirrhosis of the liver and liver cancer patients. The combination of supervised principal component analysis (PCA) with ion mobility distributions of glycan isomers and conformers of 10 different glycans was sufficient to discriminate cirrhosis and liver cancer from healthy patients (Isailovic et al., 2012). The same IMS–MS and PCA approach of serum N-glycan analysis was then used for distinguishing phenotypes of Barrett's esophagus, high-grade dysplasia, and esophageal adenocarcinoma from normal control. Composite ion mobility distributions constructed from 11 glycans revealed 46 features that allowed unambiguous differentiation of the esophageal adenocarcinoma samples from the normal control samples; the authors noted, however, that improvements in IMS separation efficiency are required to observe individual isomer contributions (Gaye et al., 2012).

Abnormal or incomplete O-glycosylation has been recognized as a biomarker for cancer and is commonly observed in carcinomas. Mucins and glycoproteins in tumor cells feature truncation of O-glycans to GalNAc-α-Ser/Thr, also known as the Tn antigen, which does not appear in normal tissues. It has been reported that 70–90% of cancers of the colon, lung, bladder, cervix, ovary, stomach, and prostate express this O-glycan truncation, and in other cancers, the expression correlates with poor prognosis and metastasis

(Ju et al., 2013). O-GlcNAc has also been implicated in the cause and progression of cancer (Fardini, Dehennaut, Lefebvre, & Issad, 2013). The targeted enrichment of O-linked GalNAc or GlcNAc has been demonstrated recently using the *Helix pomotia* agglutinin lectin with human breast cancer cell lines (Rambaruth, Greenwell, & Dwek, 2012), and O-GlcNAc enrichment by reversible hydrazide chemistry, mentioned previously, has also been developed (Nishikaze et al., 2013). The identification of O-GlcNAc sites by MS/MS has been facilitated appreciably by ETD fragmentation, as demonstrated by several recent studies (Chalkley, Thalhammer, Schoepfer, & Burlingame, 2009; Myers, Daou, Affar, & Burlingame, 2013; Wang et al., 2010). Current reviews have covered O-GlcNAcylation and its role in disease and cancer in detail (Copeland et al., 2013; Fardini et al., 2013; Hart, Slawson, Ramirez-Correa, & Lagerlof, 2011).

Targeted analysis of glycoprotein cancer biomarker candidates has been performed using SRM and MRM approaches. Yoo and coworkers have investigated aberrant glycoforms of tissue inhibitor of metalloproteinase 1 (TIMP1), a colorectal cancer (CRC) biomarker candidate, in serum and from CRC cell lines that combine L-PHA lectin enrichment, stable isotope standard and capture by antipeptide antibody of a target tryptic peptide, and quantitative MRM analysis (Ahn et al., 2010, 2009). Abundance of the aberrant glycoform TIMP1 in CRC cells was shown to be 11.7-fold greater than control (Ahn et al., 2010). The same researchers recently used AAL enrichment and MRM quantitation for aberrant protein glycosylation analysis of α-1-antitrypsin and α-2*HS*-glycoprotein in HCC plasma, observing a 4.7- and 2.2-fold increase, respectively, in the aberrant glycoforms (Ahn, Shin, et al., 2013). Sanda et al. used MRM to study aberrant hyperfucosylation of a haptoglobin N-glycosite in HCC and cirrhosis patient plasma, monitoring oxonium ions for quantitation. Relative quantitation among HCC, cirrhosis, and control revealed that glycoforms displaying multiple outer arm fucose were considerably elevated in HCC and cirrhosis (Sanda et al., 2013).

Cancer has become a primary focus of MS-based glycoproteomics and glycomics biomarker studies. A comprehensive summary of the highly active field is beyond the scope of this review, and interested readers are encouraged to consult the numerous published reviews covering recent research progress (Adamczyk et al., 2013; Fardini et al., 2013; Kim & Misek, 2011; Kuzmanov et al., 2013; Lin et al., 2012; Meany & Chan, 2011; Mechref et al., 2012; Ruhaak, Miyamoto, & Lebrilla, 2013; Tan, Lee, & Chung, 2012; Ueda, 2013).

3.2. Neurodegenerative disease research

Irregularities in glycoprotein expression and glycosylation patterns have been observed in several neurological disorders, such as Alzheimer's disease (AD), Parkinson's disease, and amyotrophic lateral sclerosis, and Creutzfeldt–Jakob disease (CJD), a human prion disease (Botella-López et al., 2006; Rudd, Merry, Wormald, & Dwek, 2002; Shan, Vocadlo, & Krieger, 2012; Silveyra et al., 2006). CSF is an ideal source for biomarker discovery for diseases of the central nervous system due to its proximity to the brain. Therefore, changes in glycoforms or in abundance of glycoproteins in the CSF may reflect neurological disease conditions. A number of CSF glycoproteins with altered glycosylation have been implicated in AD, such as amyloid precursor protein (APP), acetylcholinesterase, butyrylcholinesterase, transferrin, tau, nicastrin, and reelin, which have been discussed in a recent review of protein glycosylation in AD (Schedin-Weiss, Winblad, & Tjernberg, 2013). The pathological indicators of AD in the brain are β-amyloid (Aβ) plaques originating from abnormal cleavage of APP by α- and γ-secretase and the presence of neurofibrillar tangles made up of oligomerized hyperphosphorylated tau proteins (Barone, Sturiale, Palmigiano, & Zappia, 2012). Site-specific O-glycopeptide characterization of endogenous APP/Aβ peptides from AD and non-AD patient CSF with analysis by LC–MS^n using ECD and CID was recently reported by Halim et al. Interestingly, they discovered 27 glycopeptides with Tyr linked sialylated O-glycans, which were most abundant in AD, alongside 37 glycopeptides with Ser-/Thr-linked O-glycans (Halim et al., 2011). A follow-up collaborative study by Brinkmalm et al. using a similar method and instrumentation confirmed the previously identified endogenous Aβ peptides, glycopeptides, and Ser/Thr/Tyr O-glycans, as well as discovering several new peptides (Brinkmalm et al., 2012). Butterfield and coworkers enriched glycoproteins from AD and mild cognitive impairment (MCI) inferior parietal lobule and hippocampus brain tissue via lectin affinity—WGA in one study, and Con A in another—followed by MALDI-TOF MS or LC-ESI Orbitrap MS/MS analysis to reveal altered levels of proteins involved in metabolism, cytoskeletal integrity, synaptic function, cell signaling, protein translation, and chaperoning, all of which are relevant to the disease (Di Domenico et al., 2010; Owen et al., 2009). A recent study of IgG–Fc N-glycosylation in plasma from MCI and AD patients was conducted by Zubarev and coworkers via LC–MS/MS using ETD and HCD fragmentation on a Thermo Velos Orbitrap followed by PCA assessment. A total of 19 glycoforms of two tryptic N-linked glycopeptides were quantified in a label-free manner across samples to reveal significant differences in glycosylation profile of one peptide in the AD samples, displaying increased core fucosylation and decreased biantennary-linked galactosylated and sialylated structures. The investigators connected the anti-inflammatory effect of sialylation of IgG–Fc receptor function to the observed inverse correlation between galactosylated and sialylated IgG glycoforms and AD. Additionally, it was determined that glycan truncation in females correlated with disease progress, but in the males, glycan complexity increased prior to the onset of AD, suggesting different inflammation response between genders (Lundström et al., 2014). Luider and coworkers explored the serum proteome of amnestic MCI patients by LC–MS/MS on a Thermo LTQ Orbitrap XL followed by MRM quantitative analysis of select proteins on an AB Sciex 400 QTrap. Two of the eight selected proteins were downregulated glycoproteins, galectin-3-binding protein, which has not previously been associated with neurodegenerative disease, and serum amyloid P-component (SAP), which has been previously suggested as an AD biomarker (Ijsselstijn et al., 2013).

Several studies have shown that O-GlcNAcylation deregulation of tau, APP, and nicastrin is concomitant with AD. The pathophysiological impairment of glucose uptake and metabolism observed in the AD brain reduces O-GlcNAcylation of proteins (Liu, Iqbal, Grundke-Iqbal, Hart, & Gong, 2004), and in the case of tau, which contains more than a dozen O-GlcNAcylation sites, this leads to hyperphosphorylation and the onset of neurofibrillary degeneration (Liu et al., 2004, 2009). In the mouse model, treatment with an O-GlcNAcase inhibitor, thiamet-G, effectively blocks this tau-driven neurodegeneration (Yuzwa et al., 2012). Similarly, reduced O-GlcNAcylation of APP results in increased Aβ production, through abnormal APP processing, and consequently amyloid plaque formation. Inhibition of O-GlcNAcase in human neuroblastoma cells by PUGNAc increases O-GlcNAcylated APP and leads to a decrease in Aβ through partially restored normal APP processing (Jacobsen & Iverfeldt, 2011). Related evidence was shown both *in vitro* and *in vivo* using the O-GlcNAcase inhibitor NButGT to decrease Aβ production, instead by inhibition of γ-secretase activity via O-GlcNAcylation of nicastrin, a component of γ-secretase (Kim et al., 2013). Because the use of O-GlcNAcase inhibitors has widespread impacts on many O-GlcNAcylated proteins, it is possible, or perhaps even likely, that more than one protein is responsible for the observed effects. Still, the results of these recent studies indicate O-GlcNAcylation as a potential therapeutic target for AD.

Prion diseases, or transmissible spongiform encephalopathies, are caused by abnormal folding of normal cellular prion protein (PrP^{C}) in the brain into a protease-resistant isoform (PrP^{Sc}) which is prone to aggregation. As a glycoprotein itself, PrP^{C} contains two N-linked glycosites and can carry one, two, or no glycans, for 52 different glycoform variants (Browning et al., 2011). Comparatively, PrP^{Sc} glycoforms have fewer bisecting GlcNAc and more tri- and tetraantennary glycans, suggesting a decrease in N-acetylglucosaminyltransferase III enzyme activity (Rudd et al., 1999). Several potential CSF biomarkers for CJD have been determined, which include 14-3-3 protein, total tau, S100B, neuron-specific enolase, ERK1/2, and transferrin (Hsich, Kenney, Gibbs, Lee, & Harrington, 1996; Jesse et al., 2009; Otto et al., 2002; Sanchez-Juan et al., 2006; Singh, Beveridge, & Singh, 2011; Steinacker et al., 2010). Recently, Wei et al. reported a quantitative MS-based glycoproteomics approach for premortem prion biomarker discovery in plasma from prion-infected mice using LAC, isotopic formaldehyde labeling, and multidimensional separation. Out of 708 proteins identified from plasma collected at three time points postinoculation, 53 proteins had a greater than twofold increase, and 58 proteins had a greater than twofold decrease in at least one time point. The glycoprotein SAP, which has previously been linked with amyloid deposits *in vivo* and accompanying neurofibrils in other neurodegenerative disorders, was detected in all three time points and was significantly elevated in the earliest one. The glycosylated form of SAP was also detected at significantly greater abundance compared to the unglycosylated form at the same time point, and showed significant elevation compared to the control sample. These results suggest that glycosylation of SAP plays a role in prion disease progression and that glycosylated SAP may be a useful preclinical biomarker for prion disease (Wei, Herbst, Ma, Aiken, & Li, 2011).

The use of MS-based glycoproteomics strategies in neurodegenerative disease research is still expanding, and a few current general glycoproteomics and glycomics reviews are available (Barone et al., 2012; Hwang et al., 2010), as is a recent comprehensive review of general MS-based quantitative neuroproteomics (Craft, Chen, & Nairn, 2013).

4. CONCLUDING REMARKS

MS-based glycoproteomics and glycomics have become a vital part of biomedical research for discovery and characterization of disease biomarkers due to rapid technological advances in affinity enrichment, chromatographic separation, quantitation, MS methodology and instrumentation, and bioinformatics. Quantitative comparative analyses of low-abundant, disease-related glycoforms in complex biological samples, enhanced by efficient separation and highly sensitive and accurate mass measurement, are providing deeper understanding into the onset and progression of disease. With this knowledge, improvements in diagnostic methods and therapeutic treatments are sure to follow. The glycoproteomics and glycomics strategies that were previously isolated from each other are increasingly converging in sophisticated workflows that better provide quantitative, site-specific structural characterization of glycoforms that potentially serve as the real biomarkers of a disease state. As the field moves forward, this unification is critical to take full advantage of new technological developments to effectively characterize disease phenotypes, identify drug targets, and develop treatments.

ACKNOWLEDGMENTS

Preparation of this manuscript was supported in part by the National Institutes of Health Grant R01 NS071513. L. L. acknowledges an H. I. Romnes Faculty Fellowship.

REFERENCES

Adamczyk, B., Struwe, W. B., Ercan, A., Nigrovic, P. A., & Rudd, P. M. (2013). Characterization of fibrinogen glycosylation and its importance for serum/plasma N-glycome analysis. *Journal of Proteome Research*, *12*(1), 444–454. http://dx.doi.org/10.1021/pr300813h.

Adamson, J. T., & Håkansson, K. (2006). Infrared multiphoton dissociation and electron capture dissociation of high-mannose type glycopeptides. *Journal of Proteome Research*, *5*(3), 493–501. http://dx.doi.org/10.1021/pr0504081.

Ahn, Y. H., Kim, Y.-S., Ji, E. S., Lee, J. Y., Jung, J.-A., Ko, J. H., et al. (2010). Comparative quantitation of aberrant glycoforms by lectin-based glycoprotein enrichment coupled with multiple-reaction monitoring mass spectrometry. *Analytical Chemistry*, *82*(11), 4441–4447. http://dx.doi.org/10.1021/ac1001965.

Ahn, Y. H., Lee, J. Y., Lee, J. Y., Kim, Y.-S., Ko, J. H., & Yoo, J. S. (2009). Quantitative analysis of an aberrant glycoform of TIMP1 from colon cancer serum by L-PHA-enrichment and SISCAPA with MRM mass spectrometry. *Journal of Proteome Research*, *8*(9), 4216–4224. http://dx.doi.org/10.1021/pr900269s.

Ahn, Y. H., Shin, P. M., Kim, Y.-S., Oh, N. R., Ji, E. S., Kim, K. H., et al. (2013). Quantitative analysis of aberrant protein glycosylation in liver cancer plasma by AAL-enrichment and MRM mass spectrometry. *The Analyst*, *138*(21), 6454–6462. http://dx.doi.org/10.1039/c3an01126g.

Ahn, J.-M., Sung, H.-J., Yoon, Y.-H., Kim, B.-G., Yang, W. S., Lee, C., et al. (2013). Integrated glycoproteomics demonstrates fucosylated serum paraoxonase 1 alterations in small cell lung cancer. *Molecular & Cellular Proteomics*. http://dx.doi.org/10.1074/mcp.M113.028621.

Alley, W. R., Mann, B. F., & Novotny, M. V. (2013). High-sensitivity analytical approaches for the structural characterization of glycoproteins. *Chemical Reviews*, *113*(4), 2668–2732. http://dx.doi.org/10.1021/cr3003714.

Alley, W. R., Mechref, Y., & Novotny, M. V. (2009a). Characterization of glycopeptides by combining collision-induced dissociation and electron-transfer dissociation mass spectrometry data. *Rapid Communications in Mass Spectrometry*, *23*(1), 161–170. http://dx.doi.org/10.1002/rcm.3850.

Alley, W. R., Mechref, Y., & Novotny, M. V. (2009b). Use of activated graphitized carbon chips for liquid chromatography/mass spectrometric and tandem mass spectrometric analysis of tryptic glycopeptides. *Rapid Communications in Mass Spectrometry*, *23*(4), 495–505. http://dx.doi.org/10.1002/rcm.3899.

Alvarez-Manilla, G., Atwood, J., Guo, Y., Warren, N. L., Orlando, R., & Pierce, M. (2006). Tools for glycoproteomic analysis: Size exclusion chromatography facilitates identification of tryptic glycopeptides with N-linked glycosylation sites. *Journal of Proteome Research*, *5*(3), 701–708. http://dx.doi.org/10.1021/pr050275j.

Alvarez-Manilla, G., Warren, N. L., Abney, T., Atwood, J., Azadi, P., York, W. S., et al. (2007). Tools for glycomics: Relative quantitation of glycans by isotopic permethylation using 13CH3I. *Glycobiology*, *17*(7), 677–687. http://dx.doi.org/10.1093/glycob/cwm033.

Alvarez-Manilla, G., Warren, N. L., Atwood, J., III, Orlando, R., Dalton, S., & Pierce, M. (2010). Glycoproteomic analysis of embryonic stem cells: Identification of potential glycobiomarkers using lectin affinity chromatography of glycopeptides. *Journal of Proteome Research*, *9*(5), 2062–2075. http://dx.doi.org/10.1021/pr8007489.

An, H. J., Froehlich, J. W., & Lebrilla, C. B. (2009). Determination of glycosylation sites and site-specific heterogeneity in glycoproteins. *Current Opinion in Chemical Biology*, *13*(4), 421–426. http://dx.doi.org/10.1016/j.cbpa.2009.07.022.

Anderson, N. L., & Anderson, N. G. (2002). The human plasma proteome: History, character, and diagnostic prospects. *Molecular & Cellular Proteomics*, *1*(11), 845–867. http://dx.doi.org/10.1074/mcp.R200007.

Apweiler, R., Hermjakob, H., & Sharon, N. (1999). On the frequency of protein glycosylation, as deduced from analysis of the SWISS-PROT database. *Biochimica et Biophysica Acta*, *1473*(1), 4–8. http://dx.doi.org/10.1016/s0304-4165(99)00165-8.

Atwood, J. A., Cheng, L., Alvarez-Manilla, G., Warren, N. L., York, W. S., & Orlando, R. (2008). Quantitation by isobaric labeling: Applications to glycomics. *Journal of Proteome Research*, 7(1), 367–374. http://dx.doi.org/10.1021/pr070476i.

Atwood, J. A., Sahoo, S. S., Alvarez-Manilla, G., Weatherly, D. B., Kolli, K., Orlando, R., et al. (2005). Simple modification of a protein database for mass spectral identification of N-linked glycopeptides. *Rapid Communications in Mass Spectrometry*, *19*(21), 3002–3006. http://dx.doi.org/10.1002/rcm.2162.

Barone, R., Sturiale, L., Palmigiano, A., & Zappia, M. (2012). Glycomics of pediatric and adulthood diseases of the central nervous system. *Journal of Proteomics*, *75*(17), 5123–5139. http://dx.doi.org/10.1016/j.jprot.2012.07.007.

Berven, F. S., Ahmad, R., Clauser, K. R., & Carr, S. A. (2010). Optimizing performance of glycopeptide capture for plasma proteomics. *Journal of Proteome Research*, *9*(4), 1706–1715. http://dx.doi.org/10.1021/pr900845m.

Boja, E. S., & Rodriguez, H. (2012). Mass spectrometry-based targeted quantitative proteomics: Achieving sensitive and reproducible detection of proteins. *Proteomics*, *12*(8), 1093–1110. http://dx.doi.org/10.1002/pmic.201100387.

Botella-López, A., Burgaya, F., Gavín, R., García-Ayllón, M. S., Gómez-Tortosa, E., Peña-Casanova, J., et al. (2006). Reelin expression and glycosylation patterns are altered in Alzheimer's disease. *Proceedings of the National Academy of Sciences of the United States of America*, *103*(14), 5573–5578. http://dx.doi.org/10.1073/pnas.0601279103.

Brinkmalm, G., Portelius, E., Öhrfelt, A., Mattsson, N., Persson, R., Gustavsson, M. K., et al. (2012). An online nano-LC-ESI-FTICR-MS method for comprehensive characterization of endogenous fragments from amyloid β and amyloid precursor protein in human and cat cerebrospinal fluid. *Journal of Mass Spectrometry*, *47*(5), 591–603. http://dx.doi.org/10.1002/jms.2987.

Browning, S., Baker, C. A., Smith, E., Mahal, S. P., Herva, M. E., Demczyk, C. A., et al. (2011). Abrogation of complex glycosylation by swainsonine results in strain- and cell-specific inhibition of prion replication. *Journal of Biological Chemistry*, *286*(47), 40962–40973. http://dx.doi.org/10.1074/jbc.M111.283978.

Carlson, D. M. (1968). Structures and immunochemical properties of oligosaccharides isolated from pig submaxillary mucins. *Journal of Biological Chemistry*, *243*(3), 616–626.

Chalkley, R. J., Thalhammer, A., Schoepfer, R., & Burlingame, A. L. (2009). Identification of protein O-GlcNAcylation sites using electron transfer dissociation mass spectrometry on native peptides. *Proceedings of the National Academy of Sciences of the United States of America*, *106*(22), 8894–8899. http://dx.doi.org/10.1073/pnas.0900288106.

Chen, R., Jiang, X., Sun, D., Han, G., Wang, F., Ye, M., et al. (2009). Glycoproteomics analysis of human liver tissue by combination of multiple enzyme digestion and hydrazide chemistry. *Journal of Proteome Research*, *8*(2), 651–661. http://dx.doi.org/10.1021/pr8008012.

Chen, J., Shah, P., & Zhang, H. (2013). Solid phase extraction of N-linked glycopeptides using hydrazide tip. *Analytical Chemistry*, *85*(22), 10670–10674. http://dx.doi.org/10.1021/ac401812b.

Chen, C. C., Su, W. C., Huang, B. Y., Chen, Y. J., & Tai, H. C. (2014). Interaction modes and approaches to glycopeptide and glycoprotein enrichment. *The Analyst*, *139*(4), 688–704. http://dx.doi.org/10.1039/c3an01813j.

Cho, W., Jung, K., & Regnier, F. E. (2008). Use of glycan targeting antibodies to identify cancer-associated glycoproteins in plasma of breast cancer patients. *Analytical Chemistry*, *80*(14), 5286–5292. http://dx.doi.org/10.1021/ac8008675.

Choi, E., Loo, D., Dennis, J. W., O'Leary, C. A., & Hill, M. M. (2011). High-throughput lectin magnetic bead array-coupled tandem mass spectrometry for glycoprotein biomarker discovery. *Electrophoresis*, *32*(24), 3564–3575. http://dx.doi.org/10.1002/elps.201100341.

Ciucanu, I., & Kerek, F. (1984). A simple and rapid method for the permethylation of carbohydrates. *Carbohydrate Research*, *131*, 209–217. http://dx.doi.org/10.1016/0008-6215(84)85242-8.

Clowers, B. H., Dodds, E. D., Seipert, R. R., & Lebrilla, C. B. (2007). Site determination of protein glycosylation based on digestion with immobilized nonspecific proteases and Fourier transform ion cyclotron resonance mass spectrometry. *Journal of Proteome Research*, *6*(10), 4032–4040. http://dx.doi.org/10.1021/pr070317z.

Comer, F. I., Vosseller, K., Wells, L., Accavitti, M. A., & Hart, G. W. (2001). Characterization of a mouse monoclonal antibody specific for O-linked N-acetylglucosamine. *Analytical Biochemistry*, *293*(2), 169–177. http://dx.doi.org/10.1006/abio.2001.5132.

Copeland, R. J., Han, G., & Hart, G. W. (2013). O-GlcNAcomics—Revealing roles of O-GlcNAcylation in disease mechanisms and development of potential diagnostics. *Proteomics. Clinical Applications*, 7, 597–606. http://dx.doi.org/10.1002/prca.201300001.

Craft, G. E., Chen, A., & Nairn, A. C. (2013). Recent advances in quantitative neuroproteomics. *Methods*, *61*(3), 186–218. http://dx.doi.org/10.1016/j.ymeth.2013.04.008.

Creese, A. J., & Cooper, H. J. (2012). Separation and identification of isomeric glycopeptides by high field asymmetric waveform ion mobility spectrometry. *Analytical Chemistry*, *84*(5), 2597–2601. http://dx.doi.org/10.1021/ac203321y.

Cummings, R. D., & Kornfeld, S. (1982). Fractionation of asparagine-linked oligosaccharides by serial lectin-Agarose affinity chromatography. A rapid, sensitive, and specific technique. *Journal of Biological Chemistry*, *257*(19), 11235–11240.

Dallas, D. C., Martin, W. F., Hua, S., & German, J. B. (2013). Automated glycopeptide analysis—Review of current state and future directions. *Briefings in Bioinformatics*, *14*(3), 361–374. http://dx.doi.org/10.1093/bib/bbs045.

Di Domenico, F., Owen, J. B., Sultana, R., Sowell, R. A., Perluigi, M., Cini, C., et al. (2010). The wheat germ agglutinin-fractionated proteome of subjects with Alzheimer's disease and mild cognitive impairment hippocampus and inferior parietal lobule: Implications for disease pathogenesis and progression. *Journal of Neuroscience Research*, *88*(16), 3566–3577. http://dx.doi.org/10.1002/jnr.22528.

Di Palma, S., Raijmakers, R., Heck, A. J. R., & Mohammed, S. (2011). Evaluation of the deuterium isotope effect in zwitterionic hydrophilic interaction liquid chromatography separations for implementation in a quantitative proteomic approach. *Analytical Chemistry*, *83*(21), 8352–8356. http://dx.doi.org/10.1021/ac2018074.

Dodds, E. D. (2012). Gas-phase dissociation of glycosylated peptide ions. *Mass Spectrometry Reviews*, *31*(6), 666–682. http://dx.doi.org/10.1002/mas.21344.

Dodds, E. D., Seipert, R. R., Clowers, B. H., German, J. B., & Lebrilla, C. B. (2009). Analytical performance of immobilized pronase for glycopeptide footprinting and implications for surpassing reductionist glycoproteomics. *Journal of Proteome Research*, *8*(2), 502–512. http://dx.doi.org/10.1021/pr800708h.

Drake, R. R., Cazares, L. H., Jones, E. E., Fuller, T. W., Semmes, O. J., & Laronga, C. (2011). Challenges to developing proteomic-based breast cancer diagnostics. *OMICS: A Journal of Integrative Biology*, *15*(5), 251–259. http://dx.doi.org/10.1089/omi.2010.0120.

Drake, P. M., Schilling, B., Niles, R. K., Prakobphol, A., Li, B., Jung, K., et al. (2012). Lectin chromatography/mass spectrometry discovery workflow identifies putative biomarkers of aggressive breast cancers. *Journal of Proteome Research*, *11*(4), 2508–2520. http://dx.doi.org/10.1021/pr201206w.

Dube, D. H., & Bertozzi, C. R. (2005). Glycans in cancer and inflammation—Potential for therapeutics and diagnostics. *Nature Reviews. Drug Discovery*, *4*(6), 477–488. http://dx.doi.org/10.1038/nrd1751.

Fanayan, S., Hincapie, M., & Hancock, W. S. (2012). Using lectins to harvest the plasma/serum glycoproteome. *Electrophoresis*, *33*(12), 1746–1754. http://dx.doi.org/10.1002/elps.201100567.

Fardini, Y., Dehennaut, V., Lefebvre, T., & Issad, T. (2013). O-GlcNAcylation: A new cancer hallmark? *Frontiers in Endocrinology*, *4*, 99. http://dx.doi.org/10.3389/fendo.2013.00099.

Fenn, L. S., & McLean, J. A. (2009). Simultaneous glycoproteomics on the basis of structure using ion mobility-mass spectrometry. *Molecular BioSystems*, *5*(11), 1298–1302. http://dx.doi.org/10.1039/b909745g.

Fenn, L. S., & McLean, J. A. (2011). Structural resolution of carbohydrate positional and structural isomers based on gas-phase ion mobility-mass spectrometry. *Physical Chemistry Chemical Physics*, *13*(6), 2196–2205. http://dx.doi.org/10.1039/c0cp01414a.

Froehlich, J. W., Barboza, M., Chu, C., Lerno, L. A., Clowers, B. H., Zivkovic, A. M., et al. (2011). Nano-LC-MS/MS of glycopeptides produced by nonspecific proteolysis enables rapid and extensive site-specific glycosylation determination. *Analytical Chemistry*, *83*(14), 5541–5547. http://dx.doi.org/10.1021/ac2003888.

Furukawa, J.-I., Fujitani, N., Araki, K., Takegawa, Y., Kodama, K., & Shinohara, Y. (2011). A versatile method for analysis of serine/threonine posttranslational modifications by β-elimination in the presence of pyrazolone analogues. *Analytical Chemistry*, *83*(23), 9060–9067. http://dx.doi.org/10.1021/ac2019848.

Fuster, M. M., & Esko, J. D. (2005). The sweet and sour of cancer: Glycans as novel therapeutic targets. *Nature Reviews. Cancer*, *5*(7), 526–542. http://dx.doi.org/10.1038/nrc1649.

Gaye, M. M., Valentine, S. J., Hu, Y., Mirjankar, N., Hammoud, Z. T., Mechref, Y., et al. (2012). Ion mobility-mass spectrometry analysis of serum N-linked glycans from esophageal adenocarcinoma phenotypes. *Journal of Proteome Research, 11*(12), 6102–6110. http://dx.doi.org/10.1021/pr300756e.

Gbormittah, F. O., Haab, B. B., Partyka, K., Garcia-Ott, C., Hincapie, M., & Hancock, W. S. (2013). Characterization of glycoproteins in pancreatic cyst fluid using a high performance multiple lectin affinity chromatography platform. *Journal of Proteome Research*. http://dx.doi.org/10.1021/pr400813u.

Gerlach, J. Q., Kilcoyne, M., Farrell, M. P., Kane, M., & Joshi, L. (2012). Differential release of high mannose structural isoforms by fungal and bacterial endo-β-N-acetylglucosaminidases. *Molecular BioSystems, 8*(5), 1472. http://dx.doi.org/10.1039/c2mb05455h.

Gillet, L. C., Navarro, P., Tate, S., Röst, H., Selevsek, N., Reiter, L., et al. (2012). Targeted data extraction of the MS/MS spectra generated by data-independent acquisition: A new concept for consistent and accurate proteome analysis. *Molecular & Cellular Proteomics. 11*(6). http://dx.doi.org/10.1074/mcp.O111.016717, O111.016717.

Gillette, M. A., & Carr, S. A. (2013). Quantitative analysis of peptides and proteins in biomedicine by targeted mass spectrometry. *Nature Methods, 10*(1), 28–34.

Goetz, J. A., Novotny, M. V., & Mechref, Y. (2009). Enzymatic/chemical release of O-glycans allowing MS analysis at high sensitivity. *Analytical Chemistry, 81*(23), 9546–9552. http://dx.doi.org/10.1021/ac901363h.

Gong, B., Hoyt, E., Lynaugh, H., Burnina, I., Moore, R., Thompson, A., et al. (2013). N-glycosylamine-mediated isotope labeling for mass spectrometry-based quantitative analysis of N-linked glycans. *Analytical and Bioanalytical Chemistry, 405*(17), 5825–5831. http://dx.doi.org/10.1007/s00216-013-6988-9.

Grass, J., Pabst, M., Chang, M., Wozny, M., & Altmann, F. (2011). Analysis of recombinant human follicle-stimulating hormone (FSH) by mass spectrometric approaches. *Analytical and Bioanalytical Chemistry, 400*, 2427–2438. http://dx.doi.org/10.1007/s00216-011-4923-5.

Guillard, M., Gloerich, J., Wessels, H. J. C. T., Morava, E., Wevers, R. A., & Lefeber, D. J. (2009). Automated measurement of permethylated serum N-glycans by MALDI-linear ion trap mass spectrometry. *Carbohydrate Research, 344*(12), 1550–1557. http://dx.doi.org/10.1016/j.carres.2009.06.010.

Gupta, G., Surolia, A., & Sampathkumar, S.-G. (2010). Lectin microarrays for glycomic analysis. *OMICS: A Journal of Integrative Biology, 14*(4), 419–436. http://dx.doi.org/10.1089/omi.2009.0150.

Gygi, S. P., Rist, B., Gerber, S. A., Turecek, F., Gelb, M. H., & Aebersold, R. (1999). Quantitative analysis of complex protein mixtures using isotope-coded affinity tags. *Nature Biotechnology, 17*(10), 994–999. http://dx.doi.org/10.1038/13690.

Hage, D. S., Anguizola, J. A., Bi, C., Li, R., Matsuda, R., Papastavros, E., et al. (2012). Pharmaceutical and biomedical applications of affinity chromatography: Recent trends and developments. *Journal of Pharmaceutical and Biomedical Analysis, 69*, 93–105. http://dx.doi.org/10.1016/j.jpba.2012.01.004.

Hägglund, P., Matthiesen, R., Elortza, F., Højrup, P., Roepstorff, P., Jensen, O. N., et al. (2007). An enzymatic deglycosylation scheme enabling identification of core fucosylated N-glycans and O-glycosylation site mapping of human plasma proteins. *Journal of Proteome Research, 6*(8), 3021–3031. http://dx.doi.org/10.1021/pr0700605.

Hahne, H., Neubert, P., Kuhn, K., Etienne, C., Bomgarden, R., Rogers, J. C., et al. (2012). Carbonyl-reactive tandem mass tags for the proteome-wide quantification of N-linked glycans. *Analytical Chemistry, 84*(8), 3716–3724. http://dx.doi.org/10.1021/ac300197c.

Håkansson, K., Cooper, H. J., Emmett, M. R., Costello, C. E., Marshall, A. G., & Nilsson, C. L. (2001). Electron capture dissociation and infrared multiphoton dissociation MS/MS of an N-glycosylated tryptic peptide to yield complementary sequence

information. *Analytical Chemistry*, *73*(18), 4530–4536. http://dx.doi.org/10.1021/ac0103470.
Halim, A., Brinkmalm, G., Rüetschi, U., Westman-Brinkmalm, A., Portelius, E., Zetterberg, H., et al. (2011). Site-specific characterization of threonine, serine, and tyrosine glycosylations of amyloid precursor protein/amyloid beta-peptides in human cerebrospinal fluid. *Proceedings of the National Academy of Sciences of the United States of America*, *108*(29), 11848–11853. http://dx.doi.org/10.1073/pnas.1102664108.
Halim, A., Nilsson, J., Rüetschi, U., Hesse, C., & Larson, G. (2012). Human urinary glycoproteomics; attachment site specific analysis of N- and O-linked glycosylations by CID and ECD. *Molecular & Cellular Proteomics*. *11*(4). http://dx.doi.org/10.1074/mcp.M111.013649, M111.013649.
Halim, A., Rüetschi, U., Larson, G., & Nilsson, J. (2013). LC-MS/MS characterization of O-glycosylation sites and glycan structures of human cerebrospinal fluid glycoproteins. *Journal of Proteome Research*, *12*(2), 573–584. http://dx.doi.org/10.1021/pr300963h.
Hart, G. W., Slawson, C., Ramirez-Correa, G., & Lagerlof, O. (2011). Cross talk between O-GlcNAcylation and phosphorylation: Roles in signaling, transcription, and chronic disease. *Annual Review of Biochemistry*, *80*, 825–858. http://dx.doi.org/10.1146/annurev-biochem-060608-102511.
Harvey, D. J. (2011). Derivatization of carbohydrates for analysis by chromatography; electrophoresis and mass spectrometry. *Journal of Chromatography. B, Analytical Technologies in the Biomedical and Life Sciences*, *879*(17–18), 1196–1225. http://dx.doi.org/10.1016/j.jchromb.2010.11.010.
Harvey, D. J., Scarff, C. A., Edgeworth, M., Crispin, M., Scanlan, C. N., Sobott, F., et al. (2013). Travelling wave ion mobility and negative ion fragmentation for the structural determination of N-linked glycans. *Electrophoresis*, *34*(16), 2368–2378. http://dx.doi.org/10.1002/elps.201200669.
Helenius, A., & Aebi, M. (2004). Roles of N-linked glycans in the endoplasmic reticulum. *Annual Review of Biochemistry*, *73*, 1019–1049. http://dx.doi.org/10.1146/annurev.biochem.73.011303.073752.
Hirabayashi, J., Yamada, M., Kuno, A., & Tateno, H. (2013). Lectin microarrays: Concept, principle and applications. *Chemical Society Reviews*, *42*(10), 4443–4458. http://dx.doi.org/10.1039/c3cs35419a.
Hong, Q., Lebrilla, C. B., Miyamoto, S., & Ruhaak, L. R. (2013). Absolute quantitation of immunoglobulin G and its glycoforms using multiple reaction monitoring. *Analytical Chemistry*, *85*(18), 8585–8593. http://dx.doi.org/10.1021/ac4009995.
Hsich, G., Kenney, K., Gibbs, C. J., Lee, K. H., & Harrington, M. G. (1996). The 14-3-3 brain protein in cerebrospinal fluid as a marker for transmissible spongiform encephalopathies. *The New England Journal of Medicine*, *335*(13), 924–930. http://dx.doi.org/10.1056/NEJM199609263351303.
Hsu, J., Huang, S., Chow, N., & Chen, S. (2003). Stable-isotope dimethyl labeling for quantitative proteomics. *Analytical Chemistry*, *75*(24), 6843–6852.
Hua, S., An, H. J., Ozcan, S., Ro, G. S., Soares, S., DeVere-White, R., et al. (2011). Comprehensive native glycan profiling with isomer separation and quantitation for the discovery of cancer biomarkers. *The Analyst*, *136*(18), 3663–3671. http://dx.doi.org/10.1039/c1an15093f.
Hua, S., Hu, C. Y., Kim, B.-J., Totten, S. M., Oh, M. J., Yun, N., et al. (2013). Glycoanalytical multispecific proteolysis (Glyco-AMP): A simple method for detailed and quantitative glycoproteomic characterization. *Journal of Proteome Research*, *12*(10), 4414–4423. http://dx.doi.org/10.1021/pr400442y.
Hua, S., Nwosu, C. C., Strum, J. S., Seipert, R. R., An, H. J., Zivkovic, A. M., et al. (2011). Site-specific protein glycosylation analysis with glycan isomer differentiation. *Analytical*

and Bioanalytical Chemistry, *403*(5), 1291–1302. http://dx.doi.org/10.1007/s00216-011-5109-x.

Hua, S., Saunders, M., Dimapasoc, L. M., Jeong, S. H., Kim, B. J., Kim, S., et al. (2014). Differentiation of cancer cell origin and molecular subtype by plasma membrane N-glycan profiling. *Journal of Proteome Research*, *13*, 961–968. http://dx.doi.org/10.1021/pr400987f.

Hua, S., Williams, C. C., Dimapasoc, L. M., Ro, G. S., Ozcan, S., Miyamoto, S., et al. (2013). Isomer-specific chromatographic profiling yields highly sensitive and specific potential N-glycan biomarkers for epithelial ovarian cancer. *Journal of Chromatography. A*, *1279*, 58–67. http://dx.doi.org/10.1016/j.chroma.2012.12.079.

Huang, G., Cheng, F., Chen, X., Peng, D., Hu, X., & Liang, G. (2013). Recent progress on the applications of multifunctional glyconanoparticles. *Current Pharmaceutical Design*, *19*(13), 2454–2458.

Hüttenhain, R., Surinova, S., Ossola, R., Sun, Z., Campbell, D., Cerciello, F., et al. (2013). N-glycoprotein SRMAtlas: A resource of mass spectrometric assays for N-glycosites enabling consistent and multiplexed protein quantification for clinical applications. *Molecular & Cellular Proteomics*, *12*(4), 1005–1016. http://dx.doi.org/10.1074/mcp.O112.026617.

Hwang, H., Zhang, J., Chung, K. A., Leverenz, J. B., Zabetian, C. P., Peskind, E. R., et al. (2010). Glycoproteomics in neurodegenerative diseases. *Mass Spectrometry Reviews*, *29*(1), 79–125. http://dx.doi.org/10.1002/mas.20221.

Ijsselstijn, L., Papma, J. M., Dekker, L. J. M., Calame, W., Stingl, C., Koudstaal, P. J., et al. (2013). Serum proteomics in amnestic mild cognitive impairment. *Proteomics*, *13*(16), 2526–2533. http://dx.doi.org/10.1002/pmic.201200190.

Iliuk, A., Galan, J., & Tao, W. A. (2008). Playing tag with quantitative proteomics. *Analytical and Bioanalytical Chemistry*, *393*(2), 503–513. http://dx.doi.org/10.1007/s00216-008-2386-0.

Isailovic, D., Plasencia, M. D., Gaye, M. M., Stokes, S. T., Kurulugama, R. T., Pungpapong, V., et al. (2012). Delineating diseases by IMS-MS profiling of serum N-linked glycans. *Journal of Proteome Research*, *11*(2), 576–585. http://dx.doi.org/10.1021/pr200777u.

Jacobsen, K. T., & Iverfeldt, K. (2011). O-GlcNAcylation increases non-amyloidogenic processing of the amyloid-β precursor protein (APP). *Biochemical and Biophysical Research Communications*, *404*(3), 882–886. http://dx.doi.org/10.1016/j.bbrc.2010.12.080.

Jesse, S., Steinacker, P., Cepek, L., von Arnim, C. A. F., Tumani, H., Lehnert, S., et al. (2009). Glial fibrillary acidic protein and protein S-100B: Different concentration pattern of glial proteins in cerebrospinal fluid of patients with Alzheimer's disease and Creutzfeldt-Jakob disease. *Journal of Alzheimer's Disease*, *17*(3), 541–551. http://dx.doi.org/10.3233/JAD-2009-1075.

Jiao, J., Zhang, H., & Reinhold, V. N. (2011). High performance IT-MS sequencing of glycans (spatial resolution of ovalbumin isomers). *International Journal of Mass Spectrometry*, *303*(2–3), 109–117. http://dx.doi.org/10.1016/j.ijms.2011.01.016.

Ju, T., Wang, Y., Aryal, R. P., Lehoux, S. D., Ding, X., Kudelka, M. R., et al. (2013). Tn and sialyl-Tn antigens, aberrant O-glycomics as human disease markers. *Proteomics. Clinical Applications*. http://dx.doi.org/10.1002/prca.201300024.

Kailemia, M. J., Ruhaak, L. R., Lebrilla, C. B., & Amster, I. J. (2014). Oligosaccharide analysis by mass Spectrometry: A review of recent developments. *Analytical Chemistry*, *86*(1), 196–212. http://dx.doi.org/10.1021/ac403969n.

Kaji, H., Ocho, M., Togayachi, A., Kuno, A., Sogabe, M., Ohkura, T., et al. (2013). Glycoproteomic discovery of serological biomarker candidates for HCV/HBV infection-associated liver fibrosis and hepatocellular carcinoma. *Journal of Proteome Research*, *12*(6), 2630–2640. http://dx.doi.org/10.1021/pr301217b.

Kaji, H., Saito, H., Yamauchi, Y., Shinkawa, T., Taoka, M., Hirabayashi, J., et al. (2003). Lectin affinity capture, isotope-coded tagging and mass spectrometry to identify N-linked glycoproteins. *Nature Biotechnology*, *21*(6), 667–672. http://dx.doi.org/10.1038/nbt829.

Kim, E. H., & Misek, D. E. (2011). Glycoproteomics-based identification of cancer biomarkers. *International Journal of Proteomics*, *2011*, 601937. http://dx.doi.org/10.1155/2011/601937.

Kim, C., Nam, D. W., Park, S. Y., Song, H., Hong, H. S., Boo, J. H., et al. (2013). O-linked β-N-acetylglucosaminidase inhibitor attenuates β-amyloid plaque and rescues memory impairment. *Neurobiology of Aging*, *34*(1), 275–285. http://dx.doi.org/10.1016/j.neurobiolaging.2012.03.001.

Kitteringham, N. R., Jenkins, R. E., Lane, C. S., Elliott, V. L., & Park, B. K. (2009). Multiple reaction monitoring for quantitative biomarker analysis in proteomics and metabolomics. *Journal of Chromatography. B, Analytical Technologies in the Biomedical and Life Sciences*, *877*(13), 1229–1239. http://dx.doi.org/10.1016/j.jchromb.2008.11.013.

Klement, E., Lipinszki, Z., Kupihár, Z., Udvardy, A., & Medzihradszky, K. F. (2010). Enrichment of O-GlcNAc modified proteins by the periodate oxidation—hydrazide resin capture approach. *Journal of Proteome Research*, *9*(5), 2200–2206. http://dx.doi.org/10.1021/pr900984h.

Kozak, R. P., Royle, L., Gardner, R. A., Fernandes, D. L., & Wuhrer, M. (2012). Suppression of peeling during the release of O-glycans by hydrazinolysis. *Analytical Biochemistry*, *423*(1), 119–128. http://dx.doi.org/10.1016/j.ab.2012.01.002.

Kullolli, M., Hancock, W. S., & Hincapie, M. (2008). Preparation of a high-performance multi-lectin affinity chromatography (HP-M-LAC) adsorbent for the analysis of human plasma glycoproteins. *Journal of Separation Science*, *31*(14), 2733–2739. http://dx.doi.org/10.1002/jssc.200800233.

Kullolli, M., Hancock, W. S., & Hincapie, M. (2010). Automated platform for fractionation of human plasma glycoproteome in clinical proteomics. *Analytical Chemistry*, *82*(1), 115–120. http://dx.doi.org/10.1021/ac9013308.

Kurogochi, M., Matsushista, T., Amano, M., Furukawa, J.-I., Shinohara, Y., Aoshima, M., et al. (2010). Sialic acid-focused quantitative mouse serum glycoproteomics by multiple reaction monitoring assay. *Molecular & Cellular Proteomics*, *9*(11), 2354–2368. http://dx.doi.org/10.1074/mcp.M110.000430.

Küster, B., & Mann, M. (1999). ^{18}O-labeling of N-glycosylation sites to improve the identification of gel-separated glycoproteins using peptide mass mapping and database searching. *Analytical Chemistry*, *71*(7), 1431–1440. http://dx.doi.org/10.1021/ac981012u.

Kuzmanov, U., Kosanam, H., & Diamandis, E. P. (2013). The sweet and sour of serological glycoprotein tumor biomarker quantification. *BMC Medicine*, *11*, 31. http://dx.doi.org/10.1186/1741-7015-11-31.

Larsen, M. R., Jensen, S. S., Jakobsen, L. A., & Heegaard, N. H. H. (2007). Exploring the sialiome using titanium dioxide chromatography and mass spectrometry. *Molecular & Cellular Proteomics*, *6*(10), 1778–1787. http://dx.doi.org/10.1074/mcp.M700086-MCP200.

Leach, F. E., Ly, M., Laremore, T. N., Wolff, J. J., Perlow, J., Linhardt, R. J., et al. (2012). Hexuronic acid stereochemistry determination in chondroitin sulfate glycosaminoglycan oligosaccharides by electron detachment dissociation. *Journal of the American Society for Mass Spectrometry*, *23*(9), 1488–1497. http://dx.doi.org/10.1007/s13361-012-0428-5.

Lemoine, J., Fortin, T., Salvador, A., Jaffuel, A., Charrier, J.-P., & Choquet-Kastylevsky, G. (2012). The current status of clinical proteomics and the use of MRM and MRM(3) for biomarker validation. *Expert Review of Molecular Diagnostics, 12*(4), 333–342. http://dx.doi.org/10.1586/erm.12.32.

Leymarie, N., & Zaia, J. (2012). Effective use of mass spectrometry for glycan and glycopeptide structural analysis. *Analytical Chemistry, 84*(7), 3040–3048. http://dx.doi.org/10.1021/ac3000573.

Li, F., Glinskii, O. V., & Glinsky, V. V. (2013). Glycobioinformatics: Current strategies and tools for data mining in MS-based glycoproteomics. *Proteomics, 13*(2), 341–354. http://dx.doi.org/10.1002/pmic.201200149.

Li, Y., Wen, T., Zhu, M., Li, L., Wei, J., Wu, X., et al. (2013). Glycoproteomic analysis of tissues from patients with colon cancer using lectin microarrays and nanoLC-MS/MS. *Molecular BioSystems, 9*(7), 1877–1887. http://dx.doi.org/10.1039/c3mb00013c.

Liebler, D. C., & Zimmerman, L. J. (2013). Targeted quantitation of proteins by mass spectrometry. *Biochemistry, 52*(22), 3797–3806. http://dx.doi.org/10.1021/bi400110b.

Liedtke, S., Geyer, H., Wuhrer, M., Geyer, R., Frank, G., Gerardy-Schahn, R., et al. (2001). Characterization of N-glycans from mouse brain neural cell adhesion molecule. *Glycobiology, 11*(5), 373–384.

Lin, Z., Lo, A., Simeone, D. M., Ruffin, M. T., & Lubman, D. M. (2012). An N-glycosylation analysis of human alpha-2-macroglobulin using an integrated approach. *Journal of Proteomics & Bioinformatics, 5*, 127–134. http://dx.doi.org/10.4172/jpb.1000224.

Liu, Y., Hüttenhain, R., Surinova, S., Gillet, L. C., Mouritsen, J., Brunner, R., et al. (2013). Quantitative measurements of N-linked glycoproteins in human plasma by SWATH-MS. *Proteomics, 13*(8), 1247–1256. http://dx.doi.org/10.1002/pmic.201200417.

Liu, F., Iqbal, K., Grundke-Iqbal, I., Hart, G. W., & Gong, C. X. (2004). O-GlcNAcylation regulates phosphorylation of tau: A mechanism involved in Alzheimer's disease. *Proceedings of the National Academy of Sciences of the United States of America, 101*(29), 10804–10809. http://dx.doi.org/10.1073/pnas.0400348101.

Liu, T., Qian, W.-J., Gritsenko, M. A., Camp, D. G., Monroe, M. E., Moore, R. J., et al. (2005). Human plasma N-glycoproteome analysis by immunoaffinity subtraction, hydrazide chemistry, and mass spectrometry. *Journal of Proteome Research, 4*(6), 2070–2080. http://dx.doi.org/10.1021/pr0502065.

Liu, F., Shi, J., Tanimukai, H., Gu, J., Gu, J., Grundke-Iqbal, I., et al. (2009). Reduced O-GlcNAcylation links lower brain glucose metabolism and tau pathology in Alzheimer's disease. *Brain: A Journal of Neurology, 132*(Pt. 7), 1820–1832. http://dx.doi.org/10.1093/brain/awp099.

Loo, D., Jones, A., & Hill, M. M. (2010). Lectin magnetic bead array for biomarker discovery. *Journal of Proteome Research, 9*(10), 5496–5500. http://dx.doi.org/10.1021/pr100472z.

Lundström, S. L., Yang, H., Lyutvinskiy, Y., Rutishauser, D., Herukka, S.-K., Soininen, H., et al. (2014). Blood plasma IgG Fc glycans are significantly altered in Alzheimer's disease and progressive mild cognitive impairment. *Journal of Alzheimer's Disease, 38*(3), 567–579. http://dx.doi.org/10.3233/JAD-131088.

Lux, A., & Nimmerjahn, F. (2011). Impact of differential glycosylation on IgG activity. *Advances in Experimental Medicine and Biology, 780*, 113–124. http://dx.doi.org/10.1007/978-1-4419-5632-3_10.

Mariño, K., Bones, J., Kattla, J. J., & Rudd, P. M. (2010). A systematic approach to protein glycosylation analysis: A path through the maze. *Nature Chemical Biology, 6*(10), 713–723. http://dx.doi.org/10.1038/nchembio.437.

Mayampurath, A., Yu, C.-Y., Song, E., Balan, J., Mechref, Y., & Tang, H. (2014). A computational framework for identification of intact glycopeptides in complex samples. *Analytical Chemistry, 86*(1), 453–463. http://dx.doi.org/10.1021/ac402338u 131126112938007.

Meany, D. L., & Chan, D. W. (2011). Aberrant glycosylation associated with enzymes as cancer biomarkers. *Clinical Proteomics*, *8*(1), 7. http://dx.doi.org/10.1186/1559-0275-8-7.

Mechref, Y., Hu, Y., Desantos-Garcia, J. L., Hussein, A., & Tang, H. (2013). Quantitative glycomics strategies. *Molecular & Cellular Proteomics*, *12*(4), 874–884. http://dx.doi.org/10.1074/mcp.R112.026310.

Mechref, Y., Hu, Y., Garcia, A., Zhou, S., Desantos-Garcia, J. L., & Hussein, A. (2012). Defining putative glycan cancer biomarkers by MS. *Bioanalysis*, *4*(20), 2457–2469. http://dx.doi.org/10.4155/bio.12.246.

Mechref, Y., & Novotny, M. V. (2002). Structural investigations of glycoconjugates at high sensitivity. *Chemical Reviews*, *102*(2), 321–370. http://dx.doi.org/10.1021/cr0103017.

Medvedev, A., Kopylov, A., Buneeva, O., Zgoda, V., & Archakov, A. (2012). Affinity-based proteomic profiling: Problems and achievements. *Proteomics*, *12*(4–5), 621–637. http://dx.doi.org/10.1002/pmic.201100373.

Meng, Z., & Veenstra, T. D. (2011). Targeted mass spectrometry approaches for protein biomarker verification. *Journal of Proteomics*, *74*(12), 2650–2659. http://dx.doi.org/10.1016/j.jprot.2011.04.011.

Miyoshi, E., Moriwaki, K., & Nakagawa, T. (2008). Biological function of fucosylation in cancer biology. *Journal of Biochemistry*, *143*(6), 725–729. http://dx.doi.org/10.1093/jb/mvn011.

Mondal, G., Chatterjee, U., Chawla, Y. K., & Chatterjee, B. P. (2011). Alterations of glycan branching and differential expression of sialic acid on alpha fetoprotein among hepatitis patients. *Glycoconjugate Journal*, *28*(1), 1–9. http://dx.doi.org/10.1007/s10719-010-9316-z.

Myers, S. A., Daou, S., Affar, E. B., & Burlingame, A. (2013). Electron transfer dissociation (ETD): The mass spectrometric breakthrough essential for O-GlcNAc protein site assignments-a study of the O-GlcNAcylated protein Host Cell Factor C1. *Proteomics*, *13*(6), 982–991. http://dx.doi.org/10.1002/pmic.201200332 (R. Zahedi & A. Sickmann, Eds.).

Nakada, H., Numata, Y., Inoue, M., Tanaka, N., Kitagawa, H., Funakoshi, I., et al. (1991). Elucidation of an essential structure recognized by an anti-GalNAc alpha-Ser(Thr) monoclonal antibody (MLS 128). *The Journal of Biological Chemistry*, *266*(19), 12402–12405.

Nanni, P., Panse, C., Gehrig, P., Mueller, S., Grossmann, J., & Schlapbach, R. (2013). PTM MarkerFinder, a software tool to detect and validate spectra from peptides carrying post-translational modifications. *Proteomics*, *13*(15), 2251–2255. http://dx.doi.org/10.1002/pmic.201300036.

Nie, H., Li, Y., & Sun, X.-L. (2012). Recent advances in sialic acid-focused glycomics. *Journal of Proteomics*, *75*(11), 3098–3112. http://dx.doi.org/10.1016/j.jprot.2012.03.050.

Nie, S., Lo, A., Zhu, J., Wu, J., Ruffin, M. T., & Lubman, D. M. (2013). Isobaric protein-level labeling strategy for serum glycoprotein quantification analysis by liquid chromatography–tandem mass spectrometry. *Analytical Chemistry*, *85*(11), 5353–5357. http://dx.doi.org/10.1021/ac400838s.

Nilsson, J., & Larson, G. (2013). Sialic acid capture-and-release and LC-MS(n) analysis of glycopeptides. *Methods in Molecular Biology (Clifton, NJ)*, *951*, 79–100. http://dx.doi.org/10.1007/978-1-62703-146-2_7.

Nilsson, J., Rüetschi, U., Halim, A., Hesse, C., Carlsohn, E., Brinkmalm, G., et al. (2009). Enrichment of glycopeptides for glycan structure and attachment site identification. *Nature Methods*, *6*(11), 809–811. http://dx.doi.org/10.1038/nmeth.1392.

Nishikaze, T., Kawabata, S.-I., Iwamoto, S., & Tanaka, K. (2013). Reversible hydrazide chemistry-based enrichment for O-GlcNAc-modified peptides and glycopeptides having non-reducing GlcNAc residues. *The Analyst*, *138*(23), 7224–7232. http://dx.doi.org/10.1039/c3an00880k.

Nwosu, C. C., Seipert, R. R., Strum, J. S., Hua, S. S., An, H. J., Zivkovic, A. M., et al. (2011). Simultaneous and extensive site-specific N- and O-glycosylation analysis in protein mixtures. *Journal of Proteome Research*, *10*(5), 2612–2624. http://dx.doi.org/10.1021/pr2001429.

Nyalwidhe, J. O., Betesh, L. R., Powers, T. W., Jones, E. E., White, K. Y., Burch, T. C., et al. (2013). Increased bisecting N-acetylglucosamine and decreased branched chain glycans of N-linked glycoproteins in expressed prostatic secretions associated with prostate cancer progression. *Proteomics. Clinical Applications*, 7, 677–689. http://dx.doi.org/10.1002/prca.201200134.

Ongay, S., Boichenko, A., Govorukhina, N., & Bischoff, R. (2012). Glycopeptide enrichment and separation for protein glycosylation analysis. *Journal of Separation Science*, *35*(18), 2341–2372. http://dx.doi.org/10.1002/jssc.201200434.

Orlando, R. (2013). Quantitative analysis of glycoprotein glycans. *Methods in Molecular Biology (Clifton, NJ)*, *951*, 197–215. http://dx.doi.org/10.1007/978-1-62703-146-2_13.

Otto, M., Wiltfang, J., Cepek, L., Neumann, M., Mollenhauer, B., Steinacker, P., et al. (2002). Tau protein and 14-3-3 protein in the differential diagnosis of Creutzfeldt-Jakob disease. *Neurology*, *58*(2), 192–197. http://dx.doi.org/10.1212/wnl.58.2.192.

Otvos, L., Urge, L., & Thurin, J. (1992). Influence of different N- and O-linked carbohydrates on the retention times of synthetic peptides in reversed-phase high-performance liquid chromatography. *Journal of Chromatography*, *599*(1–2), 43–49.

Owen, J. B., Di Domenico, F., Sultana, R., Perluigi, M., Cini, C., Pierce, W. M., et al. (2009). Proteomics-determined differences in the concanavalin-a-fractionated proteome of hippocampus and inferior parietal lobule in subjects with Alzheimer's disease and mild cognitive Impairment: Implications for progression of AD. *Journal of Proteome Research*, *8*(2), 471–482. http://dx.doi.org/10.1021/pr800667a.

Palmisano, G., Lendal, S. E., Engholm-Keller, K., Leth-Larsen, R., Parker, B. L., & Larsen, M. R. (2010). Selective enrichment of sialic acid-containing glycopeptides using titanium dioxide chromatography with analysis by HILIC and mass spectrometry. *Nature Protocols*, *5*(12), 1974–1982. http://dx.doi.org/10.1038/nprot.2010.167.

Palmisano, G., Melo-Braga, M. N., Engholm-Keller, K., Parker, B. L., & Larsen, M. R. (2012). Chemical deamidation: A common pitfall in large-scale N-linked glycoproteomic mass spectrometry-based analyses. *Journal of Proteome Research*, *11*(3), 1949–1957. http://dx.doi.org/10.1021/pr2011268.

Pan, M., Sun, Y., Zheng, J., & Yang, W. (2013). Boronic acid-functionalized core-shell-shell magnetic composite microspheres for the selective enrichment of glycoprotein. *ACS Applied Materials & Interfaces*, *5*(17), 8351–8358. http://dx.doi.org/10.1021/am401285x.

Parker, B. L., Thaysen-Andersen, M., Solis, N., Scott, N. E., Larsen, M. R., Graham, M. E., et al. (2013). Site-specific glycan-peptide analysis for determination of N-glycoproteome heterogeneity. *Journal of Proteome Research*, *12*(12), 5791–5800. http://dx.doi.org/10.1021/pr400783j.

Percy, A. J., Parker, C. E., & Borchers, C. H. (2013). Pre-analytical and analytical variability in absolute quantitative MRM-based plasma proteomic studies. *Bioanalysis*, *5*(22), 2837–2856. http://dx.doi.org/10.4155/bio.13.245.

Pernemalm, M., Lewensohn, R., & Lehtiö, J. (2009). Affinity prefractionation for MS-based plasma proteomics. *Proteomics*, *9*(6), 1420–1427. http://dx.doi.org/10.1002/pmic.200800377.

Plasencia, M. D., Isailovic, D., Merenbloom, S. I., Mechref, Y., & Clemmer, D. E. (2008). Resolving and assigning N-linked glycan structural isomers from ovalbumin by IMS-MS. *Journal of the American Society for Mass Spectrometry*, *19*(11), 1706–1715. http://dx.doi.org/10.1016/j.jasms.2008.07.020.

Plavina, T., Wakshull, E., Hancock, W. S., & Hincapie, M. (2007). Combination of abundant protein depletion and multi-lectin affinity chromatography (M-LAC) for plasma protein biomarker discovery. *Journal of Proteome Research, 6*(2), 662–671. http://dx.doi.org/10.1021/pr060413k.

Plomp, R., Hensbergen, P. J., Rombouts, Y., Zauner, G., Dragan, I., Koeleman, C. A. M., et al. (2013). Site-specific N-glycosylation analysis of human immunoglobulin E. *Journal of Proteome Research, 13*(2), 536–546. http://dx.doi.org/10.1021/pr400714w.

Pompach, P., Chandler, K. B., Lan, R., Edwards, N., & Goldman, R. (2012). Semi-automated identification of N-glycopeptides by hydrophilic interaction Chromatography, nano-reverse-phase LC–MS/MS, and glycan database search. *Journal of Proteome Research, 11*(3), 1728–1740. http://dx.doi.org/10.1021/pr201183w.

Prien, J. M., Ashline, D. J., Lapadula, A. J., Zhang, H., & Reinhold, V. N. (2009). The high mannose glycans from bovine ribonuclease B isomer characterization by ion trap MS. *Journal of the American Society for Mass Spectrometry, 20*(4), 539–556. http://dx.doi.org/10.1016/j.jasms.2008.11.012.

Rambaruth, N. D. S., Greenwell, P., & Dwek, M. V. (2012). The lectin Helix pomatia agglutinin recognizes O-GlcNAc containing glycoproteins in human breast cancer. *Glycobiology, 22*(6), 839–848. http://dx.doi.org/10.1093/glycob/cws051.

Ross, P. L., Huang, Y. N., Marchese, J. N., Williamson, B., Parker, K., Hattan, S., et al. (2004). Multiplexed protein quantitation in Saccharomyces cerevisiae using amine-reactive isobaric tagging reagents. *Molecular & Cellular Proteomics, 3*(12), 1154–1169. http://dx.doi.org/10.1074/mcp.M400129-MCP200.

Rudd, P. M., Endo, T., Colominas, C., Groth, D., Wheeler, S. F., Harvey, D. J., et al. (1999). Glycosylation differences between the normal and pathogenic prion protein isoforms. *Proceedings of the National Academy of Sciences of the United States of America, 96*(23), 13044–13049. http://dx.doi.org/10.1073/pnas.96.23.13044.

Rudd, P. M., Merry, A. H., Wormald, M. R., & Dwek, R. A. (2002). Glycosylation and prion protein. *Current Opinion in Structural Biology, 12*, 578–586. http://dx.doi.org/10.1016/S0959-440X(02)00377-9.

Ruhaak, L. R., Miyamoto, S., & Lebrilla, C. B. (2013). Developments in the identification of glycan biomarkers for the detection of cancer. *Molecular & Cellular Proteomics, 12*(4), 846–855. http://dx.doi.org/10.1074/mcp.R112.026799.

Ruhaak, L. R., Zauner, G., Huhn, C., Bruggink, C., Deelder, A. M., & Wuhrer, M. (2010). Glycan labeling strategies and their use in identification and quantification. *Analytical and Bioanalytical Chemistry, 397*(8), 3457–3481. http://dx.doi.org/10.1007/s00216-010-3532-z.

Sanchez-Juan, P., Green, A., Ladogana, A., Cuadrado-Corrales, N., Sáanchez-Valle, R., Mitrováa, E., et al. (2006). CSF tests in the differential diagnosis of Creutzfeldt-Jakob disease. *Neurology, 67*(4), 637–643. http://dx.doi.org/10.1212/01.wnl.0000230159.67128.00.

Sanda, M., Pompach, P., Brnakova, Z., Wu, J., Makambi, K., & Goldman, R. (2013). Quantitative liquid chromatography-mass spectrometry-multiple reaction monitoring (LC-MS-MRM) analysis of site-specific glycoforms of haptoglobin in liver disease. *Molecular & Cellular Proteomics, 12*(5), 1294–1305. http://dx.doi.org/10.1074/mcp.M112.023325.

Schedin-Weiss, S., Winblad, B., & Tjernberg, L. O. (2013). The role of protein glycosylation in Alzheimer disease. *The FEBS Journal*. http://dx.doi.org/10.1111/febs.12590.

Schiel, J. E., Smith, N. J., & Phinney, K. W. (2013). Universal proteolysis and MS(n) for N- and O-glycan branching analysis. *Journal of Mass Spectrometry, 48*(4), 533–538. http://dx.doi.org/10.1002/jms.3196.

Scott, N. E., Parker, B. L., Connolly, A. M., Paulech, J., Edwards, A. V. G., Crossett, B., et al. (2011). Simultaneous glycan-peptide characterization using hydrophilic interaction chromatography and parallel fragmentation by CID, higher energy collisional dissociation, and electron transfer dissociation MS applied to the N-linked glycoproteome of Campylobacter jejuni. *Molecular & Cellular Proteomics. 10*(2). http://dx.doi.org/10.1074/mcp.M000031-MCP201, M000031-MCP201.

Segu, Z. M., Hussein, A., Novotny, M. V., & Mechref, Y. (2010). Assigning N-glycosylation sites of glycoproteins using LC/MSMS in conjunction with endo-M/exoglycosidase mixture. *Journal of Proteome Research, 9*(7), 3598–3607. http://dx.doi.org/10.1021/pr100129n.

Segu, Z. M., & Mechref, Y. (2010). Characterizing protein glycosylation sites through higher-energy C-trap dissociation. *Rapid Communications in Mass Spectrometry, 24*(9), 1217–1225. http://dx.doi.org/10.1002/rcm.4485.

Selvaraju, S., & Rassi, E. Z. (2012). Tandem lectin affinity chromatography monolithic columns with surface immobilised concanavalin A, wheat germ agglutinin and Ricinus communis agglutinin-I for capturing sub-glycoproteomics from breast cancer and disease-free human sera. *Journal of Separation Science, 35*(14), 1785–1795. http://dx.doi.org/10.1002/jssc.201200230.

Selvaraju, S., & Rassi, E. Z. (2011). Liquid-phase-based separation systems for depletion, prefractionation and enrichment of proteins in biological fluids and matrices for in-depth proteomics analysis—An update covering the period 2008–2011. *Electrophoresis, 33*(1), 74–88. http://dx.doi.org/10.1002/elps.201100431.

Serang, O., Froehlich, J. W., Muntel, J., McDowell, G., Steen, H., Lee, R. S., et al. (2013). SweetSEQer, simple de novo filtering and annotation of glycoconjugate mass spectra. *Molecular & Cellular Proteomics, 12*(6), 1735–1740. http://dx.doi.org/10.1074/mcp.O112.025940.

Shan, X., Vocadlo, D. J., & Krieger, C. (2012). Reduced protein O-glycosylation in the nervous system of the mutant SOD1 transgenic mouse model of amyotrophic lateral sclerosis. *Neuroscience Letters, 516*(2), 296–301. http://dx.doi.org/10.1016/j.neulet.2012.04.018.

Silveyra, M.-X., Cuadrado-Corrales, N., Marcos, A., Barquero, M.-S., Rábano, A., Calero, M., et al. (2006). Altered glycosylation of acetylcholinesterase in Creutzfeldt-Jakob disease. *Journal of Neurochemistry, 96*(1), 97–104. http://dx.doi.org/10.1111/j.1471-4159.2005.03514.x.

Singh, A., Beveridge, A. J., & Singh, N. (2011). Decreased CSF transferrin in sCJD: A potential pre-mortem diagnostic test for prion disorders. *PLoS ONE, 6*(3), e16804. http://dx.doi.org/10.1371/journal.pone.0016804.

Singh, C., Zampronio, C. G., Creese, A. J., & Cooper, H. J. (2012). Higher energy collision dissociation (HCD) product ion-triggered electron transfer dissociation (ETD) mass spectrometry for the analysis of N-linked glycoproteins. *Journal of Proteome Research, 11*(9), 4517–4525. http://dx.doi.org/10.1021/pr300257c.

Song, E., Pyreddy, S., & Mechref, Y. (2012). Quantification of glycopeptides by multiple reaction monitoring liquid chromatography/tandem mass spectrometry. *Rapid Communications in Mass Spectrometry, 26*(17), 1941–1954. http://dx.doi.org/10.1002/rcm.6290.

Sparbier, K., Koch, S., Kessler, I., Wenzel, T., & Kostrzewa, M. (2005). Selective isolation of glycoproteins and glycopeptides for MALDI-TOF MS detection supported by magnetic particles. *Journal of Biomolecular Techniques, 16*(4), 407–413.

Sparbier, K., Wenzel, T., & Kostrzewa, M. (2006). Exploring the binding profiles of ConA, boronic acid and WGA by MALDI-TOF/TOF MS and magnetic particles. *Journal of Chromatography. B, Analytical Technologies in the Biomedical and Life Sciences, 840*(1), 29–36. http://dx.doi.org/10.1016/j.jchromb.2006.06.028.

Steinacker, P., Klafki, H., Lehnert, S., Jesse, S., Arnim, C. A. F. V., Tumani, H., et al. (2010). ERK2 is increased in cerebrospinal fluid of Creutzfeldt-Jakob disease patients. *Journal of Alzheimer's Disease, 22*(1), 119–128. http://dx.doi.org/10.3233/JAD-2010-100030.

Sun, C., Chen, P., Chen, Q., Sun, L., Kang, X., Qin, X., et al. (2012). Serum paraoxonase 1 heteroplasmon, a fucosylated, and sialylated glycoprotein in distinguishing early hepatocellular carcinoma from liver cirrhosis patients. *Acta Biochimica et Biophysica Sinica*, *44*(9), 765–773. http://dx.doi.org/10.1093/abbs/gms055.

Taga, Y., Kusubata, M., Ogawa-Goto, K., & Hattori, S. (2013). Site-specific quantitative analysis of overglycosylation of collagen in osteogenesis imperfecta using hydrazide chemistry and SILAC. *Journal of Proteome Research*, *12*(5), 2225–2232. http://dx.doi.org/10.1021/pr400079d.

Takátsy, A., Böddi, K., Nagy, L., Nagy, G., Szabó, S., Markó, L., et al. (2009). Enrichment of Amadori products derived from the nonenzymatic glycation of proteins using microscale boronate affinity chromatography. *Analytical Biochemistry*, *393*(1), 8–22. http://dx.doi.org/10.1016/j.ab.2009.06.007.

Takegawa, Y., Deguchi, K., Ito, H., Keira, T., Nakagawa, H., & Nishimura, S.-I. (2006). Simple separation of isomeric sialylated N-glycopeptides by a zwitterionic type of hydrophilic interaction chromatography. *Journal of Separation Science*, *29*(16), 2533–2540. http://dx.doi.org/10.1002/jssc.200600133.

Tan, H. T., Lee, Y. H., & Chung, M. C. M. (2012). Cancer proteomics. *Mass Spectrometry Reviews*, *31*(5), 583–605. http://dx.doi.org/10.1002/mas.20356.

Tang, J., Liu, Y., Qi, D., Yao, G., Deng, C., & Zhang, X. (2009). On-plate-selective enrichment of glycopeptides using boronic acid-modified gold nanoparticles for direct MALDI-QIT-TOF MS analysis. *Proteomics*, *9*(22), 5046–5055. http://dx.doi.org/10.1002/pmic.200900033.

Tang, J., Liu, Y., Yin, P., Yao, G., Yan, G., Deng, C., et al. (2010). Concanavalin A-immobilized magnetic nanoparticles for selective enrichment of glycoproteins and application to glycoproteomics in hepatocelluar carcinoma cell line. *Proteomics*, *10*(10), 2000–2014. http://dx.doi.org/10.1002/pmic.200900377.

Temporini, C., Perani, E., Calleri, E., Dolcini, L., Lubda, D., Caccialanza, G., et al. (2007). Pronase-immobilized enzyme Reactor: An approach for automation in glycoprotein analysis by LC/LC–ESI/MS n. *Analytical Chemistry*, *79*(1), 355–363. http://dx.doi.org/10.1021/ac0611519.

Teo, C. F., Ingale, S., Wolfert, M. A., Elsayed, G. A., Nöt, L. G., Chatham, J. C., et al. (2010). Glycopeptide-specific monoclonal antibodies suggest new roles for O-GlcNAc. *Nature Chemical Biology*, *6*(5), 338–343. http://dx.doi.org/10.1038/nchembio.338.

Tep, S., Hincapie, M., & Hancock, W. S. (2012). A general approach for the purification and quantitative glycomic analysis of human plasma. *Analytical and Bioanalytical Chemistry*, *402*(9), 2687–2700. http://dx.doi.org/10.1007/s00216-012-5712-5.

Thompson, A., Schäfer, J., Kuhn, K., Kienle, S., Schwarz, J., Schmidt, G., et al. (2003). Tandem mass tags: A novel quantification strategy for comparative analysis of complex protein mixtures by MS/MS. *Analytical Chemistry*, *75*(8), 1895–1904. http://dx.doi.org/10.1021/ac0262560.

Ueda, K. (2013). Glycoproteomic strategies: From discovery to clinical application of cancer carbohydrate biomarkers. *Proteomics. Clinical Applications*. http://dx.doi.org/10.1002/prca.201200123.

Varki, A. (1993). Biological roles of oligosaccharides: All of the theories are correct. *Glycobiology*, *3*(2), 97–130.

Walker, S. H., Budhathoki-Uprety, J., Novak, B. M., & Muddiman, D. C. (2011). Stable-isotope labeled hydrophobic hydrazide reagents for the relative quantification of N-linked glycans by electrospray ionization mass spectrometry. *Analytical Chemistry*, *83*(17), 6738–6745. http://dx.doi.org/10.1021/ac201376q.

Walker, S. H., Carlisle, B. C., & Muddiman, D. C. (2012). Systematic comparison of reverse phase and hydrophilic interaction liquid chromatography platforms for the analysis of N-linked glycans. *Analytical Chemistry*, *84*(19), 8198–8206. http://dx.doi.org/10.1021/ac3012494.

Walker, S. H., Taylor, A. D., & Muddiman, D. C. (2013). Individuality normalization when labeling with isotopic glycan hydrazide tags (INLIGHT): A novel glycan-relative quantification strategy. *Journal of the American Society for Mass Spectrometry*, *24*(9), 1376–1384. http://dx.doi.org/10.1007/s13361-013-0681-2.

Wang, L., Aryal, U. K., Dai, Z., Mason, A. C., Monroe, M. E., Tian, Z.-X., et al. (2012). Mapping N-linked glycosylation sites in the secretome and whole cells of Aspergillus niger using hydrazide chemistry and mass spectrometry. *Journal of Proteome Research*, *11*(1), 143–156. http://dx.doi.org/10.1021/pr200916k.

Wang, C., Fan, W., Zhang, P., Wang, Z., & Huang, L. (2011). One-pot nonreductive O-glycan release and labeling with 1-phenyl-3-methyl-5-pyrazolone followed by ESI-MS analysis. *Proteomics*, *11*(21), 4229–4242. http://dx.doi.org/10.1002/pmic.201000677.

Wang, Z., Pandey, A., & Hart, G. W. (2007). Dynamic interplay between O-linked N-acetylglucosaminylation and glycogen synthase kinase-3-dependent phosphorylation. *Molecular & Cellular Proteomics*, *6*(8), 1365–1379. http://dx.doi.org/10.1074/mcp.M600453-MCP200.

Wang, Z., Udeshi, N. D., O'Malley, M., Shabanowitz, J., Hunt, D. F., & Hart, G. W. (2010). Enrichment and site mapping of O-linked N-acetylglucosamine by a combination of chemical/enzymatic tagging, photochemical cleavage, and electron transfer dissociation mass spectrometry. *Molecular & Cellular Proteomics*, *9*(1), 153–160. http://dx.doi.org/10.1074/mcp.M900268-MCP200.

Wang, Y., Wu, S.-L., & Hancock, W. S. (2006). Approaches to the study of N-linked glycoproteins in human plasma using lectin affinity chromatography and nano-HPLC coupled to electrospray linear ion trap—Fourier transform mass spectrometry. *Glycobiology*, *16*(6), 514–523. http://dx.doi.org/10.1093/glycob/cwj091.

Wang, C., Wu, Z., Yuan, J., Wang, B., Zhang, P., Zhang, Y., et al. (2013). Simplified quantitative glycomics using the stable isotope label Girard's reagent P by electrospray ionization mass spectrometry. *Journal of Proteome Research*, *13*(2), 372–384. http://dx.doi.org/10.1021/pr4010647.

Wei, X., Dulberger, C., & Li, L. (2010). Characterization of murine brain membrane glycoproteins by detergent assisted lectin affinity chromatography. *Analytical Chemistry*, *82*(15), 6329–6333.

Wei, X., Herbst, A., Ma, D., Aiken, J., & Li, L. (2011). A quantitative proteomic approach to prion disease biomarker research: Delving into the glycoproteome. *Journal of Proteome Research*, *10*(6), 2687–2702. http://dx.doi.org/10.1021/pr2000495.

Wolff, J. J., Leach, F. E., Laremore, T. N., Kaplan, D. A., Easterling, M. L., Linhardt, R. J., et al. (2010). Negative electron transfer dissociation of glycosaminoglycans. *Analytical Chemistry*, *82*(9), 3460–3466. http://dx.doi.org/10.1021/ac100554a.

Woodin, C. L., Maxon, M., & Desaire, H. (2013). Software for automated interpretation of mass spectrometry data from glycans and glycopeptides. *The Analyst*, *138*(10), 2793–2803. http://dx.doi.org/10.1039/c2an36042j.

Wuhrer, M. (2012). Glycomics using mass spectrometry. *Glycoconjugate Journal*, *30*(1), 11–22. http://dx.doi.org/10.1007/s10719-012-9376-3.

Wuhrer, M., Catalina, M. I., Deelder, A. M., & Hokke, C. H. (2007). Glycoproteomics based on tandem mass spectrometry of glycopeptides. *Journal of Chromatography. B, Analytical Technologies in the Biomedical and Life Sciences*, *849*(1–2), 115–128. http://dx.doi.org/10.1016/j.jchromb.2006.09.041.

Xu, Y., Wu, Z., Zhang, L., Lu, H., Yang, P., Webley, P. A., et al. (2009). Highly specific enrichment of glycopeptides using boronic acid-functionalized mesoporous silica. *Analytical Chemistry*, *81*(1), 503–508. http://dx.doi.org/10.1021/ac801912t.

Xu, Y., Zhang, L., Lu, H., & Yang, P. (2010). On-plate enrichment of glycopeptides by using boronic acid functionalized gold-coated Si wafer. *Proteomics*, *10*, 1079–1086. http://dx.doi.org/10.1002/pmic.200900097.

Yang, G., Cui, T., Wang, Y., Sun, S., Ma, T., Wang, T., et al. (2013). Selective isolation and analysis of glycoprotein fractions and their glycomes from hepatocellular carcinoma sera. *Proteomics*, *13*(9), 1481–1498. http://dx.doi.org/10.1002/pmic.201200259.

Yang, Z., & Hancock, W. S. (2004). Approach to the comprehensive analysis of glycoproteins isolated from human serum using a multi-lectin affinity column. *Journal of Chromatography. A*, *1053*(1–2), 79–88. http://dx.doi.org/10.1016/j.chroma.2004.08.150.

Yang, Z., & Hancock, W. S. (2005). Monitoring glycosylation pattern changes of glycoproteins using multi-lectin affinity chromatography. *Journal of Chromatography. A*, *1070*(1–2), 57–64.

Yang, Z., Hancock, W. S., Chew, T. R., & Bonilla, L. (2005). A study of glycoproteins in human serum and plasma reference standards (HUPO) using multilectin affinity chromatography coupled with RPLC-MS/MS. *Proteomics*, *5*(13), 3353–3366. http://dx.doi.org/10.1002/pmic.200401190.

Yang, Z., Harris, L. E., Palmer-Toy, D. E., & Hancock, W. S. (2006). Multilectin affinity chromatography for characterization of multiple glycoprotein biomarker candidates in serum from breast cancer patients. *Clinical Chemistry*, *52*(10), 1897–1905. http://dx.doi.org/10.1373/clinchem.2005.065862.

Yang, S., & Zhang, H. (2012). Solid-phase glycan isolation for glycomics analysis. *Proteomics. Clinical Applications*, *6*(11–12), 596–608. http://dx.doi.org/10.1002/prca.201200045.

Ye, H., Boyne, M. T., II, Buhse, L. F., & Hill, J. (2013). Direct approach for qualitative and quantitative characterization of glycoproteins using tandem mass tags and an LTQ Orbitrap XL electron transfer dissociation hybrid mass spectrometer. *Analytical Chemistry*, *85*(3), 1531–1539. http://dx.doi.org/10.1021/ac3026465.

Yin, X., Bern, M., Xing, Q., Ho, J., Viner, R., & Mayr, M. (2013). Glycoproteomic analysis of the secretome of human endothelial cells. *Molecular & Cellular Proteomics*, *12*(4), 956–978. http://dx.doi.org/10.1074/mcp.M112.024018.

Yu, X., Huang, Y., Lin, C., & Costello, C. E. (2012). Energy-dependent electron activated dissociation of metal-adducted permethylated oligosaccharides. *Analytical Chemistry*, *84*(17), 7487–7494. http://dx.doi.org/10.1021/ac301589z.

Yu, X., Jiang, Y., Chen, Y., Huang, Y., Costello, C. E., & Lin, C. (2013). Detailed glycan structural characterization by electronic excitation dissociation. *Analytical Chemistry*, *85*(21), 10017–10021. http://dx.doi.org/10.1021/ac402886q.

Yue, T., & Haab, B. B. (2009). Microarrays in glycoproteomics research. *Clinics in Laboratory Medicine*, *29*(1), 15–29. http://dx.doi.org/10.1016/j.cll.2009.01.001.

Yuzwa, S. A., Shan, X., Macauley, M. S., Clark, T., Skorobogatko, Y., Vosseller, K., et al. (2012). Increasing O-GlcNAc slows neurodegeneration and stabilizes tau against aggregation. *Nature Chemical Biology*, *8*(4), 393–399. http://dx.doi.org/10.1038/nchembio.797.

Zaia, J. (2010). Mass spectrometry and glycomics. *OMICS: A Journal of Integrative Biology*, *14*(4), 401–418. http://dx.doi.org/10.1089/omi.2009.0146.

Zauner, G., Deelder, A. M., & Wuhrer, M. (2011). Recent advances in hydrophilic interaction liquid chromatography (HILIC) for structural glycomics. *Electrophoresis*, *32*(24), 3456–3466. http://dx.doi.org/10.1002/elps.201100247.

Zauner, G., Koeleman, C. A. M., Deelder, A. M., & Wuhrer, M. (2010). Protein glycosylation analysis by HILIC-LC-MS of proteinase K-generated N- and O-glycopeptides. *Journal of Separation Science*, *33*(6–7), 903–910. http://dx.doi.org/10.1002/jssc.200900850.

Zauner, G., Koeleman, C. A. M., Deelder, A. M., & Wuhrer, M. (2012). Mass spectrometric O-glycan analysis after combined O-glycan release by beta-elimination and 1-phenyl-3-methyl-5-pyrazolone labeling. *Biochimica et Biophysica Acta*, *1820*(9), 1420–1428. http://dx.doi.org/10.1016/j.bbagen.2011.07.004.

Zauner, G., Kozak, R. P., Gardner, R. A., Fernandes, D. L., Deelder, A. M., & Wuhrer, M. (2012). Protein O-glycosylation analysis. *Biological Chemistry*, *393*(8), 687–708. http://dx.doi.org/10.1515/hsz-2012-0144.

Zeng, Z., Hincapie, M., Pitteri, S. J., Hanash, S., Schalkwijk, J., Hogan, J. M., et al. (2011). A proteomics platform combining depletion, multi-lectin affinity chromatography (M-LAC), and isoelectric focusing to study the breast cancer proteome. *Analytical Chemistry*, *83*(12), 4845–4854. http://dx.doi.org/10.1021/ac2002802.

Zhang, H., Li, X.-J., Martin, D. B., & Aebersold, R. (2003). Identification and quantification of N-linked glycoproteins using hydrazide chemistry, stable isotope labeling and mass spectrometry. *Nature Biotechnology*, *21*(6), 660–666. http://dx.doi.org/10.1038/nbt827.

Zhang, L., Lu, H., & Yang, P. (2009). Specific enrichment methods for glycoproteome research. *Analytical and Bioanalytical Chemistry*, *396*(1), 199–203. http://dx.doi.org/10.1007/s00216-009-3086-0.

Zhang, Q., Schepmoes, A. A., Brock, J. W. C., Wu, S., Moore, R. J., Purvine, S. O., et al. (2008). Improved methods for the enrichment and analysis of glycated peptides. *Analytical Chemistry*, *80*(24), 9822–9829. http://dx.doi.org/10.1021/ac801704j.

Zhang, B., Sheng, Q., Li, X., Liang, Q., Yan, J., & Liang, X. (2011). Selective enrichment of glycopeptides for mass spectrometry analysis using C18 fractionation and titanium dioxide chromatography. *Journal of Separation Science*, *34*(19), 2745–2750. http://dx.doi.org/10.1002/jssc.201100427.

Zhang, Q., Tang, N., Brock, J. W. C., Mottaz, H. M., Ames, J. M., Baynes, J. W., et al. (2007). Enrichment and analysis of nonenzymatically glycated peptides: Boronate affinity chromatography coupled with electron-transfer dissociation mass spectrometry. *Journal of Proteome Research*, *6*(6), 2323–2330. http://dx.doi.org/10.1021/pr070112q.

Zhang, W., Wang, H., Zhang, L., Yao, J., & Yang, P. (2011). Large-scale assignment of N-glycosylation sites using complementary enzymatic deglycosylation. *Talanta*, *85*(1), 499–505. http://dx.doi.org/10.1016/j.talanta.2011.04.019.

Zhang, L., Xu, Y., Yao, H., Xie, L., Yao, J., Lu, H., et al. (2009). Boronic acid functionalized core-satellite composite nanoparticles for advanced enrichment of glycopeptides and glycoproteins. *Chemistry (Weinheim an der Bergstrasse, Germany)*, *15*(39), 10158–10166. http://dx.doi.org/10.1002/chem.200901347.

Zhao, Y., Jia, W., Wang, J., Ying, W., Zhang, Y., & Qian, X. (2011). Fragmentation and site-specific quantification of core fucosylated glycoprotein by multiple reaction monitoring-mass spectrometry. *Analytical Chemistry*, *83*(22), 8802–8809. http://dx.doi.org/10.1021/ac201676a.

Zhou, Y., Aebersold, R., & Zhang, H. (2007). Isolation of N-linked glycopeptides from plasma. *Analytical Chemistry*, *79*(15), 5826–5837. http://dx.doi.org/10.1021/ac0623181.

Zhou, W., Yao, N., Yao, G., Deng, C., Zhang, X., & Yang, P. (2008). Facile synthesis of aminophenylboronic acid-functionalized magnetic nanoparticles for selective separation of glycopeptides and glycoproteins. *Chemical Communications (Cambridge, England)*, (43), 5577–5579. http://dx.doi.org/10.1039/b808800d.

Zhu, J., He, J., Liu, Y., Simeone, D. M., & Lubman, D. M. (2012). Identification of glycoprotein markers for pancreatic cancer CD24+CD44+ stem-like cells using nano-LC-MS/MS and tissue microarray. *Journal of Proteome Research*, *11*(4), 2272–2281. http://dx.doi.org/10.1021/pr201059g.

Zhu, Z., Hua, D., Clark, D. F., Go, E. P., & Desaire, H. (2013). GlycoPep Detector: A tool for assigning mass spectrometry data of N-linked glycopeptides on the basis of their electron transfer dissociation spectra. *Analytical Chemistry*, *85*(10), 5023–5032. http://dx.doi.org/10.1021/ac400287n.

Zielinska, D. F., Gnad, F., Wiśniewski, J. R., & Mann, M. (2010). Precision mapping of an in vivo N-glycoproteome reveals rigid topological and sequence constraints. *Cell*, *141*(5), 897–907. http://dx.doi.org/10.1016/j.cell.2010.04.012.

CHAPTER FOUR

Proteomics and Proteogenomics Approaches for Oral Diseases

Nicola Luigi Bragazzi[*,†,‡], **Eugenia Pechkova**[*,†], **Claudio Nicolini**[*,†,§,1]
[*]Nanobiotechnology and Biophysics Laboratories (NBL), Department of Experimental Medicine (DIMES), University of Genoa, Genoa, Italy
[†]Nanoworld Institute Fondazione ELBA Nicolini (FEN), Pradalunga, Bergamo, Italy
[‡]School of Public Health, Department of Health Sciences (DISSAL), University of Genoa, Genoa, Italy
[§]Biodesign Institute, Arizona State University, Tempe, Arizona, USA
[1]Corresponding author: e-mail address: info@fondazioneelba-nicolini.org

Contents

Abstract

Design and implementation of new biocompatible materials and achievements in the field of nanogenomics and nanoproteomics as well as in other related and allied sciences in the broader framework of translational and clinical nanomedicine are paving new avenues for nanodentistry. Classical dentistry is becoming more predictive, preventive, personalized, and participatory, providing the patients with a tailored and targeted treatment and handling of their diseases. Considering the global impact of the oral pathologies, being particularly heavy in underdeveloped and developing countries, it is mandatory from an ethical perspective to ensure a global oral health. Nanobiotechnologies play a major role in this ambitious goal. In this review, we will focus on the bioinformatics, nanogenomics, and nanoproteomics aspects of contemporary nanodentistry, emphasizing the urgent need for an integrated proteogenomics approach and addressing its clinical and translational implications and new future perspectives and scenarios.

Advances in Protein Chemistry and Structural Biology, Volume 95
ISSN 1876-1623
http://dx.doi.org/10.1016/B978-0-12-800453-1.00004-X

1. INTRODUCTION

Advancements in the field of oral biomaterials (Choi, Ben-Nissan, Matinlinna, & Conway, 2013; Covani et al., 2007; Mallineni, Nuvvula, Matinlinna, Yiu, & King, 2013; Marconcini et al., 2014; Riley, Bavastrello, Covani, Barone, & Nicolini, 2005; Zandparsa, 2014), nanotechnologies (Ozak & Ozkan, 2013) and nanobiotechnologies, such as nanogenomics (Nicolini, 2006, 2010) and nanoproteomics (Kobeissy et al., 2014; Nicolini & Pechkova, 2010a,2010b) tools as fundamental components of a modern nanobiomedical approach (Nicolini et al., 2012; Nicolini, Bragazzi, & Pechkova, 2012; Nicolini, Bragazzi, & Pechkova, 2013) have enabled the birth of a new, highly interdisciplinary and rapidly growing discipline, termed as nanodentistry (Freitas, 2000; Kanaparthy & Kanaparthy, 2011; Mantri & Mantri, 2013), emerging from complementary and converging approaches.

Early diagnosing and properly monitoring oral diseases, avoiding their recurrence, providing the patients with a tailored, individualized, and targeted treatment (Bragazzi, 2013a, 2013b, 2013c) are important tasks within the field of personalized dentistry (Garcia et al., 2013; Glurich et al., 2013; Kornman & Duff, 2012; Razzouk & Termechi, 2013), that is becoming more predictive, preventive and participatory (Cafiero & Matarasso, 2013). Oral diseases have a tremendous burden and societal impact, affecting approximately 3.9 billions of people worldwide and particularly in underdeveloped and developing countries (Richards, 2013), and therefore ensuring global oral health is an ethical onus (Giannobile, 2013).

In this review, we will focus on the bioinformatics, nanogenomics, and nanoproteomics aspects of contemporary nanodentistry, addressing its clinical and translational implications and foreseeing its new future perspectives and scenarios.

2. AN INTEGRATED PROTEOGENOMICS PROTOCOL FOR PERSONALIZED DENTISTRY

2.1. Human samples

Human samples collected from a properly stratified cohort of patients offer the possibility to study and underpin the diseases-related biomarkers, connecting the symptoms, the diagnosis and the prognosis with the molecular and cellular levels, and ensuring the possibility of a targeted and individualized treatment. Many different kinds of biospecimens are available in the field of oral dentistry, each one with its own peculiarity and advantages, as well as pitfalls and drawback (Fig. 4.1). Once obtained, data can be eventually combined in order to have a robust molecular signature and a panel of selected, differentially expressed markers, that need to be replicated and validated before entering everyday clinical practice and routine. A bio-marker is indeed defined as reliable, reproducible, sensitive, and specific (Strimbu & Tavel, 2010). In the following subsections, we briefly overview the main sources of biomarkers in the field of oral pathologies, namely tissues biopsies, blood, dental plaque and oral biofilms, gingival crevicular fluid (GCF), saliva, and oral rinse.

2.1.1 Tissues

Oral cavity is a multifunctional environment made up of different components, building up a complex architecture. Its anatomy includes tissues from the mucosa (which is divided into different parts, namely the labial, buccal,

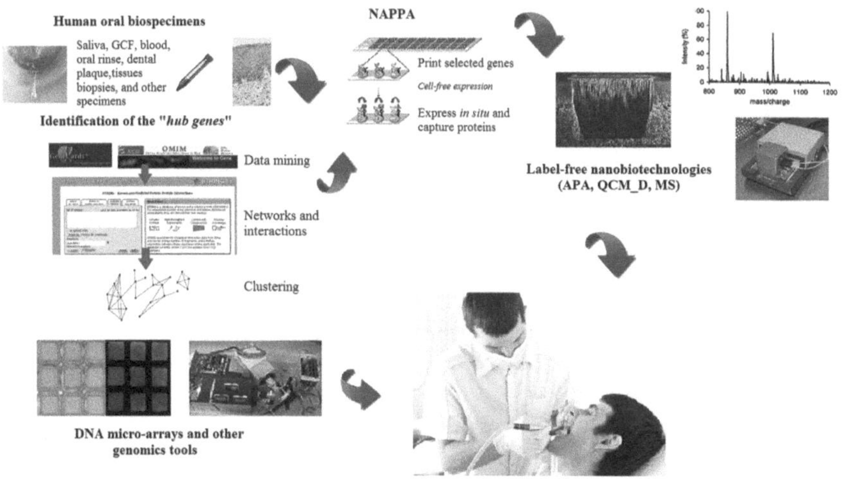

Figure 4.1 Protocol of the integrated nanoproteogenomics approach for personalized nanodentistry. After collecting the samples from the patients from one or more sources (tissue biopsies, saliva, gingival crevicular fluid or GCF, dental plaque and oral biofilms, and oral rinse), differentially expressed genes (DEG) are studied using for example the Leader Gene Algorithm (LGA). Thus, only few genes of interest are selected (genomics signature), which can be used as a panel for monitoring the diseases or can be subsequently expressed via the Nucleic Acid Programmable Protein Array (NAPPA) technology and protein–protein interactions (proteomics signature) are fully characterized via label-free nanobiotechnologies (namely Quartz Crystal Microbalance with Dissipation Factor Monitoring or QCM_D, Anodic Porous Allumina or APA, mass spectrometry or MS), overcoming the limitations and difficulties encountered in the use of labeled technologies.

lingual, palatoglossal, gingival, and palatal one), alveolar bone, periodontal ligament, and the cementum, as well as the salivary glands. These samples can be collected only in an invasive way, and even though they are quite reliable and reproducible, they are characterized by low acceptance and are therefore challenging. Few studies, indeed, give detailed pieces of information about oral tissues genomics and proteomics (Jágr et al., 2012).

2.1.2 Blood

Blood is a bodily fluid that delivers nutrients, oxygen, and other fundamental molecules for life, clearing and removing cellular waste products. It is a very common and popular, accepted sample, which is easy-to-obtain, and not difficult to store and process. It can be used as a whole blood or selected purified components.

In oral diseases, blood-derived biomarkers are associated with systemic risks and pathologies, such as cardiovascular (Meurman, Janket, Qvarnström, & Nuutinen, 2003), rheumatological (Joseph, Rajappan, Nath, & Paul, 2013; Kobayashi et al., 2014; Okada et al., 2013), gastrointestinal (Jaiswal, Deo, Bhongade, & Jaiswal, 2011), and metabolic (Pradeep, Kumari, Kalra, & Priyanka, 2013) diseases. Some studies have correlated blood with other samples like saliva (Haririan et al., 2012; Sundar, Krishnan, Krishnaraj, Hemalatha, & Alam, 2013), GCF (Fiorini et al., 2013; Gokul, Faizuddin, & Pradeep, 2012; Patel & Raju, 2013; Pradeep et al., 2011; Raghavendra et al., 2012; Sharma, Pradeep, Raghavendra, Arjun, & Kathariya, 2012; Thorat, Pradeep, & Garg, 2010), finding a positive correlation, even though in few cases not always concordant (Fiorini et al., 2013).

2.1.3 Dental plaque and oral biofilms

Dental plaque is a unique and dynamic biofilm, highly heterogeneous and poly-microbial, usually of a yellowish color, that develops naturally on the teeth's smooth surfaces (Kuboniwa et al., 2012). The oral flora colonizing the human oral cavity is also called the human oral microbiome (Dewhirst et al., 2010; Dimitrov & Hoeng, 2013), being different from an individual to another, as well as reflecting the effect of the treatment (Schwarzberg et al., 2014), and genomics/metagenomics and proteomics approaches have underpinned the different stages of dental (Peterson et al., 2011) plaque formation (dental biofilm, dental calculus or tartar) and shed light on the interaction between the host and the pathogens (Lemos & Burne, 2008; Lemos et al., 2005).

2.1.4 Gingival crevicular fluid

GCF is known also as sulcular fluid, since it is a serum transudate an inflammatory exudate produced by the sulcular epithelium of the oral mucosa in physiological conditions or from periodontal or gingival pockets in oral disorders (Lamster & Ahlo, 2007). It exerts a variety of functions, from antimicrobial activity to the lubrication of the oral cavity. It can be site-specific (from 1 of the 168 possible sampling sites) or not (Guzman et al., 2014). It is emerging as a promising sample for collecting information, thanks also to progresses in the way of obtaining it via extra-crevicular techniques (Lamster & Ahlo, 2007) and its unique transforming nature from a transudate to exudate and enrichment in disease-specific proteins, molecules, and microorganisms (Guzman et al., 2014). However, it is not-so-easy to obtain and requires specialized and trained staff (Guzman et al., 2014), moreover a complete mouth examination can be demanding, time consuming, and therefore challenging (Guzman et al., 2014).

2.1.5 Saliva

Saliva is a complex biological fluid (Huang, 2004; Ruhl, 2012; Wong, 2009; Ogawa et al., 2011; Zhang et al., 2013), produced by major salivary glands (submandibular, sublingual and parotid glands) and minor ones (scattered throughout the entire oral mucosa), and made up of water for the 99–99.5% (594–1194 mL/day) and of a mixture of microorganisms (bacteria, viruses, fungi, protozoa), (Jagtap et al., 2012; Wong, 2009) ions, enzymes and catalytic proteins, DNA and RNA, hormones, desquamated cells, food debris, and other molecules for the remaining 0.5–1% (3–12 mL/day), ranging up to 4–5% ore more (24–60 mL/day) in some clinical cases (Wong, 2009). Moreover, being the oral cavity in intimate contact with the gastrointestinal and respiratory tracts (Wong, 2009), it may contain also expectorated bronchial and nasal secretions, typical gastrointestinal or respiratory microorganisms, and some serum constituents that are derived from the local vasculature of the salivary glands and GCF, as well as from oral wounds (Deepa & Thirrunavukkarasu, 2010). Its production is finely tuned by the autonomic system, at least for the exocrine components (Wong, 2009), and plays a role in different functions, from speech and phonation, bolus formation and swallowing, starch digestion, protection, lubrication, buffering action, maintenance of tooth integrity through maintenance of an adequate level of mineralization, to perception of taste. Its proteome has unique features that makes it different from other proteomes: 73% of proteins present in saliva are absent in plasma, being exclusive to saliva (Cuevas-Córdoba & Santiago-Garcia, 2014). Moreover, it is highly heterogeneous when compared to the plasmatic one or other proteomes (Cuevas-Cordoba & Santiago-Garcia, 2014).

Even though quite attractive, being easy to obtain in noninvasive and acceptable ways and enabling also the study of other nonoral diseases (Bassim et al., 2012; Cuevas-Cordoba & Santiago-Garcia, 2014; Wong, 2009), saliva has many pitfalls and technical drawbacks that need to be addressed and overcome, in order to be a reliable source of information. Molecules are found at generally lower concentrations than in other fluids, in the nanomolar or picomolar range, proteins for example are found at concentration ranges of 150–400 mg% (Wong, 2009), and only recent advancements in the field of microseparation, purification, detection, and nanobiosensors have enabled its feasibility as source sample (Cuevas-Cordoba & Santiago-Garcia, 2014; Wong, 2009).

Moreover, many biomarkers need to be still validated (Cuevas-Cordoba & Santiago-Garcia, 2014). Saliva diagnostics represents the frontiers in oral diseases diagnostics: recently many high-throughput and OMICS technologies have focused on this fluid (Bencharit et al., 2012; Cuevas-Cordoba & Santiago-Garcia, 2014).

2.1.6 Oral rinse

Concentrated oral rinse has been used to detect the presence of oral bacteria and in particular fungi, which are difficult to find with other techniques, but can be found in abundance if collected from the dorsum of the tongue and the oral mucosa (Ghannoum et al., 2010). This procedure has different advantages: for example, it can be performed prior to sputus and expectoration, enhancing the sensitivity, and specificity of these procedures (Kalema et al., 2012). Moreover, it is relatively simple and noninvasive to collect, therefore is accepted by patients, and is safer to handle than other biospecimens (like blood; Ghannoum et al., 2010). Some drawbacks of the technique are possible contamination with extraoral environments, meal interference, and variations induced by different salivary flow rates.

2.2. Bioinformatics analysis

Bioinformatics is playing a growing role in the field of oral pathologies (Giacomelli & Covani, 2010), since it enables to collect many data and to store, retrieve them. Not only big data can be collected but also new data can be inferred and simulated through sophisticated and powerful algorithms.

2.2.1 Bioinformatics resources

Besides classical bioinformatics tools that have been developed for general purposes, the specifically designed and available bioinformatics resources for oral health are summarized in Table 4.1. Moreover, bioinformatics has been used to study complex interaction networks of microbial communities and oral biofilms, in the effort to associate particular microbial flora and interactions with cellular events, such as immune alterations (Yu, Hu, Zhou, Xia, & Amar, 2010), metabolic conditions (Mazumdar, Snitkin, Amar, & Segrè, 2009), and/or with a specific clinical trait (Duran-Pinedo, Paster, Teles, & Frias-Lopez, 2011; Hsiao et al., 2012; Zainal-Abidin et al., 2012). Pathways underlying human osseous remodeling (Sbordone et al., 2009), as well as the interaction between genetic factors, environment and microorganisms (a new field called "infectogenomics"; Kellam & Weiss, 2006; Nibali, Donos, & Henderson, 2009) and connections among the different diseases (Covani, Marconcini, Derchi, Barone, & Giacomelli, 2009) have been elucidated using a systems biology approach.

Bioinformatics has been used also in the field of reverse vaccinology (Rappuoli, 2000) for developing vaccine candidates for oral pathologies (Ross et al., 2001).

Dental informatics (Schleyer, 2003; Schleyer et al., 2011), dental bioinformatics (Giacomelli & Covani, 2010), and dental nanoinformatics (De La Iglesia et al., 2009) are growing and expanding fields, and new tools are likely to be released in the next years.

2.2.2 Leader-gene algorithm

Previously, we introduced a bio-data mining strategy for gene prioritizing, that is to say for selecting the most important, highly interconnected genes (termed as "*leader genes*" or "*hub genes*"; Bragazzi, Giacomelli, Sivozhelezov, & Nicolini, 2011; Bragazzi et al., 2011; Bragazzi & Nicolini, 2013) involved in different biological events, both at a cellular or molecular level (Giacomelli & Nicolini, 2006; Nicolini, 2006; Sivozhelezov, Giacomelli, Tripathi, & Nicolini, 2006) and more specifically in human diseases (Covani et al., 2008; Jovanovic et al., 2010; Marconcini et al., 2011; Orlando et al., 2013; Racapé et al., 2012; Sivozhelezov et al., 2008).

There are different candidate genes prioritizing computational approaches in the literature—for a broad and comprehensive review, the reader is referred to Moreau and Tranchevent (2012)—but our strategy is not limited to mono- or oligogenetic Mendelian diseases (Bragazzi et al., 2011; Bragazzi & Nicolini, 2013).

Table 4.1 A comprehensive list of bioinformatics resources available for oral diseases

Database	Features	URL	Reference
Bioinformatics Resource for Oral Pathogens (BROP)	It includes and integrates the Oral Pathogen Microarray Database, the Genome Viewer, and the Genome-wide ORF Alignment (GOAL)	http://brop.org	Chen, Abbey, Deng, and Cheng (2005)
CORE (a streamlined and phylogenetically curated database of 16S rDNA sequences that represent the core oral microbiome)	It allows to recognize microorganisms from clinical samples, exploiting next-generation sequencing technologies	http://microbiome.osu.edu/	Griffen et al. (2011)
Head and Neck and Oral Cancer Database (HNOCDB)	It enables to mine genes, miRNAs, and altered loci/chrosomes related to oral diseases	http://gyanxet.com/hno.html	Mitra et al. (2012)
Human Oral Microbiome Database (HOMD)	It is a vast, comprehensive, and authoritative database that includes information about the human oral microorganisms. It also addresses nomenclature issues	http://homd.org	Chen et al. (2010) and Wade (2013)
OralCard	Manually curated database, it combines different resources and approaches: the ecological one (the oral molecular ecosystem or OralPhysiOme),	http://bioinformatics.ua.pt/oralcard	Arrais et al. (2013)

	the oral proteome of human (OralOme), and microbial origin (MicroOralOme). It integrates both nonproteomics and proteomics resources		
Oral Cancer Gene Database (OCGD) OrCGDB	It exploits STRING database and integrates genomics and proteomics resources	http://www.actrec.gov.in/oralcancer/GeneHome.htm (version I) http://www.actrec.gov.in/OCDB/index.htm (version II)	Gadewal and Zingde (2011)
Oral Fungal Microbiome (Mycobiome)	Manually curated list of oral fungal pathogens	Available as supplementary materials, at http://www.plospathogens.org/article/info%3Adoi%2F10.1371%2Fjournal.ppat.1000713#s5	Ghannoum et al. (2010)
OralOme	Manually curated list of proteins	Available as supplementary materials, at http://www.sciencedirect.com/science/article/pii/S0003996912000027	Rosa et al. (2012)
Oral Pathogen Sequence Databases of the Los Alamos National Laboratory Bioscience Division (ORALGEN)	It provides the researchers with genomics and metagenomics tools and resources	http://www.oralgen.lanl.gov/oralgen/ http://www.oralgen.org/ (archived and mirror copy)	Xie et al. (2010)
Orca-DB	It is a manually curated database that includes molecular and other clinically relevant information about oral cancer	http://210.212.237.35/orcadb/ http://www.rgcb.res.in/orcadb	Reshmi et al. (2012)

Continued

Table 4.1 A comprehensive list of bioinformatics resources available for oral diseases—cont'd

Database	Features	URL	Reference
OrCGDB	It integrates mining tools such as PubMed/MEDLINE and OMIM	http://www.tumor-gene.org/Oral/oral.html	Levine and Steffen (2001)
Pathogenic Pathway Database for Periodontitis	It is a manually curated database containing pathogenic pathways for periodontitis, linked to causal relations, and biological entities obtained from text mining	http://bio-omix.tmd.ac.jp/disease/perio/	Suzuki et al. (2009)
Salivaomics Knowledge Base (SKB)	It enables saliva diagnostics, exploiting tools such as Saliva Ontology, and SdxMart	http://www.hspp.ucla.edu/skb.swf	Ai, Smith, and Wong (2012)
Salivary Gland Tumor Biorepository (SGTB)	It is a curated collection of salivary gland, tumor-related biospecimens, and cell lines, enabling both basic and translational research	https://research.mdacc.tmc.edu/Salivary_DB/index.html	Used in Matse et al. (2013)
A searchable database for proteomes of oral microorganisms	It is a database that includes proteomes of oral microorganisms obtained with two-dimensional electrophoresis (2DE) gel	http://www.myamagu.dent.kyushu-u.ac.jp/~bioinformatics/index.html%20 http://www.bipos.mascat.nihon-u.ac.jp/index.html	Nakano et al. (2005)

Namely, a comprehensive combination of exhaustive, recursive, and iterative search of diseases-related genes mining different databases is performed, like PubMed using the National Library of Medicine (NLM) standardized and controlled vocabulary based on medical subject headings or MESH terms (Doğan, Leaman, & Lu, 2014), the National Center for Biotechnology Information Online Mendelian Inheritance in Man (NCBI OMIM; Doğan et al., 2014; NCBI Resource Coordinators, 2014), GenBank (Benson et al., 2012), GeneCards (Stelzer et al., 2011) and MalaCards (Rappaport et al., 2013), GeneAtlas, using the standardized nomenclature (HUGO, or Human Genome Organization), and/or repositories containing DNA microarrays data, such as GEO or Gene Expression Omnibus (Edgar, Domrachev, & Lash, 2002), DDBJ, or DNA Data Bank of Japan (Kodama et al., 2010), and ArrayExpress (Rustici et al., 2013). For further information about mining and accessing genomics public repositories, the reader is referred to Huttenhower and Hofmann (2010) and references therein.

Links to other repositories and databases such as Genetic Association Database (GAD; Becker, Barnes, Bright, & Wang, 2004) are currently in progress.

The obtained hits can be integrated after quality-check, preprocessing, and statistical meta-analysis, using also available resources such as AnnotCompute (Zheng, Stoyanovich, Manduchi, Liu, & Stoeckert, 2011), MageComet (Xue et al., 2012), M(2)DB (Cheng et al., 2010), ArrayMining (Glaab, Garibaldi, & Krasnogor, 2009), or virtualArray (Heider & Alt, 2013).

Moreover, in an updated version, that is currently in progress, the user will be able to choose whether to use specifically curated databases (such as the resources and tools listed in Table 4.1), it will be possible also to interrogate cross-species databases (Le, Oltvai, & Bar-Joseph, 2010).

This step is followed by the networks and pathways reconstruction using STRING software (Search Tool for the Retrieval of Interacting Genes, Heidelberg, Germany; Franceschini et al., 2013) and finally the list of genes is clustered according to their weighted number of links (WNLs).

This measure is calculated for each gene using the program STRING, and this value is derived from the weighed sum of three types of interactions:

1. literature co-occurrence of the names of genes, and/or their products in abstracts and/or full texts of papers available on the Internet. The scores assigned are derived from a benchmarked and validated scoring system, which is based on the frequencies and distributions of gene/gene

products names in the aforementioned abstracts and/or full texts. The benchmarks themselves are set from a manual evaluation and assessment of predictions of gene and protein interactions by experts and are typically below 0.5;

2. scores derived from different databases dedicated to gene networks, containing data on induction and expression of a particular genes by other genes derived from microarray experiments, or other high-throughput omics techniques. The score of 1 is assigned if the link is already present in the databases, while putative links have lower values (typically in the range 0.6–0.8);
3. the same range of scores is assigned to gene interactions via physically observed interactions between proteins. The software used does not discriminate between *in vivo* or *in vitro* experiment-derived data. Generally, the scores are close to those of interaction type 2, but links of this type occur much rarely than of type 2.

The combined association scores S_{ij} were summed for each gene i over its neighbors (i,j), giving the final WNL for the gene i. Further, clustering methods are applied to the WNLs in order to identify the group of leader genes. Cluster analysis, also called segmentation analysis or taxonomy analysis, is a way to partition a set of objects into homogeneous and separated groups or clusters, in such a way that the profiles of objects in the same cluster are very similar and the profiles of objects in different clusters are quite distinct.

Genes belonging to the highest rank are defined as "*leader genes*" or "*hub genes*" because they may be assumed to play an important role in the analyzed processes. Leader-gene algorithm (LGA) can suggest a list of few, but strong candidate genes potentially relevant within a given cellular process or a pathology, according to the already available experimental data. Moreover, the interaction map among all the genes involved in the same process may be useful in interpreting the experimental and clinical results, and in planning new targeted experimentation. Interestingly, such experimentation may be simpler to be analyzed than mass-scale molecular genomics, whose wealth of details may raise problems and complications. This computational method gave promising results, when applied to the human T lymphocyte cell cycle (Giacomelli & Nicolini, 2006; Nicolini et al., 2006; Sivozhelezov et al., 2006) and its malignant transformation (Sivozhelezov et al., 2009), human kidney transplant with a focus on operational tolerance (Braud et al., 2008; Jovanovic et al., 2010; Racapé et al., 2012; Sivozhelezov et al., 2008), oral lichen planus (OLP; Orlando, Bragazzi, & Nicolini, 2013), and periodontitis (Covani et al., 2008; Marconcini et al., 2011). These results were also integrated with a targeted experimental analysis, to draw an overall picture of these processes (Giacomelli & Nicolini, 2006; Marconcini et al., 2011; Racapé et al., 2012), and only those related to oral diseases are reviewed in the following paragraphs.

This interactive, automatic, and user-friendly stand-alone tool has been written in house in Java, JavaScript, PHP, and HTML. The completely automated pipeline is performed via NCBI e-utilities (e-search, e-fetch, for further information the author is referred to the NCBI site), and other similar facilities.

A scheme of the algorithm together with a screen-shot of the software is given in Fig. 4.2.

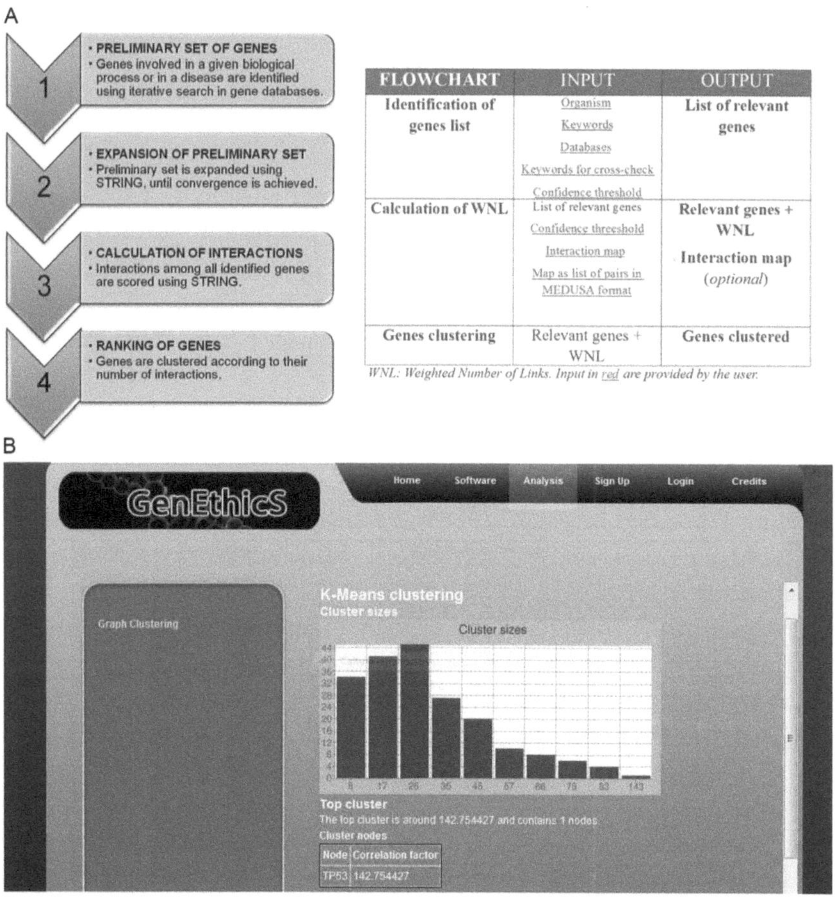

FLOWCHART	INPUT	OUTPUT
Identification of genes list	Organism Keywords Databases Keywords for cross-check Confidence threshold	List of relevant genes
Calculation of WNL	List of relevant genes Confidence threeshold Interaction map Map as list of pairs in MEDUSA format	Relevant genes + WNL Interaction map *(optional)*
Genes clustering	Relevant genes + WNL	Genes clustered

WNL: Weighted Number of Links. Input in red are provided by the user.

Figure 4.2 The algorithm on which the leader-gene tool for molecular genomics is based and a screen-shot of the software.

The clustering techniques the user can choose are Clustering K-means and Chinese whispers (which has been thought specifically for graph clustering); as far as the number of clusters is concerned, the user can choose from heuristic number or provided by the user himself.

The obtained list of Class A and Class B genes can be used for predicting further biomarkers such as miRNAs (work currently in progress) or being validated with *ad hoc* experiments, such as gene microarrays or protein arrays after being expressed and subsequently analyzed via labeled or label-free nanobiotechnologies (Fig. 4.1), which are better described in the following paragraphs.

2.3. Proteomics technologies, with a focus on the label-free tools

2.3.1 Mass spectrometry

MS is a technique widely used in the field of proteomics and is emerging as a useful technology for oral biology and dentistry (Al-Tarawneh, Border, Dibble, & Bencharit, 2011; Amado et al., 2013). However, only few studies have attempted to analyze saliva or other human bodily fluids with MS, and currently consensus on the sample collection protocol is still missing (Al-Tarawneh et al., 2011). Size of the recruited cohorts is usually low-middle and some inconsistencies among the studies have been found (Al-Tarawneh et al., 2011), even though scholars have benefit from the advancements in MS technology.

In our laboratory, we have successfully coupled the matrix assisted laser desorption ionization time of flight MS (MALDI-TOF MS) to nucleic acid programmable protein arrays (NAPPA) technology (Spera, Labaer, & Nicolini, 2011) and bioinformatics analysis (Belmonte, Spera, & Nicolini, 2013; Nicolini et al., 2013a, 2013b) for the detection of the proteins translated from the cDNA on the surface of the array. However, the development of a MALDI-TOF MS-compatible protein microarray was not a trivial and straightforward task, but was rather complex and demanding, since existing methods and techniques for obtaining protein microarrays may not be compatible with LDI MS, and therefore a specifically *ad hoc* modified support, having an electro-conductive target surface, was essential.

Moreover, one of the challenges in properly identifying the mass spectra generated from the MS coupled with the label-free NAPPA technology was their particular complexity, due to the presence of extra biological material besides the target protein, such as the BSA complex, the additional peptide chain (the GST tag), and the anti-GST capture antibody. Since this material is present in all the features of the array, as a "common background," *in house* developed bioinformatics tools can be used to better interpret the obtained results, since the available software is not fully adequate for the analysis of such complex mixtures (Belmonte et al., 2013). Namely a matching algorithm was developed to identify and discriminate between real "*protein peaks*" and "*background peaks*."

This R-script based *in house* software is termed as Spectrum Analyzer and Data Set manager, or SpADS (Belmonte et al., 2013). It is able to perform different tasks, from the data preprocessing and binning to smoothing, noise filtering, data reduction, peak extraction, and normalization, as well as peak alignment, background subtraction, and peak identification. Finally, it can be coupled to simple data mining algorithms such k-means clustering or other statistical strategies such as Principal Component Analysis (PCA) in order to identify proteins in case of failure in attempt of identifying the peak by mining MASCOT database (Matrix Sciences, Ltd, available at the URL: www.matrixscience.com; Belmonte et al., 2013).

The current NAPPA chemistry and the recent advance in MS allows us potentially to validate this label-free technology even in clinical settings through the correlation with the tremendous amount of fluorescence data already acquired over the years.

In this case, the end game is to demonstrate that we can identify proteins, in particular proteins that bind to the target proteins on the array. Toward this end we have shown that we can identify the expressed proteins, printing on gold slides four different genes (each one with $16 \times 300\ \mu$ spots)—first in a known configuration, then in an unknown configuration. MS analysis was conducted successfully by searching peptides on a database. Key to success was to do trypsin digests and get the peptides to fly and to be identified. Once dried, the array was placed on the MALDI target and analyzed. The analysis was performed with an Autoflex MALDI-TOF MS (Bruker Daltonics, Leipzig, Germany) operating in linear and reflector mode. The resulting mass accuracy for peptides was <20 ppm. MALDI-TOF mass spectra were acquired with a pulsed nitrogen laser (337 nm) in positive ion mode, using two software programs to acquire and process mass spectra: FlexControl, designed to configure and to operate TOF MS of the Bruker's flex-series, and FlexAnalysis, designed for data analysis of spectra acquired with Bruker's TOF MS. We acquired spectra for each sample (p53, JUN, CdK2, CdKN1A, A, B, C, D) and in order to identify the A, B, C, and D samples we matched, with the aid of our *in house* developed "matching algorithm" implemented by us and described in Belmonte et al. (2013), Nicolini, Adami, et al. (2012),

Nicolini, Bragazzi, and Pechkova (2012), Nicolini et al. (2013), Nicolini et al. (2013), and Spera et al. (2010), their experimental mass lists with that of the known samples (P53, JUN, CdK2, CdKN1A).

We can then conservatively conclude that the implemented chemistry and analysis for the first time demonstrate the successful use of MS for the characterization of proteins immobilized on NAPPA. Further development is in progress to bring this label-free procedure to practice as an adjunct to fluorescence NAPPA work, which has already seen significant clinical applications in the last decades (Anderson et al., 2008; Nand, Gautam, Pérez, Merino, & Zhu, 2012; Nicolini & Pechkova, 2010a,2010b; Sibani & LaBaer, 2011; Spera et al., 2013b). NAPPA approach has been indeed used for investigating cancer (Anderson et al., 2011, 2010), type 1 diabetes (Miersch et al., 2013), rheumatological diseases (Gibson et al., 2012; Wright et al., 2012), and infections (Ceroni et al., 2010; Manzano-Román et al., 2012; Montor et al., 2009; Rolfs et al., 2008; Thanawastien, Montor, Labaer, Mekalanos, & Yoon, 2009).

The background generated by the reticulocytes lysate is, however, still significant and need to be reduced to make this approach routinely applicable in the clinics. This reduction might be instead achieved by the use of a bacterial cell-free expression system with respect to the traditional mammalian lysate, particularly required by the highly sensitive nanotechnologies being here utilized. The application of bacterial PURExpress to NAPPA (in progress) consists of a template double-stranded DNA containing the gene of interest fused to a SNAP tag and the upstream T7 promoter (Nicolini, Spera et al., 2013; Pechkova et al., 2010).

By adding the PURExpress reconstituted cell-free translation system (Houlihan, Gatti-Lafranconi, Kaltenbach, Lowe, & Hollfelder, 2014), the template DNA is transcribed into mRNA, and then translated into a fusion protein containing the N-terminal SNAP tag and the C-terminal target protein. In the same spot, the SNAP tag allows the synthesized protein to bind to its own template DNA via the BG linkage, thus immobilizes the target protein. The rest of the reaction mixture can be washed away and the immobilized target protein is allowed to interact with a mixture of query proteins. After the binding reaction, the unbound proteins are washed away and the target protein complex is released by cleaving the template DNA.

To compare the backgrounds of NAPPA between PURExpress (bacterial lysate) and RRL (rabbit reticulocyte lysate) by Mass Spectrometry (MS) and Fluorescence we are presently utilizing NEB *in vitro* system and SNAP fusion as an alternative to *in vitro* system and GST tag.

Their advantages are higher expression level and cleaner for downstream analysis, making possible and really effective Label Free quantitative analysis at the nanoscale (work in progress in cooperation with Arizona State University, ASU, and New England Biolabs, NEB; Nicolini et al., 2013; Nicolini et al., 2013).

2.3.2 Anodic porous alumina

Anodic porous alumina (APA) is a versatile material that can be used in order to design nanostructured materials, such as nanoporous membranes and arrays, as well as nanoparticles (Nicolini, Adami, et al., 2012; Nicolini, Bragazzi, & Pechkova, 2012; Nicolini et al., 2013). APA surface can be prepared following a two-step protocol (Masuda & Fukuda, 1995), by a suitable electrolytic process designed to obtain a regular honeycomb distribution of deep micrometric/nanometric holes. The task of evaporating aluminum over glass has been accomplished by avoiding its detachment during the anodization process, a typical problem due to the incompatibility of cold borosilicate glass to the vapors of aluminum. This phenomenon can be easily contrasted by means of a thin layer of chromium (deposited by sputtering) as medium element between glass and aluminum.

The dielectric properties of Al_2O_3 make this structure optimal for the realization of an electrically anisotropic system; the electrochemical reactions occurring on the bottom of the well (caused by the interaction between the biological probe molecules and the test molecules) induce variation on the electrical response to alternating voltage signal, and they can be quantified by means of scanning electrodes moving on the surface of the array, placed in a proper solution. The walls of the pore behave like insulator decoupling from the electrochemical events occurring few microns/millimeters away from the measurement place. This option constitutes the APA label-free approach to the analysis of protein arrays, since no fluorescent/marked molecule is utilized, while the alternative one still linked to APA requires a fluorescent molecule to spread the luminous signal through the APA wave guide. The potential of label-free approaches to complement and even to improve other detection technologies of proteins being expressed in NAPPA microarrays has never been higher.

NAPPA printing on APA has proven (Nicolini, Correia, et al., 2013; Nicolini, Singh, Spera and Felli, 2013):

- The ability to spot a colored fluid on the APA surface in discrete spots
- The ability to rapidly exchange that fluid with a different fluid
- The ability to repeat these manipulations as needed

- The production of the APA slides in a format compatible also with a fluorescence reader and not only with electrochemistry
- The structural integrity of APA slide format (either alone or over aluminum) has been proven acceptable for "routine" manipulation and especially during fluorescence readings, mainly relying for the stand-alone configuration also on alternative printing which exploits capillary forces.
- Last but not least the usefulness of APA substrate was shown as a candidate to isolate protein expression in a defined space followed by detection of the photoluminescence signal emitted from the complex secondary antibody anchored to Cy3 with a CCD microscope. The fluorescence of these labeled proteins was clearly evident in circular shaped arrangements on a limited surface of APA where the proteins were confined in the pores. APA surface appears to allow a label-free analysis using electric impedance spectroscopy (EIS). It is known that with EIS it is possible to detect different amounts of organic materials deposited even indirectly over conducting surfaces. After the hybridization/expression experiment, using a scanning electrode controlled by a manipulator (MPC 200 by Sutter Technologies) via a PC, different EIS measurements were performed in different spots on the surface of the array in phosphate buffered saline solution (PBS; Nicolini et al., 2013; Nicolini et al., 2013).

Moreover, this nanomaterial has proven to be efficient for osteoblast growth (Karlsson, Pålsgård, Wilshaw, & Di Silvio, 2003; Salerno et al., 2013; Salerno, Giacomelli, & Larosa, 2011; Song, Ju, Morita, & Song, 2013). Moreover, its high aspect ratio (depth/width ratio) of the pores makes this material also a natural wave guide for any fluorescent molecule present on the bottom of the pores, avoiding crosstalk of many point-light sources too close as frequently in fluorescent NAPPA (Nicolini et al., 2013; Stura et al., 2007). It has been adapted and used coupling with a laser desorption/ionization (LSI) MS (Shenar, Martinez, & Enjalbal, 2008), realizing a variant of the desorption/ionization on porous silicon (DIOS) MS device, using an aqueous suspension of porous alumina (pore size of 90 Å). Other scholars have coupled APA with Localized Surface Plasmon Resonance (LSPR; Kim et al., 2008). The utilization of an APA-based substrate is indeed a promising approach for proteomics (Wang, Xia, & Guo, 2005).

2.3.3 Nucleic acid programmable protein arrays

NAPPA is a cell-free technology (Ramachandran et al., 2004). Different other protein arrays exist, like DNA-array to Protein-array (DAPA), PISA

(Reddy et al., 2011), but we focused on NAPPA since it offers many advantages in comparison with classical technologies.

Recently, we have developed a new device that couples NAPPA with the quartz crystal microbalance with dissipation factor monitoring (QCM_D; Nicolini, Adami, et al., 2012; Nicolini, Bragazzi, & Pechkova, 2012; Spera et al., 2013b).

The QCM_D instrument was developed by Elbatech Srl. The quartz was connected to an RF gain-phase detector (Analog Devices, Inc., Norwood, MA, USA) and was driven by a precision DDS (Analog Devices, Inc., Norwood, MA, USA) around its resonance frequency, thus acquiring a conductance versus frequency curve (conductance curve) which shows a typical Gaussian behavior. The conductance curve peak was at the actual resonance frequency while the shape of the curve indicated how the viscoelastic effects of the surrounding layers affected the oscillation. The QCM_D-dedicated software, QCMAgic-Q5.3.256 (Elbatech srl, Marciana—LI, Italy) allows to acquire the conductance curve or the frequency and dissipation factor variation versus time. In order to have a stable control of the temperature, the experiments were conducted in a temperature chamber. Microarrays were produced on standard nanogravimetry quartz used as highly sensitive transducers. The QC expressing proteins consisted of 9.5 MHz, AT-cut quartz crystal of 14 mm blank diameter and 7.5 mm electrode diameter, produced by ICM (Oklahoma City, OK, USA). The electrode material was 100 Å Cr and 1000 Å Au and the quartz was embedded into glass-like structures for easy handling.

The NAPPA-QC arrays were printed with 100 spots per QC.

Quartzes gold surfaces were coated with cysteamine to allow the immobilization of the NAPPA printing mix. Briefly, quartzes were washed 3 × with ethanol, dried with Argon and incubated over night at 4 °C with 2 mM cysteamine. Quartzes were then washed 3 × with ethanol to remove any unbound cysteamine and dried with Argon. Plasmids DNA coding for GST tagged proteins were transformed into *E. coli* and DNA were purified using the NucleoPrepII anion exchange resin (Macherey Nagel). NAPPA printing mix was prepared with 1.4 μg/μl DNA, 3.75 μg/μl BSA (Sigma–Aldrich), 5 mM BS3 (Pierce, Rockford, IL, USA), and 66.5 μg polyclonal capture GST antibody (GE Healthcares). Negative controls, named master mix (hereinafter abbreviated as "MM"), were obtained replacing DNA for water in the printing mix. Samples were incubated at room temperature for 1 h with agitation and then printed on the cysteamine-coated gold quartz using the Qarray II from Genetix. In order to enhance the sensitivity, each quartz was printed with 100 identical features of 300 μ diameter each, spaced by 350 μ center-to-center.

Gene expression was performed immediately before the assay, following the protocol described in Spera et al. (2013b). Briefly, *in vitro* transcription and translation (IVTT) were performed using HeLa lysate mix (1-Step Human Coupled IVTT Kit, Thermo Fisher Scientific Inc.), prepared according to the manufacturers' instructions. The quartz, connected to the nanogravimeter inside the incubator, was incubated for 10 min at 30 °C with 40 μl of HeLa lysate mix for proteins synthesis and then, the temperature was decreased to 15 °C for a period of 5 min to facilitate the proteins binding on the capture antibody (anti-GST). After the protein expression and capture, the quartz was removed from the instrument and washed at room temperature, in 500 mM NaCl PBS for 3 ×. The protocol described above was followed identically for both negative control QC (the one with only MM, i.e., all the NAPPA chemistry except the cDNA) and protein displaying QC.

After protein expression, capture, and washing the QCs were used for the interaction studies QC displaying the expressed protein was spotted with 40 μl of the desired molecule solutions in PBS at increasing concentrations at 22 °C.

We also tested the possibility to analyze drug/small molecule–protein interactions in QC displaying multiple proteins, a task which is not possible with fluorescence based arrays (Spera et al., 2013a,2013b).

QCM_D measures were calibrated for frequency and for *D* factor shifts. The calibration curves equation (obtained with Ordinary Least Squares methods, OLS) are (Spera et al., 2013a,2013b):

$$\Delta f = -7.16 - 231.18\text{m} \quad \text{with}\, r^2 = 0.9986$$
$$D = 0.831 + 0.286\eta \quad \text{with}\, r^2 = 0.9990$$

We analyzed the conductance curves acquired in NAPPA-QCs in different steps of the expressing and capturing process. Moreover, the conductometer can be employed under both flow and static conduction.

Figure 4.3 shows the conductance curves for the NAPPA-QCs expressing p53, CDK2, and Jun (all the cDNAs were co-immobilized in the same feature). These data, combined with the data previously acquired, pointed to a unique conductance curve shape for each gene/protein and suggested the possibility to identify the expressed gene/proteins by QCM_D even when combined on the same expressing QC (Fig. 4.3).

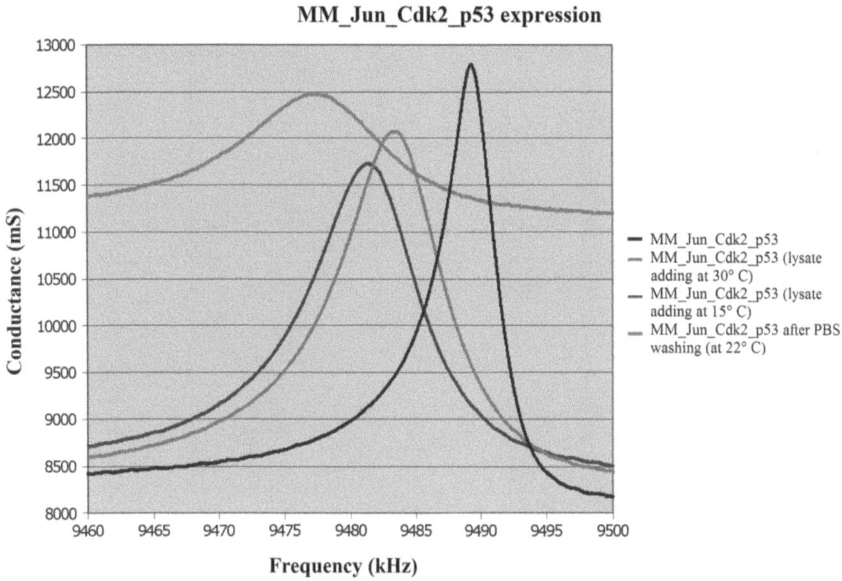

Figure 4.3 Conductance curves for the NAPPA-Quartz Crystal expressing p53, CDK2, and Jun (all the cDNAs being co-immobilized in the same feature).

The coefficients of variations yield values that are usually very low, confirming the repeatability of the experiments and the validity and portability of the technique. In our hands, NAPPA-based QCM_D proved to have an intra-assay overall C_V of 5% (range 3.3–8.0%; Spera et al., 2013b).

In conclusion, our innovative conductometer, realized by combining NAPPA technology with QCM_D, enables the study of genes and their products, the characterization of protein–protein and protein–drugs/small molecules interactions in a multiparametric way, taking advantage of the multiple information provided by the analysis of the conductance curves (i.e., conductance, viscoelasticity, and adsorbed mass). Moreover, through our conductometer it is possible to acquire detailed information about the kinetic constants of the interaction.

All these approaches can be combined and together can provide useful information (Fig. 4.4).

3. ORAL DISEASES

Oral diseases are complex pathologies, deriving from the intersection of different components: the oral microbial flora (microbiome), environmental and behavioral factors and life styles, the human genetic make-up

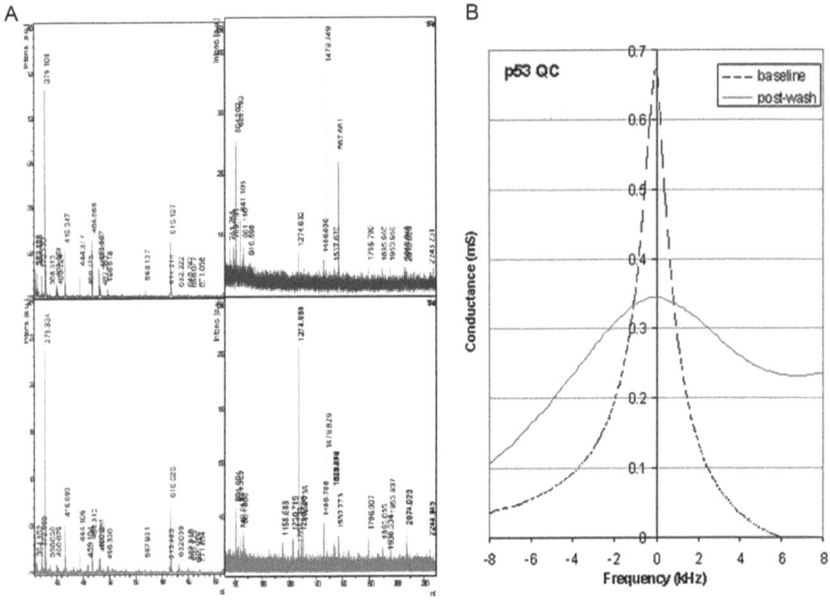

Figure 4.4 MALDI-TOF spectra of NAPPA after protein trypsin digestion, 5–20 kDa range, for p53 (upper, left) versus A (bottom, left) samples. p53 Normalized conductance curve acquired with the NAPPA-QCM_C conductomer (right). Proteomics approaches can be combined in order to get more information.

(the genome), its transcription and translation (the transcriptome, the proteome, the metabolome, or metabonome and further levels).

For this reason, all the approaches that we have overviewed in the previous sections should be coherently integrated into a proper framework.

3.1. Dental caries

Caries is a very common and expensive oral disease (Bánóczy & Rugg-Gunn, 2013).

Besides classical clinical investigations, decaying teeth have been studied with nanobiotechnologies that have enabled an unprecedented characterization down to the nanometer scale using small-angle X-ray scattering (SAXS) and synchrotron technologies (Gaiser, Deyhle, Bunk, White, & Müller, 2012).

Wang and collaborators have used the bioinformatics approach based on prioritizing candidate genes and protein–protein interaction analyses. They identified three major clusters putatively leading to dental caries: namely, the cytokine network, the matrix metalloproteinases (MMPs) family, and the transforming growth factor-beta (TGF-β) cluster (Wang et al., 2013).

Genomics and metagenomics studies have characterized the evolution and differentiation the oral microbiome (Jiang et al., 2014).

Proteomics studies have analyzed saliva fluid and superoxide dismutase (SOD), copper and zinc concentrations were found to be different between patients and healthy subjects in a statistically significant way (Hegde, Hegde, Ashok, & Shetty, 2014). In another study, the proteomics investigation of the parotid gland secretion lead to the identification of some biomarkers such as cystatin S and collagen as being upregulated, while dermcidin was downregulated (Preza, Thiede, Olsen, & Grinde, 2009).

However, some scholars have not been able to find differences in saliva proteome between subjects with and subjects without caries (Zehetbauer, Wojahn, Hiller, Schmalz, & Ruhl, 2009).

The study of the entire *Streptococcus mutans*, one of the causative microorganisms of the dental caries, has lead to the discovery of 84 uncharacterized proteins, that can be studied for identifying potential drug targets for a pharmacological intervention (Horst et al., 2012; Klein et al., 2012; Nan et al., 2009). In a series of comparative proteomics analyses, carolacton has for example shown an inhibitory effect, by disturbing the *Streptococcus* peptidoglycan biosynthesis and damaging the integrity of the cell envelope (Li, Wang, Wang, & Zeng, 2013).

Models incorporating both proteomics and genomics/metagenomics approaches have been demonstrated more predictive than models including only microbial or salivary data (Hart et al., 2011).

3.2. Periodontitis

Periodontitis is a set of inflammatory diseases affecting the periodontium, that is, the tissues that surround and support the teeth. Periodontitis involves progressive loss of the alveolar bone around the teeth, and if left untreated, can lead to the loosening and subsequent resorption and loss of teeth. Periodontitis is caused by microorganisms that adhere to and grow on the tooth's surfaces, along with an overly aggressive immune response against these microorganisms.

Until 1977, periodontitis was divided into two classes (juvenile and chronic marginal periodontitis), that have become four in 1986 (the first class has been split into subclasses, prepubertal, localized and generalized, the other classes including adult, necrotizing ulcerative gingivo-periodontitis, and refractory periodontitis). The 1989 classification has introduced new classes: (1) early onset periodontitis (that includes the former prepubertal and juvenile periodontitis, both localized and generalized, plus the rapidly progressive periodontitis), (2) adult periodontitis, (3) necrotizing ulcerative periodontitis, (4) refractory periodontitis, and (5) periodontitis with systematic disease. The 1999 classification, criticizing the concept and definition of an early onset periodontitis, has replaced it with aggressive periodontitis, has recognized the gingival diseases as a precursor of periodontitis (even though the transition from gingivitis to periodontitis is not always clear and clinically obvious), and, thinking of periodontitis as a continuum set of pathologies, has introduced further categories (abscess of the periodontium, periodontitis associated with endodontic lesions, and developmental or acquired deformities and conditions; Wiebe & Putnins, 2000).

Also for periodontitis, the gene prioritization algorithm has been applied (Zhan et al., 2014), leading to the identification of 21 putative genes involved or potentially involved in periodontitis, 9 of them have been already confirmed, while other genes, such as *CSF3, CD40, TNFSF14, IFNB1, TIRAP, IL2RA, ETS1, GADD45B, BIRC3, VAV1, COL5A1, and C3*, have not been associated with periodontitis, even though evidences from the extant literature show their involvement in bacterial infection, immune response, and inflammatory reaction.

Using bioinformatics and decision tree, modeled the risk. Laine and collaborators identified the presence of bacterial species *Tannerella forsythia, Porphyromonas gingivalis, Aggregatibacter actinomycetemcomitans*, and SNPs *TNF-857 and IL-1A-889* as discriminators between periodontitis and non-periodontitis. The model reached an accuracy of 80%, sensitivity of 85%, specificity of 73%, and AUC of 73% (Laine et al., 2013).

In a bioinformatics analysis (Covani et al., 2008), five genes (*namely, the nuclear factor-KB1 or NFKB1, the gene for transcription factor p65 or REL-A, the growth factor receptor binding protein 2 or GRB2, the Casitas B-lineage lymphoma gene or CBL, the phosphoinositide-3-kinase, regulatory subunit 1 (alpha) or PIK3R1*) were identified as "*leader genes*." Their expression in the leukocytes of 10 patients with refractory chronic periodontitis was subsequently investigated using real-time quantitative polymerase chain reaction (PCR) technology (Marconcini et al., 2011).

The authors found that the association of pathology with the genes was statistically significant for *GRB2 and CBL* ($P<0.01$), while it was not statistically significant for the other genes (Marconcini et al., 2011).

As far as the proteomics biomarkers are concerned, a recent systematic review and meta-analysis has underpinned up to 20 classes of proteins and molecules that are differentially regulated in periodontitis—4 downregulated, 15 upregulated (Guzman et al., 2014), these proteins are involved in inflammatory and immune response, cellular homeostasis, cell cycle regulation and control, catalysis, metabolism, bone mineralization and maintenance and antimicrobial activity.

Metagenomics and proteomics of periodontitis-related microorganisms have contributed to detail the molecular basis and mechanisms of pathogen–host interactions and shown that the microbial flora may act as a link between periodontitis and systemic risks (Liu et al., 2012; Pham et al., 2010; Zainal-Abidin et al., 2012; Zijnge, Kieselbach, & Oscarsson, 2012).

3.3. Oral lichen planus

OLP is a chronic T-cell-mediated autoimmune mucocutaneous disease which involves the oral cavity. Despite advancements in the field of epidemiology, its precise etiology remains unknown: OLP shows a higher incidences among females and has an overall age-standardized prevalence of 1.27% (McCartan & Healy, 2008).

According to Andreasen's classical classification, there are six recognized oral manifestations, that is, reticular, papular, plaque, atrophic, erosive (ulcerative), and bullous lesions (Andreasen, 1968). Later, this classification was modified and simplified into a new clinical one that included only the reticular, atrophic, and erosive forms (Silverman, Gorsky, & Lozada-Nur, 1985). Another, more simplified, clinical classification is that by Carbone and Gandolfo that differentiates between white and red OLP (Carbone et al., 2009; Gandolfo et al., 2004).

Histopathological criteria usually include hypergranulosis, different degrees of keratosis (hyperkeratosis, parakeratosis, orthokeratosis), acanthosis, and apoptosis with the formation of the so-called Civatte bodies, liquefaction, and hydropic degeneration of the basal cell layer, presence of irregular ridges, and band-like T-lymphohistiocytic infiltrate at the level of papillary dermis and lamina propria, absence of epithelial dysplasia.

OLP is currently considered by the World Health Organization (WHO) as a potential premalignant status, like other disorders such as leukoplakia, erythroplakia, submucous fibrosis.

Few bioinformatics analyses have been performed: one investigated the relationship between OLP and oral cancer at a genomics level (Giacomelli et al., 2009), the other applied LGA and systems biology approach in order to identify *JUN, EGFR, FOS, IL2, and ITGB4* as "*hub genes*" (Orlando et al., 2013; Fig. 4.5).

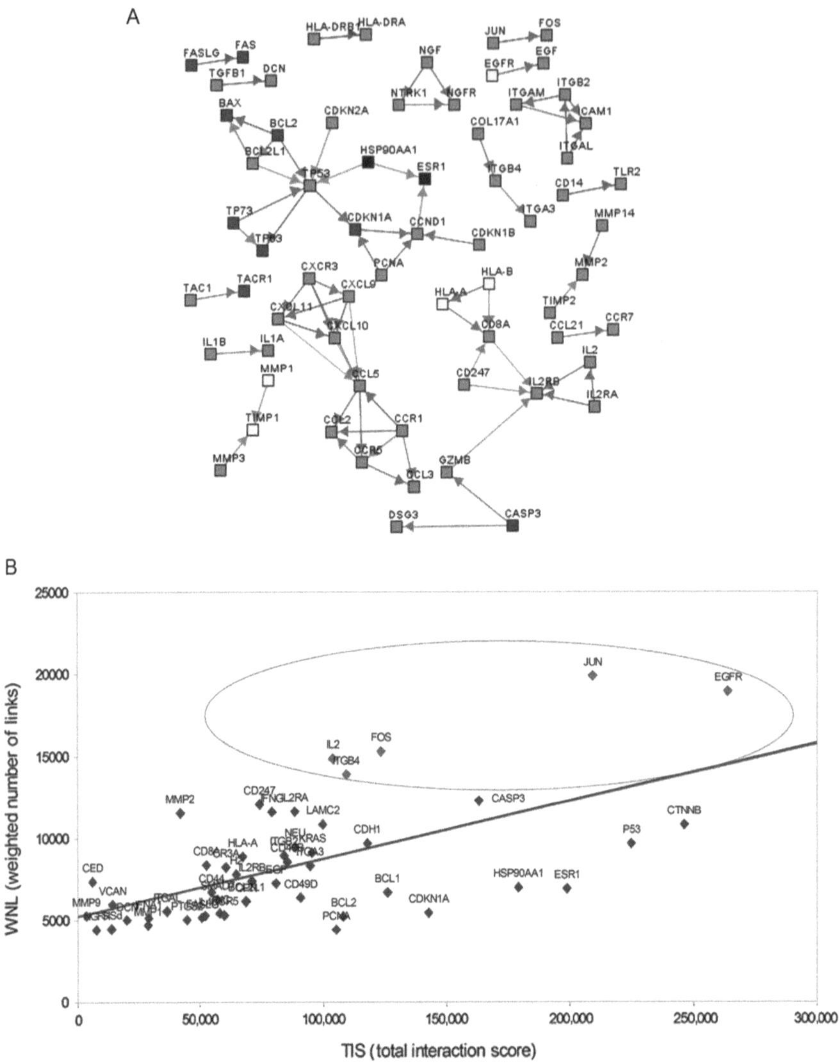

Figure 4.5 Up- and downregulated genes involved in pathogenesis of OLP. In black: genes for which there are no or little information about expression; in light grey, neutral genes in OLP disease; in grey, upregulated genes in OLP disease; in dark grey, downregulated genes in OLP disease (top). Plot of disease-related connectivities (WNL, weighted number of links) versus global connectivities (TIS, total interactions score). Calculated leader genes are above the regression tendency line (bottom).

Very few proteomics studies have been conducted on patients suffering from OLP, in a saliva-based proteomics investigation, two proteins (urinary prokallikrein, and short palate, lung and nasal epithelium carcinoma associated or PLUNC protein) were identified as potential biomarkers (Yang et al., 2006).

3.4. Oral cancer

Oral cancer is the eighth cancer in terms of prevalence, its incidence rate is in the range of 1–10 cases/100,000 people in the developed countries, whilst it is up to 12.6 cases/100,000 in South-Central Asia and in India, where is one of the three leading tumors (Petersen, 2009).

OMICS-based technologies are paving new avenues for the early diagnosis and treatment of oral cancer (Chen et al., 2013; Hu et al., 2008; Jessri & Farah, 2014; Krishna Prasad, Sharma, & Babu, 2013), even though not so many studies have been carried out (Hu & Wong, 2007; Lee et al., 2013; Marimuthu et al., 2013; Tung et al., 2013). Found protein markers belong to interleukin class, and are involved in inflammatory and immune response, showing the feasibility and clinical utility of the Salivary Transcriptome Diagnostics (STD; Li et al., 2004).

4. CONCLUDING REMARKS

Impressive progresses have been made in the last decades. New bioinformatics tools and resources have been designed, as well genomics, metagenomics and proteomics approaches that have a great added clinical value. Interestingly, integrated proteogenomics approaches have lead to models which have been proven superior to those including only data deriving from a single omics technology.

Our bioinformatics algorithm enables the prioritization and selection of few genes that can be subsequently expressed within the NAPPA array. Our conductomer appears promising in analyzing multigene and -protein interactions and seems to overcome most difficulties and hurdles of the classical techniques. Being versatile, it can be used in studying gene–gene, gene–protein, protein–protein, gene–drug, and protein–drug interactions.

However, some limitations remain, such as those due to the usually small size of the performed clinical trials and studies (Skates et al., 2013), that hinder the power of the investigations themselves and the generalizability of their findings. Efforts should be undertaken in this direction, in order to provide reliable results that can be translated into the clinical practice in order to provide the patients a tailored treatment.

REFERENCES

Ai, J. Y., Smith, B., & Wong, D. T. (2012). Bioinformatics advances in saliva diagnostics. *International Journal of Oral Science*, *4*(2), 85–87.

Al-Tarawneh, S. K., Border, M. B., Dibble, C. F., & Bencharit, S. (2011). Defining salivary biomarkers using mass spectrometry-based proteomics: A systematic review. *OMICS*, *15*(6), 353–361.

Amado, F. M., Ferreira, R. P., & Vitorino, R. (2013). One decade of salivary proteomics: Current approaches and outstanding challenges. *Clinical Biochemistry, 46*(6), 506–517.
Anderson, K. S., Ramachandran, N., Wong, J., Raphael, J. V., Hainsworth, E., Demirkan, G., et al. (2008). Application of protein microarrays for multiplexed detection of antibodies to tumor antigens in breast cancer. *Journal of Proteome Research*, 7(4), 1490–1499.
Anderson, K. S., Sibani, S., Wallstrom, G., Qiu, J., Mendoza, E. A., Raphael, J., et al. (2011). Protein microarray signature of autoantibody biomarkers for the early detection of breast cancer. *Journal of Proteome Research, 10*(1), 85–96.
Anderson, K. S., Wong, J., Vitonis, A., Crum, C. P., Sluss, P. M., Labaer, J., et al. (2010). p53 autoantibodies as potential detection and prognostic biomarkers in serous ovarian cancer. *Cancer Epidemiology, Biomarkers & Prevention, 19*(3), 859–868.
Andreasen, J. O. (1968). Oral lichen planus. 1. A clinical evaluation of 115 cases. *Oral Surgery, Oral Medicine, and Oral Pathology, 25*(1), 31–42.
Arrais, J. P., Rosa, N., Melo, J., Coelho, E. D., Amaral, D., Correia, M. J., et al. (2013). OralCard: A bioinformatic tool for the study of oral proteome. *Archives of Oral Biology, 58*(7), 762–772.
Bánóczy, J., & Rugg-Gunn, A. (2013). Epidemiology and prevention of dental caries. *Acta Medicine Academica, 42*(2), 105–107.
Bassim, C. W., Ambatipudi, K. S., Mays, J. W., Edwards, D. A., Swatkoski, S., Fassil, H., et al. (2012). Quantitative salivary proteomic differences in oral chronic graft-versus-host disease. *Journal of Clinical Immunology, 32*(6), 1390–1399.
Becker, K. G., Barnes, K. C., Bright, T. J., & Wang, S. A. (2004). The genetic association database. *Nature Genetics, 36*(5), 431–432.
Belmonte, L., Spera, R., & Nicolini, C. (2013). SpADS: An R script for mass spectrometry data preprocessing before data mining. *Journal of Computer Science and Systems Biology, 6*, 298–304.
Bencharit, S., Altarawneh, S. K., Baxter, S. S., Carlson, J., Ross, G. F., Border, M. B., et al. (2012). Elucidating role of salivary proteins in denture stomatitis using a proteomic approach. *Molecular BioSystems, 8*(12), 3216–3223.
Benson, D. A., Karsch-Mizrachi, I., Clark, K., Lipman, D. J., Ostell, J., & Sayers, E. W. (2012). GenBank. *Nucleic Acids Research, 40*, D48–D53 (Database issue).
Bragazzi, N. L. (2013a). Children, adolescents, and young adults participatory medicine: Involving them in the health care process as a strategy for facing the infertility issue. *The American Journal of Bioethics, 13*(3), 43–44.
Bragazzi, N. L. (2013b). From P0 to P6 medicine, a model of highly participatory, narrative, interactive, and "augmented" medicine: Some considerations on Salvatore Iaconesi's clinical story. *Patient Preference and Adherence*, 7, 353–359.
Bragazzi, N. L. (2013c). Rethinking psychiatry with OMICS science in the age of personalized P5 medicine: Ready for psychiatome? *Philosophy, Ethics, and Humanities in Medicine, 8*(1), 4.
Bragazzi, N., Giacomelli, L., Sivozhelezov, V., & Nicolini, C. (2011). LeaderGene: A fast data-mining tool for molecular genomics. *Journal of Proteomics & Bioinformatics*, 4(4), 083–085.
Bragazzi, N. L., & Nicolini, C. (2013). A leader genes approach-based tool for molecular genomics: From gene-ranking to gene-network systems biology and biotargets predictions. *Journal of Computer Science and Systems Biology, 6*, 165–176.
Braud, C., Baeten, D., Giral, M., Pallier, A., Ashton-Chess, J., Braudeau, C., et al. (2008). Immunosuppressive drug-free operational immune tolerancein human kidney transplant recipients: Part I. *Blood gene expression statistical analysis. Journal of Cellular Biochemistry, 103*(6), 1681–1692.

Cafiero, C., & Matarasso, S. (2013). Predictive, preventive, personalised and participatory periodontology: 'The 5Ps age' has already started. *The EPMA Journal*, *4*(1), 16. http://dx.doi.org/10.1186/1878-5085-4-16.

Carbone, M., Arduino, P. G., Carrozzo, M., Gandolfo, S., Argiolas, M. R., Bertolusso, G., et al. (2009). Course of oral lichen planus: A retrospective study of 808 northern Italian patients. *Oral Diseases*, *15*(3), 235–243.

Ceroni, A., Sibani, S., Baiker, A., Pothineni, V. R., Bailer, S. M., LaBaer, J., et al. (2010). Systematic analysis of the IgG antibody immune response against varicella zoster virus (VZV) using a self-assembled protein microarray. *Molecular BioSystems*, *6*(9), 1604–1610.

Chen, T., Abbey, K., Deng, W. J., & Cheng, M. C. (2005). The bioinformatics resource for oral pathogens. *Nucleic Acids Research*, *33*, W734–W740 (Web Server issue).

Chen, Y. T., Chong, Y. M., Cheng, C. W., Ho, C. L., Tsai, H. W., Kasten, F. H., et al. (2013). Identification of novel tumor markers for oral squamous cell carcinoma using glycoproteomic analysis. *Clinica Chimica Acta*, *420*, 45–53.

Chen, T., Yu, W. H., Izard, J., Baranova, O. V., Lakshmanan, A., & Dewhirst, F. E. (2010). The Human Oral Microbiome Database: A web accessible resource for investigating oral microbe taxonomic and genomic information. *Database (Oxford)*, 2010, baq013.

Cheng, W. C., Tsai, M. L., Chang, C. W., Huang, C. L., Chen, C. R., Shu, W. Y., et al. (2010). Microarray meta-analysis database (M(2)DB): A uniformly pre-processed, quality controlled, and manually curated human clinical microarray database. *BMC Bioinformatics*, *11*, 421.

Choi, A. H., Ben-Nissan, B., Matinlinna, J. P., & Conway, R. C. (2013). Current perspectives: Calcium phosphate nanocoatings and nanocomposite coatings in dentistry. *Journal of Dental Research*, *92*(10), 853–859.

Covani, U., Giacomelli, L., Krajewski, A., Ravaglioli, A., Spotorno, L., Loria, P., et al. (2007). Biomaterials for orthopedics: A roughness analysis by atomic force microscopy. *Journal of Biomedical Materials Research. Part A*, *82*(3), 723–730.

Covani, U., Marconcini, S., Giacomelli, L., Sivozhelevov, V., Barone, A., & Nicolini, C. (2008 Oct). Bioinformatic prediction of leader genes in human periodontitis. *Journal of Periodontology*, *79*(10), 1974–1983.

Covani, U., Marconcini, S., Derchi, G., Barone, A., & Giacomelli, L. (2009). Relationship between human periodontitis and type 2 diabetes at a genomic level: A data-mining study. *Journal of Periodontology*, *80*(8), 1265–1273.

Cuevas-Córdoba, B., & Santiago-García, J. (2014). Saliva: A fluid of study for OMICS. *Omics: A Journal of Integrative Biology*, *18*(2), 87–97.

Deepa, T., & Thirrunavukkarasu, N. (2010). Saliva as a potential diagnostic tool. *Indian Journal of Medical Sciences*, *64*(7), 293–306.

De La Iglesia, D., Chiesa, S., Kern, J., Maojo, V., Martin-Sanchez, F., Potamias, G., et al. (2009). Nanoinformatics: New challenges for biomedical informatics at the nano level. *Studies in Health Technology and Informatics*, *150*, 987–991.

Dewhirst, F. E., Chen, T., Izard, J., Paster, B. J., Tanner, A. C., Yu, W. H., et al. (2010). The human oral microbiome. *Journal of Bacteriology*, *192*(19), 5002–5017.

Dimitrov, D. V., & Hoeng, J. (2013). Systems approaches to computational modeling of the oral microbiome. *Frontiers in Physiology*, *4*, 172.

Doğan, R., Leaman, R., & Lu, Z. (2014). NCBI disease corpus: A resource for disease name recognition and concept normalization. *Journal of Biomedical Informatics*, *47*, 1–10. http://dx.doi.org/10.1016/j.jbi.2013.12.006, pii: S1532-0464(13)00197-4.

Duran-Pinedo, A. E., Paster, B., Teles, R., & Frias-Lopez, J. (2011). Correlation network analysis applied to complex biofilm communities. *PLoS One*, *6*(12), e28438.

Edgar, R., Domrachev, M., & Lash, A. E. (2002). Gene Expression Omnibus: NCBI gene expression and hybridization array data repository. *Nucleic Acids Research*, *30*(1), 207–210.

Franceschini, A., Szklarczyk, D., Frankild, S., Kuhn, M., Simonovic, M., Roth, A., et al. (2013). STRING v9.1: Protein-protein interaction networks, with increased coverage and integration. *Nucleic Acids Research*, *41*(Database issue), D808–D815.

Fiorini, T., Susin, C., da Rocha, J. M., Weidlich, P., Vianna, P., Moreira, C. H., et al. (2013). Effect of nonsurgical periodontal therapy on serum and gingival crevicular fluid cytokine levels during pregnancy and postpartum. *Journal of Periodontal Research*, *48*(1), 126–133.

Freitas, R. A., Jr. (2000). Nanodentistry. *The Journal of the American Dental Association*, *131*(11), 1559–1565.

Gadewal, N. S., & Zingde, S. M. (2011). Database and interaction network of genes involved in oral cancer: Version II. *Bioinformation*, *6*(4), 169–170.

Gaiser, S., Deyhle, H., Bunk, O., White, S. N., & Müller, B. (2012). Understanding nano-anatomy of healthy and carious human teeth: A prerequisite for nanodentistry. *Biointerphases*, 7(1–4), 4.

Gandolfo, S., Richiardi, L., Carrozzo, M., Broccoletti, R., Carbone, M., Pagano, M., et al. (2004). Risk of oral squamous cell carcinoma in 402 patients with oral lichen planus: A follow-up study in an Italian population. *Oral Oncology*, *40*(1), 77–83.

Garcia, I., Kuska, R., & Somerman, M. J. (2013). Expanding the foundation for personalized medicine: Implications and challenges for dentistry. *Journal of Dental Research*, *92*(7 Suppl), 3S–10S.

Ghannoum, M. A., Jurevic, R. J., Mukherjee, P. K., Cui, F., Sikaroodi, M., Naqvi, A., et al. (2010). Characterization of the oral fungal microbiome (mycobiome) in healthy individuals. *PLoS Pathogen*, *6*(1), e1000713.

Giacomelli, L., & Nicolini, C. (2006). Gene expression of human T lymphocytes cell cycle: Experimental and bioinformatic analysis. *Journal of Cellular Biochemistry*, *99*(5), 1326–1333.

Giacomelli, L., Oluwadara, O., Chiappe, G., Barone, A., Chiappelli, F., & Covani, U. (2009). Relationship between human oral lichen planus and oral squamous cell carcinoma at a genomic level: A datamining study. *Bioinformation*, *4*(6), 258–262.

Giacomelli, L., & Covani, U. (2010). Bioinformatics and data mining studies in oral genomics and proteomics: New trends and challenges. *The Open Dentistry Journal*, *4*, 67–71.

Giannobile, W. V. (2013). Our duty to promote global oral health. *Journal of Dental Research*, *92*(7), 573–574.

Gibson, D. S., Qiu, J., Mendoza, E. A., Barker, K., Rooney, M. E., & LaBaer, J. (2012). Circulating and synovial antibody profiling of juvenile arthritis patients by nucleic acid programmable protein arrays. *Arthritis Research & Therapy*, *14*(2), R77.

Glaab, E., Garibaldi, J. M., & Krasnogor, N. (2009). ArrayMining: A modular web-application for microarray analysis combining ensemble and consensus methods with cross-study normalization. *BMC Bioinformatics*, *10*, 358.

Glurich, I., Acharya, A., Shukla, S. K., Nycz, G. R., & Brilliant, M. H. (2013). The oral-systemic personalized medicine model at Marshfield Clinic. *Oral Diseases*, *19*(1), 1–17.

Gokul, K., Faizuddin, M., & Pradeep, A. R. (2012). Estimation of the level of tumor necrosis factor-α in gingival crevicular fluid and serum in periodontal health & disease: A biochemical study. *Indian Journal of Dental Research*, *23*(3), 348–352.

Griffen, A. L., Beall, C. J., Firestone, N. D., Gross, E. L., Difranco, J. M., Hardman, J. H., et al. (2011). CORE: A phylogenetically-curated 16S rDNA database of the core oral microbiome. *PLoS One*, *6*(4), e19051.

Guzman, Y. A., Sakellari, D., Arsenakis, M., & Floudas, C. A. (2014). Proteomics for the discovery of biomarkers and diagnosis of periodontitis: A critical review. *Expert Review of Proteomics*, *11*(1), 31–41.

Haririan, H., Bertl, K., Laky, M., Rausch, W. D., Böttcher, M., Matejka, M., et al. (2012). Salivary and serum chromogranin A and α-amylase in periodontal health and disease. *Journal of Periodontology*, *83*(10), 1314–1321.

Hart, T. C., Corby, P. M., Hauskrecht, M., Hee Ryu, O., Pelikan, R., Valko, M., et al. (2011). Identification of microbial and proteomic biomarkers in early childhood caries. *International Journal of Dentistry*, *2011*, 196721.

Hegde, M. N., Hegde, N. D., Ashok, A., & Shetty, S. (2014). Biochemical indicators of dental caries in saliva: An in vivo study. *Caries Research*, *48*(2), 170–173.

Heider, A., & Alt, R. (2013). VirtualArray: A R/bioconductor package to merge raw data from different microarray platforms. *BMC Bioinformatics*, *14*, 75.

Horst, J. A., Pieper, U., Sali, A., Zhan, L., Chopra, G., Samudrala, R., et al. (2012). Strategic protein target analysis for developing drugs to stop dental caries. *Advances in Dental Research*, *24*(2), 86–93.

Houlihan, G., Gatti-Lafranconi, P., Kaltenbach, M., Lowe, D., & Hollfelder, F. (2014). An experimental framework for improved selection of binding proteins using SNAP display. *Journal of Immunological Methods*, *405*, 47–56. http://dx.doi.org/10.1016/j.jim.2014.01.006, pii: S0022-1759(14)00016-7.

Hsiao, W. W., Li, K. L., Liu, Z., Jones, C., Fraser-Liggett, C. M., & Fouad, A. F. (2012). Microbial transformation from normal oral microbiota to acute endodontic infections. *BMC Genomics*, *13*, 345.

Hu, S., Arellano, M., Boontheung, P., Wang, J., Zhou, H., Jiang, J., et al. (2008). Salivary proteomics for oral cancer biomarker discovery. *Clinical Cancer Research*, *14*(19), 6246–6252.

Hu, S., & Wong, D. T. (2007). Oral cancer proteomics. *Current Opinion in Molecular Therapeutics*, *9*(5), 467–476.

Huang, C. M. (2004). Comparative proteomic analysis of human whole saliva. *Archives of Oral Biology*, *49*(12), 951–962.

Huttenhower, C., & Hofmann, O. (2010). A quick guide to large-scale genomic data mining. *PLoS Computational Biology*, *6*(5), e1000779.

Jágr, M., Eckhardt, A., Pataridis, S., & Mikšík, I. (2012). Comprehensive proteomic analysis of human dentin. *European Journal of Oral Sciences*, *120*(4), 259–268.

Jagtap, P., McGowan, T., Bandhakavi, S., Tu, Z. J., Seymour, S., Griffin, T. J., et al. (2012). Deep metaproteomic analysis of human salivary supernatant. *Proteomics*, *12*(7), 992–1001.

Jaiswal, G., Deo, V., Bhongade, M., & Jaiswal, S. (2011). Serum alkaline phosphatase: A potential marker in the progression of periodontal disease in cirrhosis patients. *Quintessence International*, *42*(4), 345–348.

Jessri, M., & Farah, C. S. (2014). Next generation sequencing and its application in deciphering head and neck cancer. *Oral Oncology*, *50*, 247–253. http://dx.doi.org/10.1016/j.oraloncology.2013.12.017, pii: S1368-8375(13)00805-1.

Jiang, W., Ling, Z., Lin, X., Chen, Y., Zhang, J., Yu, J., et al. (2014). Pyrosequencing analysis of oral microbiota shifting in various caries states in childhood. *Microbial Ecology*, *67*(4), 962–969.

Joseph, R., Rajappan, S., Nath, S. G., & Paul, B. J. (2013). Association between chronic periodontitis and rheumatoid arthritis: A hospital-based case-control study. *Rheumatology International*, *33*(1), 103–109.

Jovanovic, V., Giacomelli, L., Sivozhelezov, V., Degauque, N., Lair, D., Soulillou, J. P., et al. (2010). AKT1 leader gene and downstream targets are involved in a rat model of kidney allograft tolerance. *Journal of Cellular Biochemistry*, *111*(3), 709–719.

Kalema, N., Boon, S. D., Cattamanchi, A., Davis, J. L., Andama, A., Katagira, W., et al. (2012). Oral antimicrobial rinse to reduce mycobacterial culture contamination among tuberculosis suspects in Uganda: A prospective study. *PLoS One*, 7(7), e38888.

Kanaparthy, R., & Kanaparthy, A. (2011). The changing face of dentistry: nanotechnology. *International Journal of Nanomedicine*, *6*, 2799–2804. http://dx.doi.org/10.2147/IJN.S24353. Epub 2011 Nov 9.

Karlsson, M., Pålsgård, E., Wilshaw, P. R., & Di Silvio, L. (2003). Initial in vitro interaction of osteoblasts with nano-porous alumina. *Biomaterials*, *24*(18), 3039–3046.

Kellam, P., & Weiss, R. A. (2006). Infectogenomics: Insights from the host genome into infectious diseases. *Cell*, *124*(4), 695–697.

Kim, D. K., Kerman, K., Hiep, H. M., Saito, M., Yamamura, S., Takamura, Y., et al. (2008). Label-free optical detection of aptamer-protein interactions using gold-capped oxide nanostructures. *Analytical Biochemistry*, *379*(1), 1–7.

Klein, M. I., Xiao, J., Lu, B., Delahunty, C. M., Yates, J. R., 3rd., & Koo, H. (2012). Streptococcus mutans protein synthesis during mixed-species biofilm development by high-throughput quantitative proteomics. *PLoS One*, 7(9), e45795.

Kobayashi, T., Okada, M., Ito, S., Kobayashi, D., Ishida, K., Kojima, A., et al. (2014). Assessment of interleukin-6 receptor inhibition therapy on periodontal condition in patients with rheumatoid arthritis and chronic periodontitis. *Journal of Periodontology*, *85*(1), 57–67.

Kobeissy, F. H., Gulbakan, B., Alawieh, A., Karam, P., Zhang, Z., Guingab-Cagmat, J. D., et al. (2014). Post-genomics nanotechnology is gaining momentum: Nanoproteomics and applications in life sciences. *Omics: A Journal of Integrative Biology*, *18*(2), 111–131.

Kodama, Y., Kaminuma, E., Saruhashi, S., Ikeo, K., Sugawara, H., Tateno, Y., et al. (2010). Biological databases at DNA Data Bank ofJapan in the era of next-generation sequencing technologies. *Advances in Experimental Medicine and Biology*, *680*, 125–135.

Kornman, K. S., & Duff, G. W. (2012). Personalized medicine: Will dentistry ride the wave or watch from the beach? *Journal of Dental Research*, *91*(7 Suppl), 8S–11S.

Krishna Prasad, R. B., Sharma, A., & Babu, H. M. (2013). An insight into salivary markers in oral cancer. *Dental Research Journal*, *10*(3), 287–295.

Kuboniwa, M., Tribble, G. D., Hendrickson, E. L., Amano, A., Lamont, R. J., & Hackett, M. (2012). Insights into the virulence of oral biofilms: Discoveries from proteomics. *Expert Review of Proteomics*, *9*(3), 311–323.

Laine, M. L., Moustakis, V., Koumakis, L., Potamias, G., & Loos, B. G. (2013). Modeling susceptibility to periodontitis. *Journal of Dental Research*, *92*(1), 45–50.

Lamster, I. B., & Ahlo, J. K. (2007). Analysis of gingival crevicular fluid as applied to the diagnosis of oral and systemic diseases. *Annals of the New York Academy of Sciences*, *1098*, 216–229.

Le, H. S., Oltvai, Z. N., & Bar-Joseph, Z. (2010). Cross-species queries of large gene expression databases. *Bioinformatics*, *26*(19), 2416–2423.

Lee, S. Y., Park, H. R., Cho, N. H., Choi, Y. P., Rha, S. Y., Park, S. W., et al. (2013). Identifying genes related to radiation resistance in oral squamous cell carcinoma cell lines. *International Journal of Oral and Maxillofacial Surgery*, *42*(2), 169–176.

Lemos, J. A., Abranches, J., & Burne, R. A. (2005). Responses of cariogenic streptococci to environmental stresses. *Current Issues in Molecular Biology*, 7(1), 95–107.

Lemos, J. A., & Burne, R. A. (2008). A model of efficiency: Stress tolerance by Streptococcus mutans. *Microbiology*, *154*(Pt 11), 3247–3255.

Levine, A. E., & Steffen, D. L. (2001). OrCGDB: A database of genes involved in oral cancer. *Nucleic Acids Research*, *29*(1), 300–302.

Li, Y., St John, M. A., Zhou, X., Kim, Y., Sinha, U., Jordan, R. C., et al. (2004). Salivary transcriptome diagnostics for oral cancer detection. *Clinical Cancer Research*, *10*(24), 8442–8450.

Li, J., Wang, W., Wang, Y., & Zeng, A. P. (2013). Two-dimensional gel-based proteomic of the caries causative bacterium Streptococcus mutans UA159 and insight into the inhibitory effect of carolacton. *Proteomics*, *13*(23–24), 3470–3477.

Liu, B., Faller, L. L., Klitgord, N., Mazumdar, V., Ghodsi, M., Sommer, D. D., et al. (2012). Deep sequencing of the oral microbiome reveals signatures of periodontal disease. *PLoS One*, 7(6), e37919.

Mallineni, S. K., Nuvvula, S., Matinlinna, J. P., Yiu, C. K., & King, N. M. (2013). Biocompatibility of various dental materials in contemporary dentistry: A narrative insight. *Journal of Investigative and Clinical Dentistry*, *4*(1), 9–19.

Mantri, S. S., & Mantri, S. P. (2013). The nano era in dentistry. *Journal of Natural Science, Biology, and Medicine*, *4*(1), 39–44.

Manzano-Román, R., Díaz-Martín, V., González-González, M., Matarraz, S., Álvarez-Prado, A. F., LaBaer, J., et al. (2012). Self-assembled protein arrays from an Ornithodoros moubata salivary gland expression library. *Journal of Proteome Research*, *11*(12), 5972–5982.

Marconcini, S., Covani, U., Barone, A., Vittorio, O., Curcio, M., Barbuti, S., et al. (2011). Real-time quantitative polymerase chain reaction analysis of patients with refractory chronic periodontitis. *Journal of Periodontology*, *82*(7), 1018–1024.

Marconcini, S., Genovesi, A. M., Marchisio, O., Gelpi, F., Barone, A., Corega, C., et al. (2014). In vivo study of titanium healing screws surface modifications after different debridment procedure. *Minerva Stomatologica*, (Epub ahead of print).

Marimuthu, A., Chavan, S., Sathe, G., Sahasrabuddhe, N. A., Srikanth, S. M., Renuse, S., et al. (2013). Identification of head and neck squamous cell carcinoma biomarker candidates through proteomic analysis of cancer cell secretome. *Biochimica et Biophysica Acta*, *1834*(11), 2308–2316.

Masuda, H., & Fukuda, K. (1995). Ordered metal nanohole arrays made by a two-step replication of honeycomb structures of anodic alumina. *Science*, *268*, 1466–1468.

Matse, J. H., Yoshizawa, J., Wang, X., Elashoff, D., Bolscher, J. G., Veerman, E. C., et al. (2013). Discovery and prevalidation of salivary extracellular microRNA biomarkers panel for the noninvasive detection of benign and malignant parotid gland tumors. *Clinical Cancer Research*, *19*(11), 3032–3038.

Mazumdar, V., Snitkin, E. S., Amar, S., & Segrè, D. (2009). Metabolic network model of a human oral pathogen. *Journal of Bacteriology*, *191*(1), 74–90.

McCartan, B. E., & Healy, C. M. (2008). The reported prevalence of oral lichen planus: A review and critique. *Journal of Oral Pathology & Medicine*, *37*(8), 447–453.

Meurman, J. H., Janket, S. J., Qvarnström, M., & Nuutinen, P. (2003). Dental infections and serum inflammatory markers in patients with and without severe heart disease. *Oral Surgery, Oral Medicine, Oral Pathology, Oral Radiology, and Endodontics*, *96*(6), 695–700.

Miersch, S., Bian, X., Wallstrom, G., Sibani, S., Logvinenko, T., Wasserfall, C. H., et al. (2013). Serological autoantibody profiling of type 1 diabetes by protein arrays. *Journal of Proteomics*, *94*, 486–496.

Mitra, S., Das, S., Ghosal, S., & Chakrabarti, J. (2012). HNOCDB: A comprehensive database of genes and miRNAs relevant to head and neck and oral cancer. *Oral Oncology*, *48*(2), 117–119.

Montor, W. R., Huang, J., Hu, Y., Hainsworth, E., Lynch, S., Kronish, J. W., et al. (2009). Genome-wide study of Pseudomonas aeruginosa outer membrane protein immunogenicity using self-assembling protein microarrays. *Infection and Immunity*, 77(11), 4877–4886.

Moreau, Y., & Tranchevent, L. C. (2012). Computational tools for prioritizing candidate genes: Boosting disease gene discovery. *Nature Reviews. Genetics*, *13*(8), 523–536.

Nakano, Y., Shibata, Y., Kawada, M., Kojima, M., Fukamachi, H., Shibata, Y., et al. (2005). A searchable database for proteomes of oral microorganisms. *Oral Microbiology and Immunology*, *20*(6), 344–348.

Nan, J., Brostromer, E., Liu, X. Y., Kristensen, O., & Su, X. D. (2009). Bioinformatics and structural characterization of a hypothetical protein from Streptococcus mutans: Implication of antibiotic resistance. *PLoS One*, *4*(10), e7245.

Nand, A., Gautam, A., Pérez, J. B., Merino, A., & Zhu, J. (2012). Emerging technology of in situ cell free expression protein microarrays. *Protein & Cell, 3*(2), 84–88.
NCBI Resource Coordinators (2014). Database resources of the National Center for Biotechnology Information. *Nucleic Acids Research, 42*(1), D7–D17.
Nibali, L., Donos, N., & Henderson, B. (2009). Periodontal infectogenomics. *Journal of Medical Microbiology, 58*(Pt 10), 1269–1274.
Nicolini, C. (2006). Nanogenomics for medicine. *Nanomedicine (London, England), 1*(2), 147–152.
Nicolini, C., Spera, R., Stura, E., Fiordoro, S., & Giacomelli, L. (2006). Gene expression in the cell cycle of human T-lymphocytes: II Experimental determination by DNASER technology. *Journal of Cellular Biochemistry, 97*(5), 1151–1159.
Nicolini, C. (2010). Nanogenomics in medicine. *Wiley Interdisciplinary Reviews. Nanomedicine and Nanobiotechnology, 2*(1), 59–76.
Nicolini, C., Adami, M., Sartore, M., Bragazzi, N. L., Bavastrello, V., Spera, R., et al. (2012). Prototypes of newly conceived inorganic and biological sensors for health and environmental applications. *Sensors (Basel, Switzerland), 12*(12), 17112–17127.
Nicolini, C., Bragazzi, N., & Pechkova, E. (2012). Nanoproteomics enabling personalized nanomedicine. *Advanced Drug Delivery Reviews, 64*(13), 1522–1531.
Nicolini, C., Bragazzi, N., & Pechkova, E. (2013). From nanobiotechnology to organic and biological monitoring of health and environment for biosafety. *Journal of Bioanalysis & Biomedicine, 5*, 108–117.
Nicolini, C., Correia, T. B., Stura, E., Larosa, C., Spera, R., & Pechkova, E. (2013). Atomic force microscopy and anodic porous allumina of nucleic acid programmable protein arrays. *Recent Patents on Biotechnology*, 7(2), 112–121.
Nicolini, C., & Pechkova, E. (2010a). An overview of nanotechnology-based functional proteomics for cancer and cell cycle progression. *Anticancer Research, 30*(6), 2073–2080.
Nicolini, C., & Pechkova, E. (2010b). Nanoproteomics for nanomedicine. *Nanomedicine (London, England), 5*(5), 677–682.
Nicolini, C., Singh, M., Spera, R., & Felli, L. (2013). Analysis of gene expression on anodic porous alumina microarrays. *Bioengineered, 4*(5), 332–337.
Nicolini, C., Spera, R., Festa, F., Belmonte, L., Chong, S., Pechkova, E., et al. (2013). Mass spectrometry and florescence analysis of SNAP-NAPPA arrays expressed using E. coli cell_free expression system. *Journal of Nanomedicine & Nanotechnology, 4*(5), 181–195.
Ogawa, Y., Miura, Y., Harazono, A., Kanai-Azuma, M., Akimoto, Y., Kawakami, H., et al. (2011). Proteomic analysis of two types of exosomes in human whole saliva. *Biological & Pharmaceutical Bulletin, 34*(1), 13–23.
Okada, M., Kobayashi, T., Ito, S., Yokoyama, T., Abe, A., Murasawa, A., et al. (2013). Periodontal treatment decreases levels of antibodies to Porphyromonas gingivalis and citrulline in patients with rheumatoid arthritis and periodontitis. *Journal of Periodontology, 84*(12), e74–e84.
Orlando, B., Bragazzi, N., & Nicolini, C. (2013). Bioinformatics and systems biology analysis of genes network involved in OLP (Oral Lichen Planus) pathogenesis. *Archives of Oral Biology, 58*(6), 664–673.
Ozak, S. T., & Ozkan, P. (2013). Nanotechnology and dentistry. *European Journal of Dentistry*, 7(1), 145–151.
Patel, S. P., & Raju, P. A. (2013). Resistin in serum and gingival crevicular fluid as a marker of periodontal inflammation and its correlation with single-nucleotide polymorphism in human resistin gene at -420. *Contemporary Clinical Dentistry, 4*(2), 192–197.
Pechkova, E., Chong, S., Tripathi, S., & Nicolini, C. (2010). Cell free expression and APA for NAPPA and protein crystallography: Functional proteomics and nanotechnology-based microarrays. In C. Nicolini & J. LaBaer (Eds.), *Pan stanford series on nanobiotechnology 2* (pp. 121–147). London - New York - Singapore: Thomson ISI Web of Science.

Petersen, P. E. (2009). Oral cancer prevention and control—The approach of the World Health Organization. *Oral Oncology*, *45*(4–5), 454–460.

Peterson, S. N., Snesrud, E., Schork, N. J., & Bretz, W. A. (2011). Dental caries pathogenicity: A genomic and metagenomic perspective. *International Dental Journal*, *61*(Suppl 1), 11–22.

Pham, T. K., Roy, S., Noirel, J., Douglas, I., Wright, P. C., & Stafford, G. P. (2010). A quantitative proteomic analysis of biofilm adaptation by the periodontal pathogen Tannerella forsythia. *Proteomics*, *10*(17), 3130–3141.

Pradeep, A. R., Kumari, M., Kalra, N., & Priyanka, N. (2013). Correlation of MCP-4 and high-sensitivity C-reactive protein as a marker of inflammation in obesity and chronic periodontitis. *Cytokine*, *61*(3), 772–777.

Pradeep, A. R., Raghavendra, N. M., Prasad, M. V., Kathariya, R., Patel, S. P., & Sharma, A. (2011). Gingival crevicular fluid and serum visfatin concentration: Their relationship in periodontal health and disease. *Journal of Periodontology*, *82*(9), 1314–1319.

Preza, D., Thiede, B., Olsen, I., & Grinde, B. (2009). The proteome of the human parotid gland secretion in elderly with and without root caries. *Acta Odontologica Scandinavica*, *67*(3), 161–169.

Racapé, M., Bragazzi, N., Sivozhelezov, V., Danger, R., Pechkova, E., Duong Van Huyen, J. P., et al. (2012). SMILE silencing and PMA activation gene networks in HeLa cells: Comparison with kidney transplantation gene networks. *Journal of Cellular Biochemistry*, *113*(6), 1820–1832.

Raghavendra, N. M., Pradeep, A. R., Kathariya, R., Sharma, A., Rao, N. S., & Naik, S. B. (2012). Effect of non surgical periodontal therapy on gingival crevicular fluid and serum visfatin concentration in periodontal health and disease. *Disease Markers*, *32*(6), 383–388.

Ramachandran, N., Hainsworth, E., Bhullar, B., Eisenstein, S., Rosen, B., Lau, A. Y., et al. (2004). Self-assembling protein microarrays. *Science*, *305*(5680), 86–90.

Rappaport, N., Nativ, N., Stelzer, G., Twik, M., Guan-Golan, Y., Stein, T. I., et al. (2013). MalaCards: An integrated compendium for diseases and their annotation. *Database*, *2013*, bat018.

Rappuoli, R. (2000). Reverse vaccinology. *Current Opinion in Microbiology*, *3*(5), 445–450.

Razzouk, S., & Termechi, O. (2013). Host genome, epigenome, and oral microbiome interactions: Toward personalized periodontal therapy. *Journal of Periodontology*, *84*(9), 1266–1271.

Reddy, P. J., Jain, R., Paik, Y. K., Downey, R., Ptolemy, A. S., Ozdemir, V., et al. (2011). Personalized medicine in the age of pharmacoproteomics: A close up on India and need for social science engagement for responsible innovation in post-proteomic biology. *Current Pharmacogenomics and Personalized Medicine*, *9*(1), 67–75.

Reshmi, G., Charles, S., James, P., Jijith, V. S., Prathibha, R., Ramachandran, S., et al. (2012). OrCa-dB: A complete catalogue of molecular and clinical information in oral carcinogenesis. *Oral Oncology*, *48*(6), e19.

Richards, D. (2013). Oral diseases affect some 3.9 billion people. *Evidence-Based Dentistry*, *14*(2), 35.

Riley, D. J., Bavastrello, V., Covani, U., Barone, A., & Nicolini, C. (2005). An in-vitro study of the sterilization of titanium dental implants using low intensity UV-radiation. *Dental Materials*, *21*(8), 756–760.

Rolfs, A., Montor, W. R., Yoon, S. S., Hu, Y., Bhullar, B., Kelley, F., et al. (2008). Production and sequence validation of a complete full length ORF collection for the pathogenic bacterium Vibrio cholerae. *Proceedings of the National Academy of Sciences of the United States of America*, *105*(11), 4364–4369.

Rosa, N., Correia, M. J., Arrais, J. P., Lopes, P., Melo, J., Oliveira, J. L., et al. (2012). From the salivary proteome to the OralOme: Comprehensive molecular oral biology. *Archives of Oral Biology*, *57*(7), 853–864.

Ross, B. C., Czajkowski, L., Hocking, D., Margetts, M., Webb, E., Rothel, L., et al. (2001). Identification of vaccine candidate antigens from a genomic analysis of Porphyromonas gingivalis. *Vaccine, 19*(30), 4135–4142.

Ruhl, S. (2012). The scientific exploration of saliva in the post-proteomic era: From database back to basic function. *Expert Review of Proteomics, 9*(1), 85–96.

Rustici, G., Kolesnikov, N., Brandizi, M., Burdett, T., Dylag, M., Emam, I., et al. (2013). ArrayExpress update—Trends in database growth and links to data analysis tools. *Nucleic Acids Research, 41*(Database issue), D987–D990.

Salerno, M., Caneva-Soumetz, F., Pastorino, L., Patra, N., Diaspro, A., & Ruggiero, C. (2013). Adhesion and proliferation of osteoblast-like cells on anodic porous alumina substrates with different morphology. *IEEE Transactions on Nanobioscience, 12*(2), 106–111.

Salerno, M., Giacomelli, L., & Larosa, C. (2011). Biomaterials for the programming of cell growth in oral tissues: The possible role of APA. *Bioinformation, 5*(7), 291–293.

Sbordone, L., Sbordone, C., Filice, N., Menchini-Fabris, G., Baldoni, M., & Toti, P. (2009). Gene clustering analysis in human osseous remodeling. *Journal of Periodontology, 80*(12), 1998–2009.

Schleyer, T. K. (2003). Dental informatics: An emerging biomedical informatics discipline. *Journal of Dental Education, 67*(11), 1193–1200.

Schleyer, T., Mattsson, U., Ní Ríordáin, R., Brailo, V., Glick, M., Zain, R. B., et al. (2011). Advancing oral medicine through informatics and information technology: A proposed framework and strategy. *Oral Diseases, 17*(Suppl 1), 85–94.

Schwarzberg, K., Le, R., Bharti, B., Lindsay, S., Casaburi, G., Salvatore, F., et al. (2014). The personal human oral microbiome obscures the effects of treatment on periodontal disease. *PLoS One, 9*(1), e86708.

Sharma, A., Pradeep, A. R., Raghavendra, N. M., Arjun, P., & Kathariya, R. (2012). Gingival crevicular fluid and serum cystatin c levels in periodontal health and disease. *Disease Markers, 32*(2), 101–107.

Shenar, N., Martinez, J., & Enjalbal, C. (2008). Laser desorption/ionization mass spectrometry on porous silica and alumina for peptide mass fingerprinting. *Journal of the American Society for Mass Spectrometry, 19*(5), 632–644.

Sibani, S., & LaBaer, J. (2011). Immunoprofiling using NAPPA protein microarrays. *Methods in Molecular Biology, 723*, 149–161.

Silverman, S., Jr., Gorsky, M., & Lozada-Nur, F. (1985). A prospective follow-up study of 570 patients with oral lichen planus: Persistence, remission, and malignant association. *Oral Surgery, Oral Medicine, and Oral Pathology, 60*(1), 30–34.

Sivozhelezov, V., Braud, C., Giacomelli, L., Pechkova, E., Giral, M., Soulillou, J. P., et al. (2008). Immunosuppressive drug-free operational immune tolerance in human kidney transplants recipients. Part II. Non-statistical gene microarray analysis. *Journal of Cellular Biochemistry, 103*(6), 1693–1706.

Sivozhelezov, V., Giacomelli, L., Tripathi, S., & Nicolini, C. (2006). Gene expression in the cell cycle of human T lymphocytes: I. Predicted gene and protein networks. *Journal of Cellular Biochemistry, 97*(5), 1137–1150.

Sivozhelezov, V., Spera, R., Giacomelli, L., Hainsworth, E., Labaer, J., Bragazzi, N. L., et al. (2009). Bioinformatics and fluorescence DNASER for NAPPA studies on cell transformation and cell cycle. In *Functional Proteomics and Nanotechnology-Based Microarrays: Vol. 2* (pp. 31–59).

Skates, S. J., Gillette, M. A., LaBaer, J., Carr, S. A., Anderson, L., Liebler, D. C., et al. (2013). Statistical design for biospecimen cohort size in proteomics-based biomarker discovery and verification studies. *Journal of Proteome Research, 12*(12), 5383–5394.

Song, Y., Ju, Y., Morita, Y., & Song, G. (2013). Effect of the nanostructure of porous alumina on growth behavior of MG63 osteoblast-like cells. *Journal of Bioscience and Bioengineering, 116*(4), 509–515.

Spera, R., Badino, F., Hainsworth, E., Fuentes, M., Srivastava, S., LaBaer, J., et al. (2010). Label free detection of NAPPA via mass spectrometry. In J. LaBaer & C. Nicolini (Eds.), *Functional Proteomics and Nanotechnology-Based Microarrays* (pp. 61–78). Pan Stanford Publishing (Chapter).

Spera, R., Correia, T. T. B., & Nicolini, C. (2013a). NAPPA based nanogravimetric biosensor: Preliminary characterization. *Sensors and Actuators, B: Chemical, 182*, 682–688.

Spera, R., Festa, F., Bragazzi, N. L., Pechkova, E., LaBaer, J., & Nicolini, C. (2013b). Conductometric monitoring of protein-protein interactions. *Journal of Proteome Research, 12*(12), 5535–5547.

Spera, R., Labaer, J., & Nicolini, C. (2011). MALDI-TOF characterization of NAPPA-generated proteins. *Journal of Mass Spectrometry, 46*(9), 960–965.

Stelzer, G., Dalah, I., Stein, T. I., Satanower, Y., Rosen, N., Nativ, N., et al. (2011). In-silico human genomics with GeneCards. *Human Genomics, 5*(6), 709–717.

Strimbu, K., & Tavel, J. A. (2010). What are biomarkers? *Current Opinion in HIV and AIDS, 5*(6), 463–466.

Stura, E., Bruzzese, D., Valerio, F., Grasso, V., Perlo, P., & Nicolini, C. (2007). Anodic porous alumina as mechanical stability enhancer for LDL-cholesterol sensitive electrodes. *Biosensors & Bioelectronics, 23*(5), 655–660.

Sundar, N. M., Krishnan, V., Krishnaraj, S., Hemalatha, V. T., & Alam, M. N. (2013). Comparison of the salivary and the serum nitric oxide levels in chronic and aggressive periodontitis: A biochemical study. *Journal of Clinical and Diagnostic Research, 7*(6), 1223–1227.

Suzuki, A., Takai-Igarashi, T., Numabe, Y., & Tanaka, H. (2009). Development of a database and ontology for pathogenic pathways in periodontitis. In *Silico Biology, 9*(4), 233–243.

Thanawastien, A., Montor, W. R., Labaer, J., Mekalanos, J. J., & Yoon, S. S. (2009). Vibrio cholerae proteome-wide screen for immunostimulatory proteins identifies phosphatidylserine decarboxylase as a novel Toll-like receptor 4 agonist. *PLoS Pathogen, 5*(8), e1000556.

Thorat, M., Pradeep, A. R., & Garg, G. (2010). Correlation of levels of oncostatin M cytokine in crevicular fluid and serum in periodontal disease. *International Journal of Oral Science, 2*(4), 198–207.

Tung, C. L., Lin, S. T., Chou, H. C., Chen, Y. W., Lin, H. C., Tung, C. L., et al. (2013). Proteomics-based identification of plasma biomarkers in oral squamous cell carcinoma. *Journal of Pharmaceutical and Biomedical Analysis, 75*, 7–17.

Wade, W. G. (2013). The oral microbiome in health and disease. *Pharmacological Research, 69*(1), 137–143.

Wang, Y., Xia, X., & Guo, Y. (2005). Porous anodic alumina membrane as a sample support for MALDI-TOF MS analysis of salt-containing proteins. *Journal of the American Society for Mass Spectrometry, 16*(9), 1488–1492.

Wang, Q., Jia, P., Cuenco, K. T., Feingold, E., Marazita, M. L., Wang, L., et al. (2013). Multi-dimensional prioritization of dental caries candidate genes and its enriched dense network modules. *PLoS One, 8*(10), e76666.

Wiebe, C. B., & Putnins, E. E. (2000). The periodontal disease classification system of the American Academy of Periodontology—An update. *Journal of the Canadian Dental Association, 66*(11), 594–597.

Wong, D. (Ed.), (2009). *Salivary Diagnostics* (320 pages). Wiley-Blackwell. ISBN: 978-0-8138-0664-8.

Wright, C., Sibani, S., Trudgian, D., Fischer, R., Kessler, B., LaBaer, J., et al. (2012). Detection of multiple autoantibodies in patients with ankylosing spondylitis using nucleic acid programmable protein arrays. *Molecular & Cellular Proteomics, 11*(2), M9.00384.

Xie, G., Chain, P. S., Lo, C. C., Liu, K. L., Gans, J., Merritt, J., et al. (2010). Community and gene composition of a human dental plaque microbiota obtained by metagenomic sequencing. *Molecular Oral Microbiology, 25*(6), 391–405.

Xue, V., Burdett, T., Lukk, M., Taylor, J., Brazma, A., & Parkinson, H. (2012). Mage-Comet—Web application for harmonizing existing large-scale experiment descriptions. *Bioinformatics, 28*(10), 1402–1403.

Yang, L. L., Liu, X. Q., Liu, W., Cheng, B., & Li, M. T. (2006). Comparative analysis of whole saliva proteomes for the screening of biomarkers for oral lichen planus. *Inflammation Research, 55*(10), 405–407.

Yu, W. H., Hu, H., Zhou, Q., Xia, Y., & Amar, S. (2010). Bioinformatics analysis of macrophages exposed to Porphyromonas gingivalis: Implications in acute vs. chronic infections. *PLoS One, 5*(12), e15613.

Zainal-Abidin, Z., Veith, P. D., Dashper, S. G., Zhu, Y., Catmull, D. V., Chen, Y. Y., et al. (2012). Differential proteomic analysis of a polymicrobial biofilm. *Journal of Proteome Research, 11*(9), 4449–4464.

Zandparsa, R. (2014). Latest biomaterials and technology in dentistry. *Dental Clinics of North America, 58*(1), 113–134.

Zehetbauer, S., Wojahn, T., Hiller, K. A., Schmalz, G., & Ruhl, S. (2009). Resemblance of salivary protein profiles between children with early childhood caries and caries-free controls. *European Journal of Oral Sciences, 117*(4), 369–373.

Zhan, Y., Zhang, R., Lv, H., Song, X., Xu, X., Chai, L., et al. (2014). Prioritization of candidate genes for periodontitis using multiple computational tools. *Journal of Periodontology,* (Epub ahead of print).

Zhang, A., Sun, H., Wang, P., & Wang, X. (2013). Salivary proteomics in biomedical research. *Clinica Chimica Acta, 415*, 261–265.

Zheng, J., Stoyanovich, J., Manduchi, E., Liu, J., & Stoeckert, C. J., Jr. (2011). AnnotCompute: Annotation-based exploration and meta-analysis of genomics experiments. *Database, 2011,* bar045.

Zijnge, V., Kieselbach, T., & Oscarsson, J. (2012). Proteomics of protein secretion by Aggregatibacter actinomycetemcomitans. *PLoS One*, 7(7), e41662.

CHAPTER FIVE

Advances in Nanocrystallography as a Proteomic Tool

Eugenia Pechkova[*,†], **Nicola Luigi Bragazzi**[*,†,‡], **Claudio Nicolini**[*,†,§,1]
[*]Nanobiotechnology and Biophysics Laboratories (NBL), Department of Experimental Medicine (DIMES), University of Genoa, Genoa, Italy
[†]Nanoworld Institute Fondazione ELBA Nicolini (FEN), Pradalunga, Bergamo, Italy
[‡]School of Public Health, Department of Health Sciences (DISSAL), University of Genoa, Genoa, Italy
[§]Biodesign Institute, Arizona State University, Tempe, Arizona, USA
[1]Corresponding author: e-mail address: info@fondazioneelba-nicolini.org

Contents

Abstract

In order to overcome the difficulties and hurdles too much often encountered in crystallizing a protein with the conventional techniques, our group has introduced the innovative Langmuir–Blodgett (LB)-based crystallization, as a major advance in the field of both structural and functional proteomics, thus pioneering the emerging field of the so-called *nanocrystallography* or *nanobiocrystallography*. This approach uniquely combines protein crystallography and nanotechnologies within an integrated, coherent

Advances in Protein Chemistry and Structural Biology, Volume 95
ISSN 1876-1623
http://dx.doi.org/10.1016/B978-0-12-800453-1.00005-1

framework that allows one to obtain highly stable protein crystals and to fully characterize them at a nano- and subnanoscale.

A variety of experimental techniques and theoretical/semi-theoretical approaches, ranging from atomic force microscopy, circular dichroism, Raman spectroscopy and other spectroscopic methods, microbeam grazing-incidence small-angle X-ray scattering to *in silico* simulations, bioinformatics, and molecular dynamics, has been exploited in order to study the LB-films and to investigate the kinetics and the main features of LB-grown crystals.

When compared to classical hanging-drop crystallization, LB technique appears strikingly superior and yields results comparable with crystallization in microgravity environments.

Therefore, the achievement of LB-based crystallography can have a tremendous impact in the field of industrial and clinical/therapeutic applications, opening new perspectives for personalized medicine. These implications are envisaged and discussed in the present contribution.

1. INTRODUCTION

Since the birth of crystallography in 1840 (Giegé, 2013), over the past decades, a remarked progress in the field of protein structure determination has been achieved thanks to advancements in X-ray crystallography combined with the brighter and highly focused, third-generation synchrotron (Belmonte, Pechkova, Tripathi, Scudieri, & Nicolini, 2012; Riekel, 2004; Riekel, Burghammer, & Schertler, 2005). This unique method based on a combined and integrated approach is likely to remain the most important method for the determination of protein structure in the foreseeable future (Belmonte et al., 2012). In comparison with other methods, such as (cryo)electron microscopy, nuclear magnetic resonance, solution scattering, and neutron diffraction, X-ray crystallography is indeed the most utilized approach for obtaining a full atom detailed structure of the investigated protein.

However, in the literature, there is a remarked lack of both structural and functional studies devoted to the investigation and characterization of therapeutically and clinically relevant proteins, considering that at least 60% of the commercially available drugs target membrane proteins (Drews, 2000), which are instead scarcely represented on the Protein Data Bank (PDB) (Kang, Lee, & Drew, 2013; Ubarretxena-Belandia & Stokes, 2010). This is because of the initial reluctance of the protein to being crystallized and the difficulties often encountered while trying to develop standardized protocols, which would shift crystallography from an art to a science (Chayen, 2004; Chayen & Saridakis, 2008; Helliwell & Chayen, 2007). These hurdles could be in part solved by exploiting advanced techniques such as cocrystallization, structural genomics (Chayen & Saridakis, 2002), molecular modeling (Sivozhelezov, Pechkova, & Nicolini, 2006), combinatorial chemistry (Delucas et al., 2005), and the nanodroplet high-throughput crystallization approach which uses robotic platforms and interfaces, and thanks to advancements in microfluidics (Li & Ismagilov, 2010; Zheng, Gerdts, & Ismagilov, 2005), and is able to perform up to thousands of experiments per day, minimizing the necessity of doing repeated trials and extensive screening and tuning of the different parameters, and thus increasing the percentage of success (Chayen, 2007, 2009; Stevens, 2000). Biotechnology and nanobiotechnology play an important role (Nicolini & Pechkova, 2004, 2006; Pechkova, Roth, et al., 2005; Pechkova, Vasile, Spera, Fiordoro, & Nicolini, 2005), facilitating crystallization by means of nanobiomaterials such as the nanotemplates or the molecularly imprinted polymers or MIPs (Saridakis & Chayen, 2013; Saridakis et al., 2011) and other smart biomaterials, leading to universal solutions (Nicolini & Pechkova, 2004; Saridakis & Chayen, 2009) or tailored ones (Saridakis & Chayen, 2013) in the field of protein crystallography.

Another major drawback is the X-ray radiation-induced damage, which limits the quality of the collected diffraction data, causing an increase in mosaicity, Debye–Waller factor (known also as temperature factor or *B*-factor), reliability factor (termed as residual factor, *R*-factor, or *R*-value), crystal unit cell volume, and a decrease in resolution and diffracting behavior (Belmonte et al., 2012; Holton, 2009). Several solutions have been studied in order to decrease radiation-induced damage such as the usage of cryoprotectants and scavengers (Garman & Owen, 2006), a better calibration and focusing of incident X-ray intensity, computation of radiation intensity, and integration of X-ray radiation with techniques like Raman (Berweger et al., 2009), optical (Berweger et al., 2009), X-ray spectroscopy (Garman & Nave, 2009). Third-generation synchrotron monochromatic microbeams may reduce radiation damage by means of a photoelectron escape from a narrow diffraction channel (Moukhametzianov et al., 2008). Furthermore, synchrotron microbeams are characterized by less background scattering from sample environments (Riekel, Burghammer, & Popov, 2011). Despite these advancements, radiation damage is still an important issue to address and cope with.

Indeed, crystallization is a highly demanding and time-consuming task, being a real bottleneck in the nowadays basic research. Several efforts have been made in order to understand which factors and parameters can influence this process, from the pH, ionic strength, salt, and protein concentration, as well as additives, detergent, and other molecules concentration and temperature. However, a complete explanation of the processes leading to crystal formation is still lacking.

Our group has introduced Langmuir–Blodgett (LB)-based crystallization as a major advance within both structural and functional proteomics (Pechkova & Nicolini, 2004), pioneering the emerging field of the so-called *nanocrystallography* or *nanobiocrystallography* (Pechkova & Nicolini, 2004; Pechkova, Roth, et al., 2005; Pechkova, Vasile, et al., 2005). This approach integrates protein crystallography and nanotechnologies within a unique, coherent framework that allows to obtain highly stable protein crystals and to fully characterize them down to atomic resolution (Pechkova, Roth, et al., 2005; Pechkova, Vasile, et al., 2005). It is noteworthy to underline that the prefix *nano-* is here used to refer to the use of cutting-edge nanobiotechnologies for the task of protein crystallization, and not just or not only to the size and dimension of the protein crystals (Chen & Millane, 2013).

2. LANGMUIR–BLODGETT (LB)-BASED CRYSTALLIZATION

LB nanotemplate crystallization method has proved to give prominent results in target proteins crystallization, such as thaumatin, a 207-amino acid, 22.2-kDa sweet-tasting protein, obtained from the African plant *Thaumatococcus daniellii* Bennett (Gebhardt, Pechkova, Riekel, & Nicolini, 2010; Pechkova, Gebhardt, Riekel, & Nicolini, 2010; Pechkova, Scudieri, Belmonte, & Nicolini, 2012; Pechkova, Sivozhelezov, Belmonte, & Nicolini, 2012), which can be used as a low-calorie sweetener and flavor modifier. Other proteins that have been tested for crystallization are lysozyme, EC 3.2.1.17, a 14.7-kDa protein, of *Gallus gallus* (Hen egg white lysozyme) (Pechkova & Nicolini, 2002; Pechkova, Roth, et al., 2005; Pechkova, Vasile, et al., 2005; Pechkova, Sartore, Giacomelli, & Nicolini, 2007; Pechkova, Sivozhelezov, & Nicolini, 2007), ribonuclease A, EC 3.1.27.5, a small 124-amino acid, 13.7-kDa RNAse from *Bos taurus* (Pechkova, Scudieri, et al., 2012; Pechkova, Sivozhelezov, et al., 2012), thermolysin, EC 3.4.24.27, a 34.6-kDa enzyme of *Bacillus thermoproteolyticus* (Pechkova, Scudieri, et al., 2012; Pechkova, Sivozhelezov, et al., 2012), and proteinase K, EC 3.4.21.64, of *Engyodontium album*, formerly known as *Tritirachium album* (Pechkova, Tripathi, & Nicolini, 2009; Pechkova,

Tripathi, Ravelli, McSweeney, & Nicolini, 2009; Pechkova, Scudieri, et al., 2012; Pechkova, Sivozhelezov, et al., 2012) (Fig. 5.1).

LB technology enables to make highly ordered thin films, which can be used for a variety of bioengineering tasks such as the building and implementation of biosensors (Bragazzi et al., 2012; Nicolini, Adami, et al., 2012; Nicolini, Bezerra, & Pechkova, 2012; Nicolini, Belmonte, Maksimov, Brazhe, & Pechkova, 2013; Nicolini, Bragazzi, & Pechkova, 2013; Nicolini, Bruzzese, Cambria, Bragazzi, &Pechkova, 2013; Nicolini, Correia, et al., 2013). In the field of macromolecular crystallography, LB-nanostructured templates act as a catalyst for crystal nucleation and growth (Gebhardt et al., 2010; Pechkova et al., 2010). Briefly, an LB trough is a laboratory apparatus that is used to compress and to spread monolayers or multilayers of molecules, with Hamilton syringes without the use of any dispersant, by small droplet on the surface of a given subphase (usually distilled

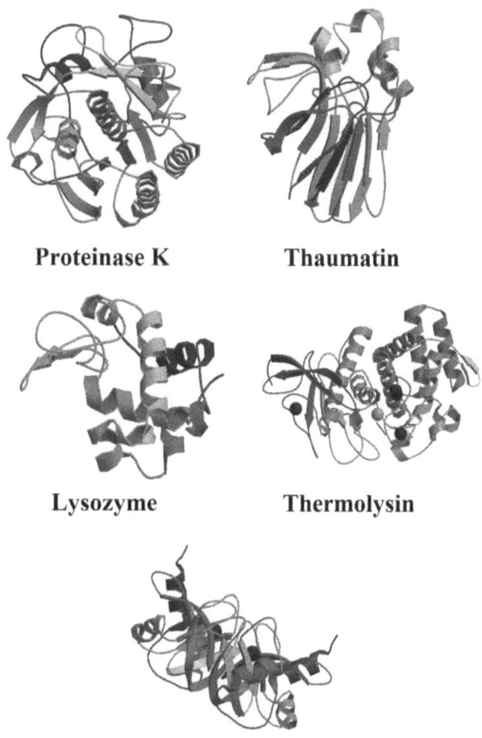

Figure 5.1 Langmuir–Blodgett (LB)-grown crystals of target proteins (namely, proteinase K, thaumatin, lysozyme, thermolysin, and ribonuclease A).

water, purified and filtered with Milli-Q system, 18.2 MΩ cm). Later, utilizing the Langmuir–Schaefer (LS) method (horizontal lift), the floating protein thin film is transferred onto the surface of a solid substrate (such as siliconized circular glass or glass wafers, washed in water, and dried in a gaseous nitrogen flux). LS technique consists of making the prepared substrate horizontally touching the monolayer or multilayer, in such a way that the layer transfers itself onto the substrate surface.

In case of multilayer deposition, the regularity and uniformity can be controlled using nanogravimetric measurements and verifying by computing the area per molecule of the proper packing density.

The trough is connected to a personal computer that, being equipped with dedicated software, enables the measuring and recording of surface phenomena due to Teflon barriers movement and compression, after properly stabilizing the paper Wilhelmy plate.

The advantages of using this nanobiotechnology are different: they include the accelerated nucleation and growth of protein crystals (Pechkova & Nicolini, 2002), their higher quality both in terms of X-ray diffraction and radiation stability when employing the high energy X-ray source and focused beans, such as the third-generation synchrotrons and the microdiffraction beamlines (Belmonte et al., 2012; Pechkova et al., 2004). This could be due to enhanced mechanical properties of the protein crystals, as we review and describe in detail in the following sections. Moreover, LB-grown crystals are larger, have a more regular and clearly defined shape, and more perfect domains (Pechkova & Nicolini, 2010; Riekel et al., 2011) (Fig. 5.2).

3. COMPARISON OF LB-BASED CRYSTALLIZATION WITH OTHER TECHNIQUES

Different crystallization techniques have been proposed during the decades (Giegé, 2013; Manuel García-Ruiz, 2003), like the "classical" hanging-drop vapor diffusion method and its variant (sitting drop and sandwiched drop), capillaries, gel, dialysis (conventional, microdiffusion, and meso- and microdialysis), cryotemperature (or cryocooling) (Garman & Owen, 2006), (micro-)batch (and its Jakoby variant) (Chernov, 2003), liquid–liquid diffusion, or free interface diffusion (FID), counterion diffusion or CID, and variants such as the liquid bridge technique, hybrid approaches such as the diffusion/dialysis one, heterogeneous nucleation approaches (such as seeding), and even in space or in simulated (Wakayama et al., 2006) microgravity environments (using techniques like gels, dialysis, FID

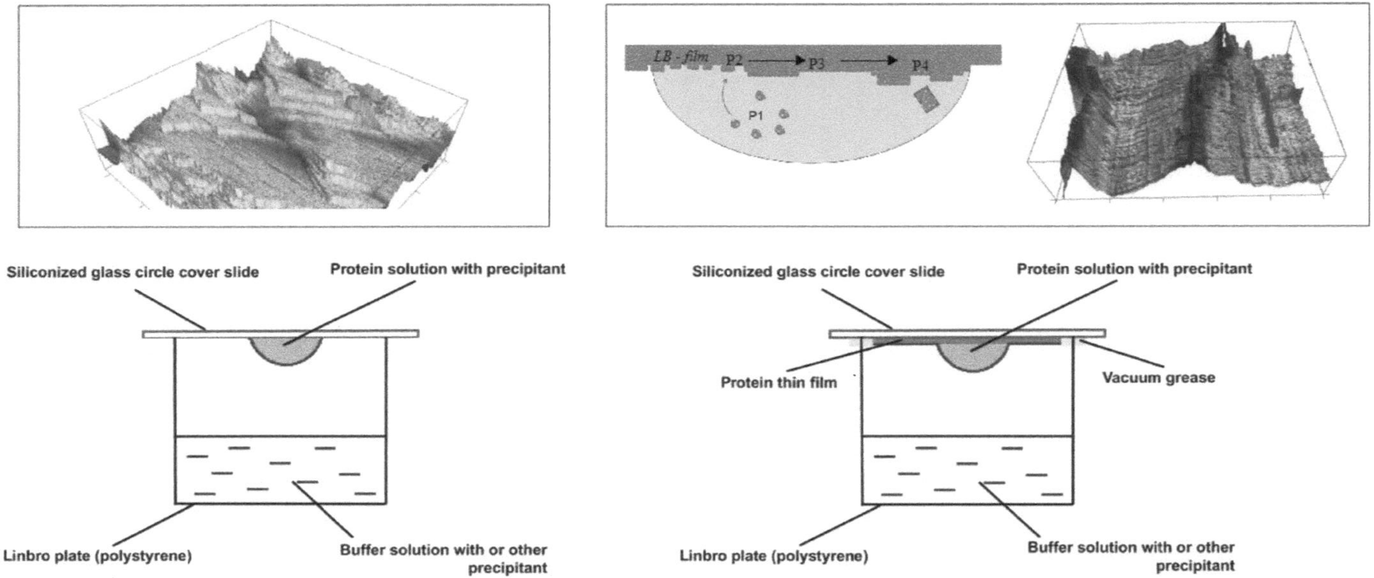

Figure 5.2 Classical hanging-drop vapor diffusion crystallization technique and atomic force microscopy (AFM) characterization of crystal domains (left); Langmuir–Blodgett (LB)-assisted crystal growth and AFM characterization of crystal domains and a pictorial explanation of the mechanisms leading to LB crystal nucleation and growth (right).

and CID, and microbatch) (Judge, Snell, & van der Woerd, 2005; Khurshid & Chayen, 2006; Otálora, Gavira, Ng, & García-Ruiz, 2009; Vergara, Lorber, Sauter, Giegé, & Zagari, 2005). Very recently, also seeding and heterogenous nucleations have been applied in space (Schöpe & Wette, 2011).

Differences in protein crystal formation between classical hanging-drop and LB have been established by a variety of methods including both *ex situ* and *in situ* GISAXS (Gebhardt et al., 2010; Nicolini & Pechkova, 2006; Nicolini, Belmonte, et al., 2013; Nicolini, Bragazzi, et al., 2013; Nicolini, Bruzzese, et al., 2013; Nicolini, Correia, et al., 2013; Nicolini, Belmonte, Riekel, Koenig, & Pechkova, 2014; Nicolini, Bragazzi, Pechkova, & Lazzari, 2014; Pechkova & Nicolini, 2006, 2011; Pechkova, Roth, et al., 2005; Pechkova, Vasile, et al., 2005; Pechkova, Tripathi, & Nicolini, 2009; Pechkova, Tripathi, Ravelli, et al., 2009; Pechkova et al., 2010), laser microdissection combined with nano- and microfocus beamlines (Nicolini, Belmonte, et al., 2014; Nicolini, Bragazzi, et al., 2014; Riekel et al., 2011), Raman spectroscopy (Nicolini, Belmonte, et al., 2013; Nicolini, Bragazzi, et al., 2013; Nicolini, Bruzzese, et al., 2013; Nicolini, Correia, et al., 2013), and atomic force microscopy (Pechkova, Sartore, et al., 2007; Pechkova, Sivozhelezov, et al., 2007). In the following paragraphs and sections, we overview the main conclusions originated from these experiments.

4. FOURIER TRANSFORM INFRARED (FTIR) SPECTROSCOPY FOR INVESTIGATING LB-FILMS

Introduction of LB film affects protein stability and preservation of secondary structure as experimentally proved in the case of lysozyme, which turned out to be quite thermally stable up to 200 °C, as shown by means of Fourier transform infrared (FTIR) spectroscopy and circular dichroism (Pechkova, Sartore, et al., 2007; Pechkova, Sivozhelezov, et al., 2007). Above 200 °C, the thermostability of lysozyme multilayer film ceases, and the protein begins to aggregate.

FTIR is a label-free technology that is widely used to characterize protein structure with a focus on folding and misfolding/aggregation dynamics (Miller, Bourassa, & Smith, 2013) as well as protein–protein interactions (Haris, 2013) and drug delivery and release (Kazarian & Ewing, 2013), together with its recent variants such as attenuated total reflection FTIR and surface-enhanced infrared spectroscopy (Glassford et al., 2013).

5. RAMAN SPECTROSCOPY

Raman spectroscopy (named after the Indian physicist Sir Chandrasekhara Venkata Raman) is an advanced spectroscopic technique that enables to observe and measure vibrational, rotational, and other low-frequency modes of a given system, relying on inelastic scattering or Raman scattering.

Raman spectroscopy is attractive as a potential diagnostic technology because it requires no extrinsic labeling, is not limited by masking water contributions, and is inherently a multiplexing technique. Raman-based measurements of biological samples have already been exploited for the identification of molecular-specific markers for disease detection and monitoring (Turzhitsky et al., 2014). The use of fiber optic technology coupled with Raman spectroscopy is ideal for application to aqueous solutions, either with or without LB nanotemplate. Raman is expected to evaluate the protein concentration changes in vapor diffusion hanging-drop method with and without LB nanotemplate. Acquired Raman spectra have previously been subjected to quantitative infrared partial least squares (PLS) models with remarkable success. The PLS model generated correlates the spectral region from 2700 to 3600 cm^{-1} with the concentration (g/ml) of lysozyme. This spectral region encompasses vibrations due to the protein C–H stretches centered at 2950 cm^{-1} and the water O–H stretches centered at 3230 cm^{-1}.

On the basis of our Raman spectroscopic analysis (Nicolini, Belmonte, et al., 2013; Nicolini, Bragazzi, et al., 2013; Nicolini, Bruzzese, et al., 2013; Nicolini, Correia, et al., 2013), we suggest that LB-assisted crystal growth with time is accompanied by:

1. Formation of disulfide bonds S6–S127/S30–S115 that brings Trp123 into new position and facilitates vibrations of its rings. It could also promote the formation of hydrogen bond between Trp indole ring and the nearby amino acids;
2. Formation of disulfide bonds S6–S127/S30–S115 that brings the whole C-end closer to Phe31 and Phe38 residues. This affects phenylalanine aromatic rings vibrations;
3. Formation of S—S bonds in C-terminal that affects the conformation of the C-terminal and, possibly, the whole lysozyme. C-terminal is more rigid in LB crystals than in classic crystals, and in larger LB crystals than in smaller ones. This can have an impact on the mechanical properties of LB-grown crystals, producing more rigid and stable crystals.

We suggest that the main difference in lysozyme conformation in LB and classical crystals is caused by higher amount of disulfide bonds, probably in C-end of protein, resulting in the higher stiffness of lysozyme molecules and LB crystal in a whole. Development of LB crystal in time and increase in its size is also accompanied by the formation of S—S bonds.

6. LASER-INDUCED MICRODISSECTION AND MICROFRAGMENTATION

Laser-induced microdissection of LB nanotemplate-facilitated protein crystals in glycerol solution results in distinct, coherently diffracting domains (Nicolini, Belmonte, et al., 2014; Nicolini, Bragazzi, et al., 2014). Laser-microdissection is indeed very useful in order to obtain pieces of crystals of very small dimensions in conjuction with X-ray nanodiffraction techniques capable to overcome the very common problem of twinned, defect, aggregated and mosaic crystals. Microdissected crystals can separatel into smaller fragments due to effects such as cavitations at domain boundaries and solvent interpenetration. Only crystals produced according to the LB nanotemplate technique reveal in all four proteins being tested (lysozyme, insulin, thaumatin, and ribonuclease) domains that are highly radiation resistant, while the crystals produced by the standard hanging-drop crystallization method do not. Actually, the very same laser exposure causes the disappearance of these "classical" protein crystals during the same time frame of 40 min needed for the laser cutting in all four proteins being tested. The microdiffraction of microcrystals prepared by the combination of LB and laser technologies (Schlichting & Miao, 2012; Smith, Fischetti, & Yamamoto, 2012) proves that not only the lysozyme survives the process, as shown recently by nanodifraction, but also all three other model proteins appear to behave similarly well, namely, insulin, thaumatin, and ribonuclease. The result confirms the emerging of a new biophysical technique uniquely useful for synchrotron radiation studies based on small protein microcrystals uniquely radiation resistant when prepared by LB nanotemplate and subsequently fragmented by laser.

7. MICROGRAZING-INCIDENCE X-RAY SCATTERING ANGLE (μ-GISAXS)

GISAXS is an advanced scattering technique that can be used to investigate large-scale structures in thin films, including biofilms (Gebhardt &

Kulozik, 2014; Gebhardt, Vendrely, & Kulozik, 2011; Metwalli et al., 2013). A combination of this technique with synchrotron radiation microbeams (μ-GISAXS) (Müller-Buschbaum, 2003; Pechkova, Roth, et al., 2005; Pechkova, Vasile, et al., 2005) has been used for studying surface gradients or confined surfaces (Bass, Berman, Singh, Konovalov, & Freger, 2010; Uhlmann et al., 2011), supported islands and buried structures, quantum dots (Buljan et al., 2012; Urban, Talapin, Shevchenko, & Murray, 2006), nanoparticles (Al-Hussein et al., 2013; Richard, Schülli, Renaud, Zhong, & Bauer, 2011), enabling the investigation and characterization of nanobiomaterials down to the nanoscale, measuring important features and parameters like the lateral correlation, the surface roughness, the size and dimension of substructures and domains.

Experiments were performed at the ID13 microfocus beamline at the European Synchrotron Radiation Facility in Grenoble, France. A monochromatic beam was focused by crossed Fresnel lenses on a spot at the sample position with a high photon flux and energy. The μ-GISAXS pattern was recorded by a two-dimensional charge-coupled device detector (MAR CCD). Specular scattering is observed for $Q_x = Q_y = 0$, $Q_z > 0$, and diffuse scattering for Q_x, $Q_y \neq 0$. Correlations vertical to the sample surface can be probed along Q_z at $Q_y = 0$. Critical angles of each specific investigated protein and glass for the used X-ray energy were calculated on the basis of their chemical formula and densities. The Fit2D software package was used for data reduction. The direct beam and the specular beam are striking, invariant features and both generate small-angle scattering. This scattering broadens the specular beam, causing scattered intensity, with a curve that can be modeled by fitting two Gaussian profiles around the direct beam. The Yoneda peak consists of a contribution from glass and a contribution from protein. Compared to the glass signal, the scattering contribution from the protein is weak. The peak height of the Yoneda peak of glass is affected by changes in the protein Yoneda peak. Changes in the Yoneda region can be referred to the interplay between specular and diffuse scatterings.

Nanotemplate-assisted crystallization experiments were acquired with *ex situ* μ-GISAXS for penicillin G acylase, urease (Pechkova, Tripathi, & Nicolini, 2009; Pechkova, Tripathi, Ravelli, et al., 2009), cytochrome P450scc (Nicolini & Pechkova, 2006), and lysozyme (Pechkova & Nicolini, 2006).

In the case of the former two enzymes, GISAXS experiments shed light on the effect of temperature on the protein reorganization taking place in an LS-multilayered enzyme film. Merging of layers is likely to occur during the heating (up to 423 K) and cooling process (down to room temperature), leading to a loss of correlation between the interfaces of the layers and to the establishment of long-range order.

Since the data generated by these experiments were extremely complex to interpret and needed to be studied in parallel with microscopy characterization, using either the classical method or the nanotemplate hanging-drop method, we planned further *ad hoc in situ* experiments.

For this purpose, we used an experimental setup, specifically designed for the *in situ* GISAXS acquisition: a modified hanging-drop crystallization cell, connected via Teflon tubes to two Harvard syringe pump for buffer exchange. This particular layout enabled to study the kinetics of crystallization by monitoring and measuring it directly on the interface of the LB film crystallization nanotemplate (Gebhardt et al., 2010; Pechkova et al., 2010).

Using dynamic mathematical modeling based on first-order differential equations, we found that the biofilm reorganizes itself and leads to crystal formation. The model assumes a reservoir of an oversaturated protein solution, P_1, which corresponds to the protein in the hanging drop. The change in the reservoir concentration results from a process that leads to a protein association on the LB film. The association depends on the amount of oversaturated thaumatin concentration in the hanging-drop reservoir, P_1, on the amounts of the LB film states, P_2 and P_3, and association rates k_1 and k_3, respectively. P_1 association leads to the association states P_{12} and P_{13} of the LB film. Besides the protein inflows, the concentration of the associated states depends also on their rates of conversion (k_2 and k_4) into the end states P_{12*} and P_{13*}. Both conversions lead to a decrease of protein in the LB film due to dissociation reactions. The change in the associated state P_{12} corresponds to the thaumatin crystalline state in the LB film, which leads to increased specular intensity due to a smoother surface and a crystalline structure factor peak in the Q_y cut. Its fast formation and dissociation takes place at high thaumatin concentration in the reservoir. Moreover, the formation of P_{12} starts from the intensity level of the LB film, which acts as a seed for crystal growth. After reaching the maximum at $t = \sim 100$ min, the process drops down to zero, which means that the crystalline state becomes completely depopulated in favor of the crystalline state P_{12*}. This state corresponds to a completely dissociated crystal, leaving a hole in the LB film. The formation of associated state P_{13} runs more slowly. In contrast to the temporal change of P_{12}, this process first starts at zero and in a second stage proceeds at comparatively low thaumatin concentrations in the hanging-drop reservoir. The former means that the state P_{13} is not populated at the beginning and not comparable to the ordered thaumatin state in the LB film. Hence, we identify this state as a less-ordered thaumatin protein that associates at low free thaumatin concentrations on the LB film. The degradation of the less-ordered state at $t > 500$ min can have different causes. Normal protein degradation or an onset of LB film disruption due to the long duration of the experiment at room temperature (295 K) and the radiation damage have to be taken into consideration.

Similar findings were obtained with lysozyme (Pechkova & Nicolini, 2011). Interestingly, while the two peaks in the Yoneda region appear to be already present at the start of plating with the LB nanotemplate, without the LB nanotemplate they are absent at the start and present only at the end. The data on LB intensity fluctuations in the GISAXS pattern versus time appear to be associated with rapid seed formation and crystal growth and damage, while the classic continuous shift of intensity in the Yoneda region is compatible with slow crystal growth and significantly larger damage, apparent by light microscopy both in the hatch and in the parallel experimentation carried out on the bench, quite compatible with the radiation damage being assessed by LB crystal diffractions.

In conclusion, taking both *ex situ* and *in situ* data, the evolution of the μ-GISAXS patterns suggests that:

1. Lateral spatial correlations of the film lead to a contribution of diffuse scattering to the overall GISAXS scattering distribution and can be studied by means of Q_y cuts;
2. The specular intensity increases considerably between $t = 100$ and $t = 300$ min and the film roughness decreases in this period;
3. The smooth decay in intensity of the cuts with increasing Q_y can be well approximated by two Gaussian profiles. When comparing the cuts, a peak at $Q_y = 0.1\ \text{nm}^{-1}$ measured after $t = 100$ min becomes conspicuous. Such a feature in the Q_y cut indicates highly ordered crystalline regions in the LB film, like a bidimensional paracrystal with a mean distance between adjacent structures on the LB film of $D = 58$ nm;
4. The increase in intensity in the Yoneda region is due to protein incorporation into the LB film. The intensity variation suggests several steps, namely, a first, fast, process, attributed to the crystal growth and its detachment from the LB film, and a second, slower process, attributed to an unordered association and conversion of protein on the LB film.

8. *IN SILICO* SIMULATIONS

Crystal growth has been simulated exploiting the 2D lattice Monte Carlo algorithm and the coarse-grained hydrophobic-polar approximation, using monomer and tetramer (aggregate) units models. These simulation lead to the conclusion that lysozyme tetramers LB-based crystal is expected to be slightly accelerated when compared to its monomer-based counterpart (Siódmiak, Gadomski, Pechkova, & Nicolini, 2006).

Previously acquired *in situ* GISAXS spectra (Gebhardt et al., 2010; Pechkova & Nicolini, 2011; Pechkova et al., 2010) were analyzed using IsGISAXS software developed by Rémi Lazzari, a tool which is dedicated to simulation of scattering from supported nanostructures. The scattering cross section is expressed in terms of island form factor and interference function and the specificity of the grazing-incidence geometry is stressed, in particular in the evaluation of the island form factor in the distorted-wave Born approximation. A full account of size and possible shape distributions is given in the decoupling approximation, where sizes and positions are not correlated, and in the local monodisperse approximation. Two types of island repartitions on the substrate can be implemented: disordered systems characterized by their particle–particle pair correlation functions, and bidimensional crystalline or paracrystalline systems of particles.

Proteins have been modeled as cylinders, LB film thickness, found with the best fit, was fixed at 7.4 nm for thaumatin and at 6.4 nm for lysozyme, while the wavelength was experimentally known (0.0991 nm). Critical incident angle for thaumatin and for lysozyme was computed to be 0.71°. The delta refraction coefficients were 3.336×10^{-6} for glass and 2.19×10^{-6} for proteins. The beta absorption coefficients were about 0 for proteins and 1.68×10^{-8} for glass. Curves have been fitted with a χ^2 Levenberg–Marquardt minimization procedure, which is an iterative technique commonly used for solving nonlinear least squares problems, with constant standard error bars of $\sigma R/R = 0.005$ by means of the IsGISAXS software.

At 100 min, the particle radius of thaumatin is about 5.62 nm, while the LB film layer thickness is about 2.25 nm, with a height ratio of 1 nm. At 900 min from the start of the experiment, the particle radius has increased up to 40.89 nm, while on the contrary, the LB film layer thickness has decreased down to 0 nm, with a decreased height ratio of 6.30×10^{-2} nm. Thus, we confirmed the working hypothesis that the protein appears to transfer directly from the nanobiostructured film into the drop to directly trigger the formation of the crystal, therefore highlighting the physical interpretation of the mechanism for nanobiotemplate-facilitated protein crystallization.

9. BIOINFORMATICS

On the basis of a mass-scale analysis of crystal structures by mining the PDB repository and confirming the hypothesis with a targeted *ad hoc* experiment using thermostable thioredoxin from *Alicyclobacillus acidocaldarius* versus the mesophilic *Escherichia coli* counterpart, we have established a role of the aqueous surroundings of a protein in its thermal stability (Pechkova, Sartore, et al., 2007; Pechkova, Sivozhelezov, et al., 2007), and in particular of the inner bounded water shell.

The introduction of LB film indeed affects the aqueous environment of the protein leading to smaller numbers of water molecules.

This explanation has been recently confirmed by the water characteristics in all model proteins (Fig. 5.1). The shape of the frequency distribution of volumes occupied by water molecules is found to be different between "classical" samples of different proteins, but surprisingly quite similar for LB samples. LB film leads to the appearance of water molecules close to the protein surface but occupying large volumes. The data suggest a "quite Gaussian distribution" for LB and a "quite periodic distribution" for classical as shown by the kurtosis and skewness analysis (Belmonte et al., 2012; Pechkova, Scudieri, et al., 2012; Pechkova, Sivozhelezov, et al., 2012).

In another study, we applied clustering algorithm and protein alignment, showing how LB-based crystals can be compared with those obtained in space. Proteins were downloaded from the PDB database (http://www.rcsb.org/pdb/home/home.do), then we iteratively refined the choice excluding proteins belonging to other *taxa* or radiation-damaged structures, and finally, we subdivided the structures using the crystallization procedure as variable (Pechkova, Bragazzi, Bozdaganyan, Belmonte, & Nicolini, 2014). The most accurate method of three-dimensional (3D) protein structure alignment algorithm based on the TM-align topological score was used, root mean square deviation (RMSD) for C-α atoms was used as the similarity measure for all structures: All the calculations were performed using the web-based protein structure comparison tool ProCKSI.

The clustering algorithm was used selecting the Ward distance, both as all-against-all and as all-against-target options.

We found that for lysozyme and human insulin structures were absolutely comparable (Fig. 5.3, left), while similar evidence was collected also for thaumatin and proteinase K.

From clustering algorithm, bioinformatics, and biostatistics, we can conclude that:

1. Bioinformatics and clustering algorithms can be applied to the study and modeling of proteins crystallized according to different crystallization techniques;
2. According to the clustering algorithm and statistical analysis of parameters like resolution, *B*-factor, and solvent content, LB-based and microgravity proteins are comparable and different from proteins crystallized with other techniques.

10. MOLECULAR DYNAMICS

In silico simulation of molecular systems dynamics (Cheng & Ivanov, 2012) is widely used in molecular physics, biotechnology, medicine, and chemistry (Kerrigan, 2013) to predict physical and mechanical properties of new molecular complexes and materials.

In the molecular dynamics (MD) method, there is a polyatomic molecular system in which all atoms are interacting like material points, and the behavior of the atoms is described by the equations of classical mechanics (CM). This method allows doing simulations of the system of the order of 10^6 atoms in the time range up to 1 μs. Despite some limitations such as approximations, it may be useful in describing the dynamics of macromolecules at the atomic level. However, this method does not take into account the chemical reactions and the formation or breaking of chemical bonds. For these purposes, there are more advanced and sophisticated approaches that combine classical Newtonian and quantum mechanics simulations (hybrid CM/QM approaches).

We downloaded our protein crystals from PDB and all other structures belonging to the same protein family and crystallized under the same experimental conditions. All the proteins were solvated and neutralized adding Cl^- or Na^+ depending on the charge of the protein. OPLS-aa (optimized potentials for liquid simulations all atom) force field were used for our MD simulation. For the long-range electrostatic interactions, we used particle-mesh Ewald method, which calculates direct space interactions within a finite distance using a modified Coulomb's law, and in reciprocal space using a Fourier transform to build a "mesh" of charges, interpolated onto a grid. It is from this charge interpolation that long-range forces can be calculated and

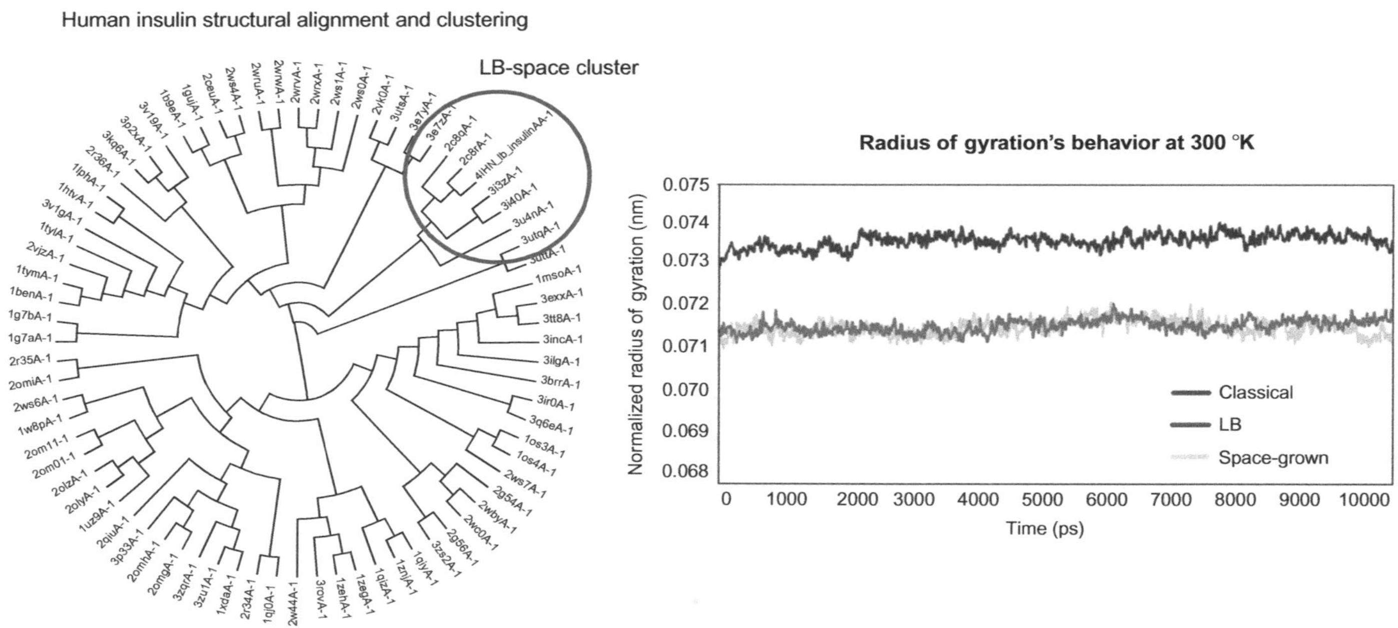

Figure 5.3 Theoretical/semi-theoretical approaches for investigating Langmuir–Blodgett (LB)-grown crystals features and behaviors in comparison with other crystallization techniques. Bioinformatics analysis is shown at left, while at right, the radius of gyration simulated with molecular dynamics (MD). LB crystal and those grown in space tend to behave in a similar way, strikingly different from the behavior of classical crystals.

incorporated into the nonbonded interactions in a simulated system. For the van der Waals interactions, a typical 12 Å cutoff was used. A standard equilibration procedure was used and adopted for all systems.

From our simulations (Bozdaganyan, Bragazzi, Pechkova, Shaytan, & Nicolini, 2014), we found that LB-grown crystals follow the trends of microgravity crystals, that is, to say higher resolution, lower content of water (in term of the Matthew coefficient), lower *B*-factor (i.e., thermal noise and atom displacement), and higher number of reflections (i.e., the crystals diffract better).

One-way ANOVA (analysis of variance) in fact showed statistically significant differences among the groups for *B*-factor (p value <0.05, with $p=0.024$) and for water content (p value <0.05, with $p=0.032$).

From MD, we can observe that (Fig. 5.3, right):

1. The structure of all proteins were very stable with the RMSD values remaining almost constant around 1–1.3 Å at 300 K for all protein crystals, namely LB, classical, and space. Such small changes in structure are in line with those expected for a protein effectively transferred from a crystal to a solution environment. The RMSD increased more significantly after 2 ns when proteins in water were simulated at 500 K. Classical proteins exhibited higher RMSD values than LB and space: the difference is around 2–3 Å. The above results indicate that, comparing to HD-proteins, LB and space proteins seem to be more stable at high temperature;
2. The same is true for the gyration radius at different temperatures. As we can see, LB and space-grown proteins have smaller meaning of R_g than the classical ones in both cases—for 300 and 500 K, indicating that the LB and space proteins are more compact than classical.

According to all data provided above, we can conclude that secondary and tertiary structures of the proteins are not destroyed during our simulation even at 500 K. Also, LB crystals are seemed more compact and resistant to high temperatures.

Large-scale atomistic MD simulations can be applied to the study and modeling of proteins crystallized according to different crystallization techniques. According to RMSD analysis, space-grown and LB proteins are more stable to unfolding at 500 K than classical ones.

11. CLINICALLY RELEVANT PROTEINS

11.1. GroEL

Phage growth λ *Escherichia coli* large (GroEL) ribosomal protein, expressed in the *Escherichia coli* cytoplasm, is a chaperonin involved in proper protein folding, and it was the first member of heat-shock protein 60 (hsp60) family to be identified, being recognized as a chromosomally encoded product whose deficiency resulted in defective morphogenesis of bacteriophage T4 head structures and T5 tail structures (Pechkova, Tripathi, Spera, & Nicolini, 2008). Despite a huge body of studies and investigations, the precise molecular mechanisms of substrate recognition and protein folding are not exactly known (Azia, Unger, & Horovitz, 2012).

GroEL is a protein of high interest in the field of biosensors for gut microbiota sensing and has a clinical relevance, being utilized in sublingual vaccination against atherosclerosis (Hagiwara et al., 2014), and against diarrhea and colitis (Péchiné, Hennequin, Boursier, Hoys, Collignon, 2013). Moreover, GroEL is also involved in the suppression of amyloidogenesis (Yagi-Utsumi et al., 2013) and this could have significant therapeutic implications in the field of protein disorder and aggregation-induced disorders, also fostering the discovery of new drugs and inhibitors (Johnson et al., 2014).

It could be anticipated that LB-based crystallization of GroEL could open new perspectives in the chaperonin biochemistry and physiopathology.

11.2. Casein kinase 2

CK2 (casein kinase II or casein kinase 2) is one of the most acidophilic, pleiotropic, versatile, and multifunctional serine/threonine protein kinases (Bian et al., 2013), first discovered in 1954 by Burnett and Kennedy (Cozza, Pinna, & Moro, 2012; Sarno & Pinna, 2008) even though its full atom structure was revealed only recently (Pechkova, Zanotti, & Nicolini, 2003), prompting the discovery of new-generation CK2 drug inhibitors (Cozza et al., 2012). It is involved in a variety of functions and biological processes, ranging from transcription, signaling, proliferation, and in various steps of cell development (Bian et al., 2013). Its abnormally elevated levels are correlated to most tumors (Trembley, Wang, Unger, Slaton, & Ahmed, 2009): from multiple myeloma (Piazza, Manni, & Semenzato, 2013), leukemia (Dovat, Song, Payne, & Li, 2011), and lymphoma, pancreatic cancer (Giroux, Dagorn, & Iovanna, 2009), breast tumor, colorectal cancer, prostate, kidney, and lung, as well as autoimmune disorders and infectious diseases (Cozza et al., 2012). It constitutes an important disease biomarker (Sarno & Pinna, 2008).

11.3. Cytochrome P-450 side-chain cleavage

Cytochrome P-450 side-chain cleavage (CYP450scc), chemically characterized by a three beta-hydroxyl and a delta 5-ring configuration, is an extremely selective enzyme, interacting with few substrates (Lambeth, 1986). It is involved in steroid synthesis, catalyzing the conversion of cholesterol into pregnenolone and its deficiency leads to adrenal insufficiency, disrupting adrenal and gonadal steroidogenesis, and clinically and hormonally mimicking congenital lipoid adrenal hyperplasia (Tee et al., 2013). CYP450scc dysregulation has been linked also to other forms of adrenal failure (Gucev, Tee, Chitayat, Wherrett, & Miller, 2013) and polycystic ovary syndrome (Wickenheisser et al., 2012).

Properly immobilized CYP450scc (e.g., via LB technology) can be used for cholesterol sensing (Arya, Datta, & Malhotra, 2008), eventually coupling it with advanced molecular modeling (Sivozhelezov & Nicolini, 2005) and new label-free techniques (Spera et al., 2013).

11.4. Rhodopsin

Rhodopsin has a molecular weight of 40 kDa and consists of an apoprotein, opsin (348 amino acid residues), a chromophore, 11-*cis*-retinal, covalently bound to Lys296 via a protonated Schiff base (PSB), and two oligosaccharide chains (Shichi & Rafferty, 1980).

It crosses the membrane with seven α-helices which constitute as much as 60% of its secondary structure and which appear oriented mostly perpendicular to the plane of the disk membrane (Unger & Schertler, 1995; Unger, Hargrave, Baldwin, & Schertler, 1997).

The rhodopsin chromophore, 11-*cis*-retinal, is located in a hydrophobic pocket between the helices (Palczewski, 2012); this covalent bond of the chromophore contributes to the tightly held rhodopsin in a nonsignaling conformation. The extracellular domain of rhodopsin is relatively rigid, which may help to reduce spontaneous activation of the receptor in the absence of light (Smith, 2010).

The visual rhodopsin is a typical representative not only of the retinal-containing proteins but also concurrently of a large family of G-protein-coupled receptors (class A G-protein-coupled receptors or GPCRs) (Lodowski, Angel, & Palczewski, 2009).

A better molecular understanding of rhodopsin structure could open new perspectives in the treatment of ophthalmological diseases, such as retinitis pigmentosa (RP) and other disorders related to retinal degeneration (Hollingsworth & Gross, 2012). Mutations related to RP cause protein misfolding and could be properly targeted by chaperones and derivative molecules (Mendes, Zaccarini, & Cheetham, 2010).

Bacteriorhodopsin can be used for nanobioelectronics (Wagner, Greco, Ranaghan, & Birge, 2013), for building and implementing 3D optical memories, real-time holographic processors, and artificial retinas, while octopus rhodopsin could be used as a new biomaterial characterized by significant photoreversibility, photostability, and photochromic properties (Paternolli et al., 2009; Sivozhelezov & Nicolini, 2006). Crystallization of octopus rhodopsin is currently in progress.

11.5. Globins

Globins are heme-containing proteins involved in binding, transporting, and delivering oxygen and other nutrients. They are typically composed of eight α-helices that fold themselves into a three-over-three α-helical sandwich structure. In particular, Hell's gate globin I (HGbI), a single-domain protein with 133 residues, was identified from the genome of *Methylacidiphilum infernorum* (Teh et al., 2011), an aerobic, acidophilic, and thermophilic obligate methanotroph that grows optimally at 608 °C and pH 2.0.

HGbI is structurally homologous to mammalian neuroglobins. Its particular features are:

1. High affinity and avidity for the oxygen;
2. Negligible auto-oxidation in the pH range of 5.2–8.6 and temperature range of 25–50 °C;
3. Unique resistance to the extreme acidity and hostile environments.

These features make globin quite interesting and attractive for building and implementing bioelectronic devices and biosensors (Chan, 2001) such as oxygen (Perutz, Paoli, & Lesk, 1999), nitric oxide sensor (Xu, Wu, & Zhao, 2013), useful for a vast array of applications in the field of molecular nephrology (Palm, Nordquist, & Buerk, 2007) to anesthesiology (Collison & Meyerhoff, 1990).

11.6. Insulin

Insulin is a 51-amino acid dimer protein, with a molecular weight of 5.8 kDa. The two chains are linked together by disulfide bonds. It is a clinically relevant peptide hormone, produced by beta cells of the pancreas, and plays a major role in the regulation of carbohydrate and fat metabolism in the body. Its dysregulation causes diabetes mellitus, metabolic syndrome, and other metabolic disorders. Insulin is required for the management of patients with diabetes and the discovery of more active biosimilar insulins represents a therapeutically important advancement (Heinemann, 2012). Our solved LB-grown insulin structure has one of the best and lowest resolutions among all the insulin crystals deposited in PDB.

12. CONCLUSIONS

LB-based crystallography has proved successful in solving the structure of both target proteins (Fig. 5.1) and proteins difficult to crystallize with the conventional techniques (Fig. 5.4).

LB-grown crystals have a lot of interesting features and properties: from resistance to radiation to better domains and regular shape.

A future step of LB-based crystallography will be the structure determination of further membrane proteins (Moraes, Evans, Sanchez-Weatherby, Newstead, & Stewart, 2014) and cytochromes (Nicolini & Pechkova, 2006; Paternolli, Ghisellini, & Nicolini, 2007; Sivozhelezov et al., 2006; Spera et al., 2013) that play a pivotal role in nanomedicine and above all personalized medicine, being the targets of commonly used drugs.

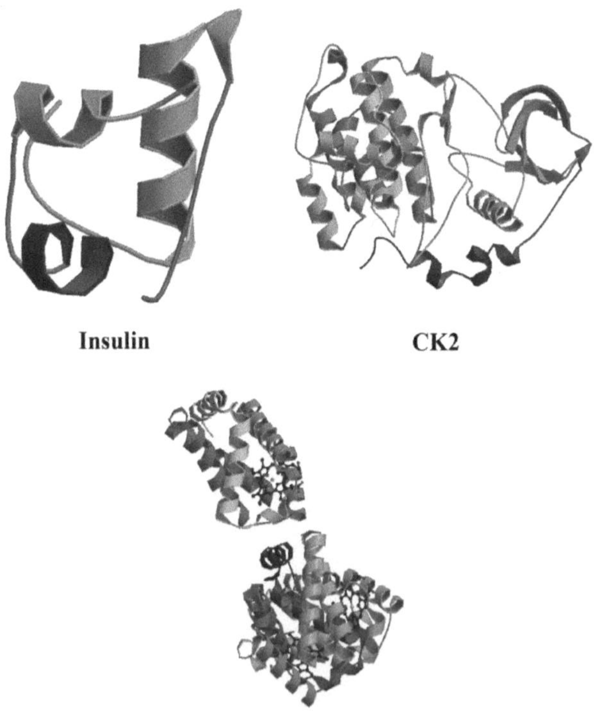

Figure 5.4 Langmuir–Blodgett (LB)-grown crystals of proteins difficult to crystallize using classical hanging-drop vapor diffusion crystallization approach (namely, insulin, casein kinase 2 or CK2, and oxygen-bound Hell's gate globin I).

In addition, nanocrystallography could also be useful for creating arrays at a nanoscale, which, until now, have been based on lithographic techniques, using protein crystals for the construction of next-generation electronic and photonic devices (Nicolini & Pechkova, 2010a,2010b; Nicolini, Adami, et al., 2012; Nicolini, Bezerra, et al., 2012, Nicolini, Bragazzi, & Pechkova, 2012; Pechkova & Nicolini, 2004). LB-immobilized enzymes may indeed have important industrial applications in the field of biocatalysis (Nicolini, Bruzzese, Sivozhelezov, & Pechkova, 2008; Nicolini, Adami, et al., 2012; Nicolini, Bezerra, et al., 2012), as well as diagnostics tools (Nicolini, Adami, et al., 2012; Nicolini, Bezerra, et al., 2012).

REFERENCES

Al-Hussein, M., Schindler, M., Ruderer, M. A., Perlich, J., Schwartzkopf, M., Herzog, G., et al. (2013). In situ X-ray study of the structural evolution of gold nano-domains by spray deposition on thin conductive P3HT films. *Langmuir*, *29*(8), 2490–2497.

Arya, S. K., Datta, M., & Malhotra, B. D. (2008). Recent advances in cholesterol biosensor. *Biosensors & Bioelectronics*, *23*(7), 1083–1100.

Azia, A., Unger, R., & Horovitz, A. (2012). What distinguishes GroEL substrates from other Escherichia coli proteins? *The FEBS Journal*, *279*(4), 543–550.

Bass, M., Berman, A., Singh, A., Konovalov, O., & Freger, V. (2010). Surface structure of Nafion in vapor and liquid. *The Journal of Physical Chemistry. B*, *114*(11), 3784–3790.

Belmonte, L., Pechkova, E., Tripathi, S., Scudieri, D., & Nicolini, C. (2012). Langmuir-Blodgett nanotemplate and radiation resistance in protein crystals: State of the art. *Critical Reviews in Eukaryotic Gene Expression*, *22*(3), 219–232.

Berweger, S., Neacsu, C. C., Mao, Y., Zhou, H., Wong, S. S., & Raschke, M. B. (2009). Optical nanocrystallography with tip-enhanced phonon Raman spectroscopy. *Nature Nanotechnology*, *4*(8), 496–499.

Bian, Y., Ye, M., Wang, C., Cheng, K., Song, C., Dong, M., et al. (2013). Global screening of CK2 kinase substrates by an integrated phosphoproteomics workflow. *Scientific Reports*, *3*, 3460.

Bozdaganyan, M., Bragazzi, N. L., Pechkova, E., Shaytan, K., & Nicolini, C. (2014). Identification of best protein crystallization methods by molecular dynamics. *Critical Reviews in Eukaryotic Gene Expression*, in press.

Bragazzi, N. L., Pechkova, E., Scudieri, D., Terencio, T. B., Adami, M., & Nicolini, C. (2012). Recombinant laccase: II. Medical biosensor. *Critical Reviews in Eukaryotic Gene Expression*, *22*(3), 197–203.

Buljan, M., Radić, N., Bernstorff, S., Dražić, G., Bogdanović-Radović, I., & Holý, V. (2012). Grazing-incidence small-angle X-ray scattering: Application to the study of quantum dot lattices. *Acta Crystallographica. Section A*, *68*(Pt. 1), 124–138.

Chan, M. K. (2001). Recent advances in heme-protein sensors. *Current Opinion in Chemical Biology*, *5*(2), 216–222.

Chayen, N. E. (2003). Protein crystallization for genomics: Throughput versus output. *Journal of Structural and Functional Genomics*, *4*(2–3), 115–120.

Chayen, N. E. (2004). Turning protein crystallisation from an art into a science. *Current Opinion in Structural Biology*, *14*(5), 577–583.

Chayen, N. E. (2007). Optimization techniques for automation and high throughput. *Methods in Molecular Biology*, *363*, 175–190.

Chayen, N. E. (2009). High-throughput protein crystallization. *Advances in Protein Chemistry and Structural Biology*, 77, 1–22.

Chayen, N. E., & Saridakis, E. (2002). Protein crystallization for genomics: Towards high-throughput optimization techniques. *Acta Crystallographica Section D, Biological Crystallography*, *58*(Pts. 2,6), 921–927.

Chayen, N. E., & Saridakis, E. (2008). Protein crystallization: From purified protein to diffraction-quality crystal. *Nature Methods*, *5*(2), 147–153.

Chen, J. P., & Millane, R. P. (2013). Diffraction by nanocrystals. *Journal of the Optical Society of America. A, Optics, Image Science, and Vision*, *30*(12), 2627–2634.

Cheng, X., & Ivanov, I. (2012). Molecular dynamics. *Methods in Molecular Biology*, *929*, 243–285.

Chernov, A. A. (2003). Protein crystals and their growth. *Journal of Structural Biology*, *142*(1), 3–21.

Collison, M. E., & Meyerhoff, M. E. (1990). Chemical sensors for bedside monitoring of critically ill patients. *Analytical Chemistry*, *62*(7), 425A–437A.

Cozza, G., Pinna, L. A., & Moro, S. (2012). Protein kinase CK2 inhibitors: A patent review. *Expert Opinion on Therapeutic Patents*, *22*(9), 1081–1097.

Delucas, L. J., Hamrick, D., Cosenza, L., Nagy, L., McCombs, D., Bray, T., et al. (2005). Protein crystallization: Virtual screening and optimization. *Progress in Biophysics and Molecular Biology*, *88*(3), 285–309.

Drews, J. (2000). Drug discovery: A historical perspective. *Science*, *287*(5460), 1960–1964.

Dovat, S., Song, C., Payne, K. J., & Li, Z. (2011). Ikaros, CK2 kinase, and the road to leukemia. *Molecular and Cellular Biochemistry*, *356*(1–2), 201–207.

Garman, E. F., & Nave, C. (2009). Radiation damage in protein crystals examined under various conditions by different methods. *Journal of Synchrotron Radiation*, *16*(Pt 2), 129–132.

Garman, E. F., & Owen, R. L. (2006). Cryocooling and radiation damage in macromolecular crystallography. *Acta Crystallographica. Section D, Biological Crystallography*, *62*(Pt. 1), 32–47.

Gebhardt, R., & Kulozik, U. (2014). Simulation of the shape and size of casein micelles in a film state. *Food & Function*, *5*(4), 780–785.

Gebhardt, R., Pechkova, E., Riekel, C., & Nicolini, C. (2010). In situ muGISAXS: II. Thaumatin crystal growth kinetic. *Biophysical Journal*, *99*(4), 1262–1267.

Gebhardt, R., Vendrely, C., & Kulozik, U. (2011). Structural characterization of casein micelles: Shape changes during film formation. *Journal of Physics. Condensed Matter*, *23*(44), 444201.

Giegé, R. (2013). A historical perspective on protein crystallization from 1840 to the present day. *The FEBS Journal*, *280*(24), 6456–6497.

Giroux, V., Dagorn, J. C., & Iovanna, J. L. (2009). A review of kinases implicated in pancreatic cancer. *Pancreatology*, *9*(6), 738–754.

Glassford, S. E., Byrne, B., & Kazarian, S. G. (2013). Recent applications of ATR FTIR spectroscopy and imaging to proteins. *Biochimica et Biophysica Acta*, *1834*(12), 2849–2858.

Gucev, Z. S., Tee, M. K., Chitayat, D., Wherrett, D. K., & Miller, W. L. (2013). Distinguishing deficiencies in the steroidogenic acute regulatory protein and the cholesterol side chain cleavage enzyme causing neonatal adrenal failure. *The Journal of Pediatrics*, *162*(4), 819–822.

Hagiwara, M., Kurita-Ochiai, T., Kobayashi, R., Hashizume-Takizawa, T., Yamazaki, K., & Yamamoto, M. (2014). Sublingual vaccine with GroEL attenuates atherosclerosis. *Journal of Dental Research*, *93*(4), 382–387.

Haris, P. I. (2013). Probing protein-protein interaction in biomembranes using Fourier transform infrared spectroscopy. *Biochimica et Biophysica Acta*, *1828*(10), 2265–2271.

Heinemann, L. (2012). Biosimilar insulins. *Expert Opinion on Biological Therapy*, *12*(8), 1009–1016.

Helliwell, J. R., & Chayen, N. E. (2007). Crystallography: A down-to-earth approach. *Nature, 448*(7154), 658–659.

Hollingsworth, T. J., & Gross, A. K. (2012). Defective trafficking of rhodopsin and its role in retinal degenerations. *International Review of Cell and Molecular Biology, 293*, 1–44.

Holton, J. M. (2009). A beginner's guide to radiation damage. *Journal of Synchrotron Radiation, 16*(Pt. 2), 133–142.

Johnson, S. M., Sharif, O., Mak, P. A., Wang, H. T., Engels, I. H., Brinker, A., et al. (2014). A biochemical screen for GroEL/GroES inhibitors. *Bioorganic & Medicinal Chemistry Letters, 24*(3), 786–789.

Judge, R. A., Snell, E. H., & van der Woerd, M. J. (2005). Extracting trends from two decades of microgravity macromolecular crystallization history. *Acta Crystallographica. Section D, Biological Crystallography, 61*(Pt. 6), 763–771.

Kang, H. J., Lee, C., & Drew, D. (2013). Breaking the barriers in membrane protein crystallography. *The International Journal of Biochemistry & Cell Biology, 45*(3), 636–644.

Kazarian, S. G., & Ewing, A. V. (2013). Applications of Fourier transform infrared spectroscopic imaging to tablet dissolution and drug release. *Expert Opinion on Drug Delivery, 10*(9), 1207–1221.

Kerrigan, J. E. (2013). Molecular dynamics simulations in drug design. *Methods in Molecular Biology, 993*, 95–113.

Khurshid, S., & Chayen, N. E. (2006). Upside-down protein crystallization: Designing microbatch experiments for microgravity. *Annals of the New York Academy of Sciences, 1077*, 208–213.

Lambeth, J. D. (1986). Cytochrome P-450scc a review of the specificity and properties of the cholesterol binding site. *Endocrine Research, 12*(4), 371–392.

Li, L., & Ismagilov, R. F. (2010). Protein crystallization using microfluidic technologies based on valves, droplets, and SlipChip. *Annual Review of Biophysics, 39*, 139–158.

Lodowski, D. T., Angel, T. E., & Palczewski, K. (2009). Comparative analysis of GPCR crystal structures. *Photochemistry and Photobiology, 85*(2), 425–430.

Manuel García-Ruiz, J. (2003). Nucleation of protein crystals. *Journal of Structural Biology, 142*(1), 22–31.

Mendes, H. F., Zaccarini, R., & Cheetham, M. E. (2010). Pharmacological manipulation of rhodopsin retinitis pigmentosa. *Advances in Experimental Medicine and Biology, 664*, 317–323.

Metwalli, E., Körstgens, V., Schlage, K., Meier, R., Kaune, G., Buffet, A., et al. (2013). Cobalt nanoparticles growth on a block copolymer thin film: A time-resolved GISAXS study. *Langmuir, 29*(21), 6331–6340.

Miller, L. M., Bourassa, M. W., & Smith, R. J. (2013). FTIR spectroscopic imaging of protein aggregation in living cells. *Biochimica et Biophysica Acta, 1828*(10), 2339–2346.

Moraes, I., Evans, G., Sanchez-Weatherby, J., Newstead, S., & Stewart, P. D. (2014). Membrane protein structure determination—The next generation. *Biochimica et Biophysica Acta, 1838*(1 Pt. A), 78–87.

Moukhametzianov, R., Burghammer, M., Edwards, P. C., Petitdemange, S., Popov, D., Fransen, M., et al. (2008). Protein crystallography with a micrometre-sized synchrotron-radiation beam. *Acta Crystallographica. Section D, Biological Crystallography, 64*(Pt 2), 158–166.

Müller-Buschbaum, P. (2003). Grazing incidence small-angle X-ray scattering: An advanced scattering technique for the investigation of nanostructured polymer films. *Analytical and Bioanalytical Chemistry, 376*(1), 3–10.

Nicolini, C., Adami, M., Sartore, M., Bragazzi, N. L., Bavastrello, V., Spera, R., et al. (2012). Prototypes of newly conceived inorganic and biological sensors for health and environmental applications. *Sensors (Basel, Switzerland), 12*(12), 17112–17127.

Nicolini, C., Belmonte, L., Maksimov, G., Brazhe, N., & Pechkova, E. (2013). In situ monitoring by raman spectroscopy of lysozyme conformation during "nanotemplate" induced crystallization. *Journal of Microbial & Biochemical Technology*, *6*, 009–016.

Nicolini, C., Belmonte, L., Riekel, C., Koenig, C., & Pechkova, E. (2014). Langmuir-Blodgett nanotemplate crystallization combined to laser micro-fragmentation uniquely characterize protein crystals by synchrotron micro-diffraction. *American Journal of Biochemistry and Biotechnology*, *10*(1), 22–30.

Nicolini, C., Bezerra, T., & Pechkova, E. (2012). Protein nanotechnology for the new design and development of biocrystals and biosensors. *Nanomedicine (London, England)*, 7(8), 1117–1120.

Nicolini, C., Bragazzi, N., & Pechkova, E. (2012). Nanoproteomics enabling personalized nanomedicine. *Advanced Drug Delivery Reviews*, *64*(13), 1522–1531.

Nicolini, C., Bragazzi, N., & Pechkova, E. (2013). From nanobiotechnology to organic and biological monitoring of health and environment for biosafety. *Journal of Bioanalysis and Biomedicine*, *5*, 108–117.

Nicolini, C., Bragazzi, N. L., Pechkova, E., & Lazzari, R. (2014). Ab initio semi-quantitative analysis of micro-beam grazing-incidence small-angle X-ray scattering (M-GISAXS) during protein crystal nucleation and growth. *Journal of Proteomics & Bioinformatics*, 7, 064–070.

Nicolini, C., Bruzzese, D., Cambria, M. T., Bragazzi, N. L., & Pechkova, E. (2013). Recombinant laccase: I. Enzyme cloning and characterization. *Journal of Cellular Biochemistry*, *114*(3), 599–605.

Nicolini, C., Bruzzese, D., Sivozhelezov, V., & Pechkova, E. (2008). Langmuir-Blodgett based lipase nanofilms of unique structure-function relationship. *Biosystems*, *94*(3), 228–232.

Nicolini, C., Correia, T. B., Stura, E., Larosa, C., Spera, R., & Pechkova, E. (2013). Atomic force microscopy and anodic porous allumina of nucleic acid programmable protein arrays. *Recent Patents on Biotechnology*, 7(2), 112–121.

Nicolini, C., & Pechkova, E. (2004). Nanocrystallography: An emerging technology for structural proteomics. *Expert Review of Proteomics*, *1*(3), 253–256.

Nicolini, C., & Pechkova, E. (2006). Nanostructured biofilms and biocrystals. *Journal of Nanoscience and Nanotechnology*, *6*(8), 2209–2236.

Nicolini, C., & Pechkova, E. (2010a). Nanoproteomics for nanomedicine. *Nanomedicine (London, England)*, *5*(5), 677–682.

Nicolini, C., & Pechkova, E. (2010b). An overview of nanotechnology-based functional proteomics for cancer and cell cycle progression. *Anticancer Research*, *30*(6), 2073–2080.

Otálora, F., Gavira, J. A., Ng, J. D., & García-Ruiz, J. M. (2009). Counterdiffusion methods applied to protein crystallization. *Progress in Biophysics and Molecular Biology*, *101*(1–3), 26–37.

Palczewski, K. (2012). Chemistry and biology of vision. *The Journal of Biological Chemistry*, *287*(3), 1612–1619.

Palm, F., Nordquist, L., & Buerk, D. G. (2007). Nitric oxide in the kidney; direct measurements of bioavailable renal nitric oxide. *Advances in Experimental Medicine and Biology*, *599*, 117–123.

Paternolli, C., Ghisellini, P., & Nicolini, C. (2007). Nanostructuring of heme-proteins for biodevice applications. *IET Nanobiotechnology*, *1*(2), 22–26.

Paternolli, C., Neebe, M., Stura, E., Barbieri, F., Ghisellini, P., Hampp, N., et al. (2009). Photoreversibility and photostability in films of octopus rhodopsin isolated from octopus photoreceptor membranes. *Journal of Biomedical Materials Research. Part A*, *88*(4), 947–951.

Péchiné, S., Hennequin, C., Boursier, C., Hoys, S., & Collignon, A. (2013). Immunization using GroEL decreases Clostridium difficile intestinal colonization. *PLoS One*, *8*(11), e81112.

Pechkova, E., Bragazzi, N. L., Bozdaganyan, M., Belmonte, L., & Nicolini, C. (2014). A review of the strategies for obtaining high quality crystals utilizing nanotechnologies and space. *Critical Reviews in Eukaryotic Gene Expression*, in press.

Pechkova, E., Gebhardt, R., Riekel, C., & Nicolini, C. (2010). In situ muGISAXS: I. Experimental setup for submicron study of protein nucleation and growth. *Biophysical Journal, 99*(4), 1256–1261.

Pechkova, E., & Nicolini, C. (2002). Protein nucleation and crystallization by homologous protein thin film template. *Journal of Cellular Biochemistry, 85*(2), 243–251.

Pechkova, E., & Nicolini, C. (2004). Protein nanocrystallography: A new approach to structural proteomics. *Trends in Biotechnology, 22*(3), 117–122.

Pechkova, E., & Nicolini, C. (2006). Structure and growth of ultrasmall protein microcrystals by synchrotron radiation: II. microGISAX and microscopy of lysozyme. *Journal of Cellular Biochemistry, 97*(3), 553–560.

Pechkova, E., & Nicolini, C. (2011). In situ study of nanotemplate-induced growth of lysozyme microcrystals by submicrometer GISAXS. *Journal of Synchrotron Radiation, 18*(Pt. 2), 287–292.

Pechkova, E., & Nicolini, C. (2010). Domain organization and properties of LB lysozyme crystals down to submicron size. *Anticancer Research, 30*(7), 2745–2748.

Pechkova, E., Roth, S. V., Burghammer, M., Fontani, D., Riekel, C., & Nicolini, C. (2005). microGISAXS and protein nanotemplate crystallization: Methods and instrumentation. *Journal of Synchrotron Radiation, 12*(Pt. 6), 713–716.

Pechkova, E., Sartore, M., Giacomelli, L., & Nicolini, C. (2007). Atomic force microscopy of protein films and crystals. *The Review of Scientific Instruments, 78*(9), 093704.

Pechkova, E., Scudieri, D., Belmonte, L., & Nicolini, C. (2012). Oxygen-bound Hell's gate globin I by classical versus LB nanotemplate method. *Journal of Cellular Biochemistry, 113*(7), 2543–2548.

Pechkova, E., Sivozhelezov, V., Belmonte, L., & Nicolini, C. (2012). Unique water distribution of Langmuir-Blodgett versus classical crystals. *Journal of Structural Biology, 180*(1), 57–64.

Pechkova, E., Sivozhelezov, V., & Nicolini, C. (2007). Protein thermal stability: The role of protein structure and aqueous environment. *Archives of Biochemistry and Biophysics, 466*(1), 40–48.

Pechkova, E., Tripathi, S., & Nicolini, C. (2009). MicroGISAXS of Langmuir-Blodgett protein films: Effect of temperature on long-range order. *Journal of Synchrotron Radiation, 16*(Pt. 3), 330–335.

Pechkova, E., Tripathi, S., Ravelli, R. B., McSweeney, S., & Nicolini, C. (2009). Radiation stability of proteinase K crystals grown by LB nanotemplate method. *Journal of Structural Biology, 168*(3), 409–418.

Pechkova, E., Tripathi, S., Spera, R., & Nicolini, C. (2008). Groel crystal growth and characterization. *Biosystems, 94*(3), 223–227.

Pechkova, E., Tropiano, G., Riekel, C., & Nicolini, C. (2004). Radiation stability of protein crystals grown by nanostructured templates: Synchrotron microfocus analysis. *Spectrochimica Acta, Part B: Atomic Spectroscopy, 59*(10–11), 1687–1693.

Pechkova, E., Vasile, F., Spera, R., Fiordoro, S., & Nicolini, C. (2005). Protein nanocrystallography: Growth mechanism and atomic structure of crystals induced by nanotemplates. *Journal of Synchrotron Radiation, 12*(Pt. 6), 772–778.

Pechkova, E., Zanotti, G., & Nicolini, C. (2003). Three-dimensional atomic structure of a catalytic subunit mutant of human protein kinase CK2. *Acta Crystallographica. Section D, Biological Crystallography, 59*(Pt. 12), 2133–2139.

Perutz, M. F., Paoli, M., & Lesk, A. M. (1999). Fix L, a haemoglobin that acts as an oxygen sensor: Signalling mechanism and structural basis of its homology with PAS domains. *Chemistry & Biology, 6*(11), R291–R297.

Piazza, F., Manni, S., & Semenzato, G. (2013). Novel players in multiple myeloma pathogenesis: Role of protein kinases CK2 and GSK3. *Leukemia Research, 37*(2), 221–227.

Richard, M. I., Schülli, T. U., Renaud, G., Zhong, Z. Z., & Bauer, G. (2011). A combined in situ grazing incidence small angle X-ray scattering and grazing incidence X-ray diffraction study of the growth of Ge Islands on pit-patterned Si(001) substrates. *Journal of Nanoscience and Nanotechnology*, *11*(10), 9123–9128.

Riekel, C. (2004). Recent developments in micro-diffraction on protein crystals. *Journal of Synchrotron Radiation*, *11*(Pt. 1), 4–6.

Riekel, C., Burghammer, M., & Popov, D. (2011). Protein micro- and nanocrystallography using synchrotron radiation. In E. Pechkova & C. Riekel (Eds.), *Synchrotron radiation and structural proteomics* (pp. 3–30): London–New York–Singapore: Pan Stanford Publishing. ISBN: 978-981-4267-38-0 eBook ISBN:978-981-4267-93-9.

Riekel, C., Burghammer, M., & Schertler, G. (2005). Protein crystallography microdiffraction. *Current Opinion in Structural Biology*, *15*(5), 556–562.

Saridakis, E., & Chayen, N. E. (2009). Towards a 'universal' nucleant for protein crystallization. *Trends in Biotechnology*, *27*(2), 99–106.

Saridakis, E., & Chayen, N. E. (2013). Imprinted polymers assisting protein crystallization. *Trends in Biotechnology*, *31*(9), 515–520.

Saridakis, E., Khurshid, S., Govada, L., Phan, Q., Hawkins, D., Crichlow, G. V., et al. (2011). Protein crystallization facilitated by molecularly imprinted polymers. *Proceedings of the National Academy of Sciences of the United States of America*, *108*(27), 11081–11086.

Sarno, S., & Pinna, L. A. (2008). Protein kinase CK2 as a druggable target. *Molecular BioSystems*, *4*(9), 889–894.

Schlichting, I., & Miao, J. (2012). Emerging opportunities in structural biology with X-ray free-electron lasers. *Current Opinion in Structural Biology*, *22*(5), 613–626.

Schöpe, H. J., & Wette, P. (2011). Seed- and wall-induced heterogeneous nucleation in charged colloidal model systems under microgravity. *Physical Review. E, Statistical, Nonlinear, and Soft Matter Physics*, *83*(5 Pt. 1), 051405.

Shichi, H., & Rafferty, C. N. (1980). The molecular aspects of visual photoreceptors. *Photochemistry and Photobiology*, *31*(6), 631–639.

Siódmiak, J., Gadomski, A., Pechkova, E., & Nicolini, C. (2006). Computer model of a lysozyme crystal growth with/without nanotemplate—A comparison. *International Journal of Modern Physics. C, Physics and Computers*, *17*(09), 1359–1366.

Sivozhelezov, V., & Nicolini, C. (2005). Homology modeling of cytochrome P450scc and the mutations for optimal amperometric sensor. *Journal of Theoretical Biology*, *234*(4), 479–485.

Sivozhelezov, V., & Nicolini, C. (2006). Theoretical framework for octopus rhodopsin crystallization. *Journal of Theoretical Biology*, *240*(2), 260–269.

Sivozhelezov, V., Pechkova, E., & Nicolini, C. (2006). Mapping electrostatic potential of a protein on its hydrophobic surface: Implications for crystallization of Cytochrome P450scc. *Journal of Theoretical Biology*, *241*(1), 73–80.

Smith, S. O. (2010). Structure and activation of the visual pigment rhodopsin. *Annual Review of Biophysics*, *39*, 309–328.

Smith, J. L., Fischetti, R. F., & Yamamoto, M. (2012). Micro-crystallography comes of age. *Current Opinion in Structural Biology*, *22*(5), 602–612.

Spera, R., Festa, F., Bragazzi, N. L., Pechkova, E., LaBaer, J., & Nicolini, C. (2013). Conductometric monitoring of protein-protein interactions. *Journal of Proteome Research*, *12*(12), 5535–5547.

Stevens, R. C. (2000). High-throughput protein crystallization. *Current Opinion in Structural Biology*, *10*(5), 558–563.

Tee, M. K., Abramsohn, M., Loewenthal, N., Harris, M., Siwach, S., Kaplinsky, A., et al. (2013). Varied clinical presentations of seven patients with mutations in CYP11A1 encoding the cholesterol side-chain cleavage enzyme, P450scc. *The Journal of Clinical Endocrinology and Metabolism*, *98*(2), 713–720.

Teh, A. H., Saito, J. A., Baharuddin, A., Tuckerman, J. R., Newhouse, J. S., Kanbe, M., et al. (2011). Hell's Gate globin I: An acid and thermostable bacterial hemoglobin resembling mammalian neuroglobin. *FEBS Letters*, *585*(20), 3250–3258.

Trembley, J. H., Wang, G., Unger, G., Slaton, J., & Ahmed, K. (2009). Protein kinase CK2 in health and disease: CK2: A key player in cancer biology. *Cellular and Molecular Life Sciences*, *66*(11–12), 1858–1867.

Turzhitsky, V., Qiu, L., Itzkan, I., Novikov, A. A., Kotelev, M. S., Getmanskiy, M., et al. (2014). Spectroscopy of scattered light for the characterization of micro and nanoscale objects in biology and medicine. *Applied Spectroscopy*, *68*(2), 133–154.

Ubarretxena-Belandia, I., & Stokes, D. L. (2010). Present and future of membrane protein structure determination by electron crystallography. *Advances in Protein Chemistry and Structural Biology*, *81*, 33–60.

Uhlmann, P., Varnik, F., Truman, P., Zikos, G., Moulin, J. F., Müller-Buschbaum, P., et al. (2011). Microfluidic emulsion separation-simultaneous separation and sensing by multilayer nanofilm structures. *Journal of Physics. Condensed Matter*, *23*(18), 184123.

Unger, V. M., Hargrave, P. A., Baldwin, J. M., & Schertler, G. F. (1997). Arrangement of rhodopsin transmembrane alpha-helices. *Nature*, *389*(6647), 203–206.

Unger, V. M., & Schertler, G. F. (1995). Low resolution structure of bovine rhodopsin determined by electron cryo-microscopy. *Biophysical Journal*, *68*(5), 1776–1786.

Urban, J. J., Talapin, D. V., Shevchenko, E. V., & Murray, C. B. (2006). Self-assembly of PbTe quantum dots into nanocrystal superlattices and glassy films. *Journal of the American Chemical Society*, *128*(10), 3248–3255.

Vergara, A., Lorber, B., Sauter, C., Giegé, R., & Zagari, A. (2005). Lessons from crystals grown in the advanced protein crystallisation facility for conventional crystallisation applied to structural biology. *Biophysical Chemistry*, *118*(2–3), 102–112.

Wagner, N. L., Greco, J. A., Ranaghan, M. J., & Birge, R. R. (2013). Directed evolution of bacteriorhodopsin for applications in bioelectronics. *Journal of the Royal Society, Interface*, *10*(84), 20130197.

Wakayama, N. I., Yin, D. C., Harata, K., Kiyoshi, T., Fujiwara, M., & Tanimoto, Y. (2006). Macromolecular crystallization in microgravity generated by a superconducting magnet. *Annals of the New York Academy of Sciences*, *1077*, 184–193.

Wickenheisser, J. K., Biegler, J. M., Nelson-Degrave, V. L., Legro, R. S., Strauss, J. F., 3rd., & McAllister, J. M. (2012). Cholesterol side-chain cleavage gene expression in theca cells: Augmented transcriptional regulation and mRNA stability in polycystic ovary syndrome. *PLoS One*, 7(11), e48963.

Xu, M. Q., Wu, J. F., & Zhao, G. C. (2013). Direct electrochemistry of hemoglobin at a graphene gold nanoparticle composite film for nitric oxide biosensing. *Sensors (Basel, Switzerland)*, *13*(6), 7492–7504.

Yagi-Utsumi, M., Kunihara, T., Nakamura, T., Uekusa, Y., Makabe, K., Kuwajima, K., et al. (2013). NMR characterization of the interaction of GroEL with amyloid β as a model ligand. *FEBS Letters*, *587*(11), 1605–1609.

Zheng, B., Gerdts, C. J., & Ismagilov, R. F. (2005). Using nanoliter plugs in microfluidics to facilitate and understand protein crystallization. *Current Opinion in Structural Biology*, *15*(5), 548–555.

CHAPTER SIX

Modern Mass Spectrometry-Based Structural Proteomics

Evgeniy V. Petrotchenko[*], **Christoph H. Borchers**[*,†,1]
[*]University of Victoria—Genome British Columbia Proteomics Centre, Victoria, British Columbia, Canada
[†]Department of Biochemistry and Microbiology, University of Victoria, Petch Building Room 207, Victoria, British Columbia, Canada
[1]Corresponding author: e-mail address: christoph@proteincentre.com

Contents

Abstract

Recent developments in the modern mass spectrometry of proteins and peptides have resulted in significant progress in structural proteomics techniques for studying protein structure. A variety of protein structural questions, ranging from defining protein interaction networks to the study of conformational changes and the structure of single proteins, can be addressed using multiple mass spectrometry-based structural proteomics approaches. Each technique provides specific structural information which can be used as experimental structural constraints in protein structure modeling. Here, we describe recent developments in limited proteolysis, surface modification, hydrogen–deuterium exchange, ion mobility, and cross-linking—all combined with modern mass spectrometric techniques—for the studying protein structure.

Advances in Protein Chemistry and Structural Biology, Volume 95
ISSN 1876-1623
http://dx.doi.org/10.1016/B978-0-12-800453-1.00006-3

1. INTRODUCTION

The knowledge of protein structures is crucial for understanding of the functioning of biological systems in healthy and disease states. The recent revolution in mass spectrometry-based proteomics has revived an interest in the traditional protein chemistry methods for studying protein structure. The combination of methods such as limited proteolysis, chemical surface modification, hydrogen–deuterium exchange (HDX), cross-linking, and affinity labeling with mass spectrometry has led to the new field of structural proteomics. Each of these methods provides specific and unique information on the protein system studied. For example, limited proteolysis can indicate portions of folded proteins that are accessible to a large probe (a proteolytic enzyme) and must therefore be exposed to the solvent. Similarly, chemical surface modification can provide similar information on the accessibility of a particular amino acid residue to a smaller probe (a modification reagent). HDX of the amide protons in the peptide bonds can provide data on the hydrogen bonding status and accessibility, and, therefore, the presence of secondary structure elements in the protein sequence. Chemical cross-linkers of different lengths form a "molecular ruler" which can provide distances between the cross-linked residues (Fasold, Klappenberger, Meyer, & Remold, 1971; Green, Reisler, & Houk, 2001; Peters & Richards, 1977). Taken together, these methods provide a set of structural constraints on the folded protein or protein complex being studied. In this review, we describe the current status of these various structural proteomics methodologies, and their application to the elucidation of the structures of proteins and protein complexes, using the example of our recent study of prion protein conversion and aggregation (Serpa, Patterson, et al., 2013).

2. THE CONCEPT OF STRUCTURAL PROTEOMICS

The concept of structural proteomics is to obtain multiple types of experimental data to characterize the structure of the protein system studied (Konermann, Vahidi, & Sowole, 2014; Serpa et al., 2012). Traditionally, as for proteomics in general, structural proteomics implies the use of mass spectrometry as the major instrumental tool. The recent revolutionary developments in the mass spectrometric analysis of peptides and proteins, which made proteomics possible, have also had an effect on advancement of structural proteomics. This new level of technology has opened the way for the combination of traditional protein chemistry methods with novel mass spectrometric methods, allowing the characterization of the protein structure with amino acid residue resolution.

Structural proteomics can be applied to the study of protein systems of varying levels of complexity. Certain types of protein structural information can be obtained for each level of organization of the sample, that is, single proteins, binary protein complexes, multisubunit protein assemblies, proteome-wide protein interaction networks, organelles, cells, and tissues. Thus, for a single protein, structural proteomics can supply information at the amino acid residue level which is often impossible to obtain by other methods, including distance information between functional groups, the degree of a specific amino acid residue's exposure to the solvent, and its involvement in hydrogen bonding and secondary structure elements. These data can be used as constraints in the molecular modeling process for an unknown protein structure, or can be used to assess the dynamic and conformational changes of the protein. In the case of known interacting proteins, information pertaining to the spatial proximity of interacting groups and to the changes in the exposure of protein surfaces upon complex formation can be useful for elucidating details of the protein interaction interfaces. This information can also be used for establishing protein complex topologies in multisubunit protein assemblies. For entire proteome-scale applications, the identities and structural details of the interacting proteins and protein domains within protein complexes can be determined. In this review, we will primarily focus on our studies of the structures of single proteins and known protein complexes.

3. LIMITED PROTEOLYSIS

The limited proteolysis method is based on short controlled exposures of the protein to a proteolytic enzyme. The first cleavage of the protein occurs while the tertiary and quaternary structure of the protein complex should still be preserved, so the initial cleavage sites should be restricted to the outermost regions of the protein subunit surfaces—that is, those are accessible to the active site of the proteolytic enzyme. Because most enzymes are globular proteins with molecular weights of at least 15–20 kDa, the location of the cleavage sites will reflect their accessibility to a nearly spherical probe, whose diameter corresponds to the size of the enzyme used.

Several mass spectrometric approaches can be used for the determination of the cleavage sites. The most common approach is to first characterize the limited proteolysis reaction by SDS-PAGE. Serial time points for short (e.g., 1–5 min) exposures of the protein to the diluted proteolytic enzyme (e.g., 1:100 enzyme:substrate ratio) are used, and the reactions quenched and the products are separated by SDS-PAGE. The process of proteolysis can be visualized by time-wise appearance of the proteolytic fragments resulting from the enzymatic cleavage. The fragments that are first to appear can then be identified by in-gel digestion followed by peptide mapping, which will indicate the sites of cleavage (Fig. 6.1A). Alternatively, cleavage sites can be deduced from measuring the exact mass of the entire fragment by Orbitrap/FTICR mass spectrometry and top-down MS/MS analysis.

Using the former approach, we have characterized conformational changes occurring in course of native (PrPC) to aggregated state (PrPβ) conversion of prion proteins. We demonstrated increased protection of the K110 cleavage site in PrPβ compared to PrPC, suggesting intra- and/or inter-protein interactions; and increased cleavage at sites in the region of residues 149–156 in PrPβ, but not in PrPC, which indicates increased exposure of the hydrophobic residues in this region (Fig. 6.1B). These changes indicate unfolding or rearrangement of the C-terminal portion of the amphipathic helix 1 (H1) during the PrPC to PrPβ conversion process (Serpa, Patterson, et al., 2013).

4. SURFACE MODIFICATION

Surface modification provides information similar to that obtained from limited proteolysis approach (i.e., protein surface accessibility), but the determined accessibilities are to a smaller probe, in this case, a modification reagent. The basis of this method is a chemical reaction of the protein with a water-soluble modification reagent. Chemical modification of the protein surface thus allows the determination of which regions of the proteins are exposed to the solvent. Although the microenvironment can have a significant influence on the reactivity of amino acid residues, it is mainly those functional groups that are solvent exposed (i.e., which are located on the surface of the protein molecules) which will be modified with amino acid specific reagents. Identification of these modification sites thus indicates which amino acid residues are on the protein surfaces and are in contact with the solvent. The regions of the protein that are internal or are involved in formation of interprotein contacts are shielded from the modification reagent and consequently remain largely unmodified. Thus, analysis of the distribution of the chemical modification sites for a multisubunit protein

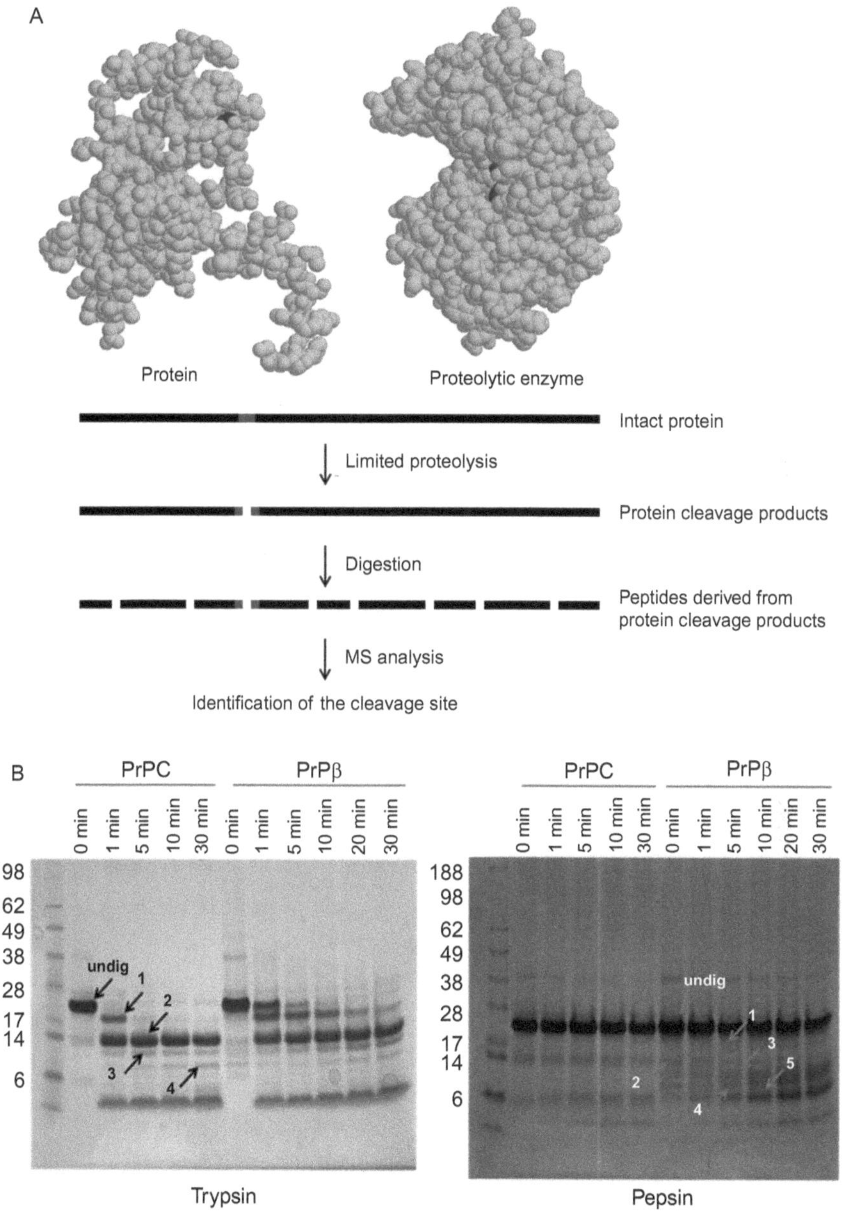

Figure 6.1 Limited proteolysis. (A) Principle of limited proteolysis cleavage-site determination by peptide mapping of the cleavage products. Following a short controlled exposure to the proteolytic enzyme, cleavage products are separated by SDS-PAGE and are in-gel digested. Peptide mapping of the cleavage products indicates the cleavage site. (B) An example of a differential limited proteolysis study of the native (PrPC) and pathological (PrPβ) states of the prion protein. Different patterns of limited proteolysis products were observed for the two states of the protein. Peptide mapping analysis of the cleavage products (indicated by the arrows) revealed that K110 and the aa149–156 region are differentially accessible in the two forms of the prion protein. *Reprinted from Serpa, Patterson, et al. (2013), with permission.*

complex indicates which regions of protein surfaces are solvent accessible and which are "shielded" or "protected" because they are either buried or are involved in protein interaction interfaces of the protein complexes.

This approach can be particularly informative for comparing two states of the protein, and can indicate changes in protection resulting from conformational changes and/or complex formation. In this type of differential experiment, surface modification is performed in parallel for two states of the protein and differences in reactivity of the particular amino acid residues, which can be quantified by mass spectrometry, will reflect their involvement in the protein's structural changes. Differential chemical surface modification can benefit greatly from the use of isotopically coded reagents which behave identically during mass spectrometric analysis. Light and heavy isotopic forms of the modification reagents, chemically identical, but different in mass, will produce modification products of different masses because of the mass differences in the stable isotopes employed. If the light form is used for one conformational state, and the heavy form is used for the other, combining the two samples before mass spectrometric analysis provides a convenient method for relative quantitation of the modification reaction yields for both samples in the same mass spectrum and under the same instrumental conditions (Fig. 6.2A).

We have also applied this approach for the characterization of the PrPC to PrPβ conversion. The isotopically coded water-soluble amine-reactive modification reagent PCASS-H4/-D4 (Fig. 6.2B), developed by our group, was used to quantitatively determine differences in specific amino acid reactivities between PrPC and PrPβ. Each form was modified with either the light or the heavy isotopic forms of the reagent so that differences in residue reactivities between the two prion isoforms could be determined from the ratios of the signal intensities of the light (H4) and heavy (D4) forms of the modified peptides. Several residues were found to be preferentially modified in the PrPC form: K110 (located on N-terminal portion of the protein), S132, S135, (located on the β1–H1 loop, residues 128–142), and K220, Y225, Y226, S231 on the C-terminal portion of H3 (residues 200–232). Changes in the reactivity of these residues, as a result of PrPC to PrPβ conversion, may indicate involvement of these regions in intra- and/or interprotein interactions within β-oligomers. The increased modification of residues on the H1–β2/H2–H3 interface (Y128, Y149, Y150, Y157, Y163, Y169) in PrPβ as compared to PrPC, also suggests a conformational change/rearrangement of this region, which was in good agreement with the limited proteolysis data (Serpa, Patterson, et al., 2013).

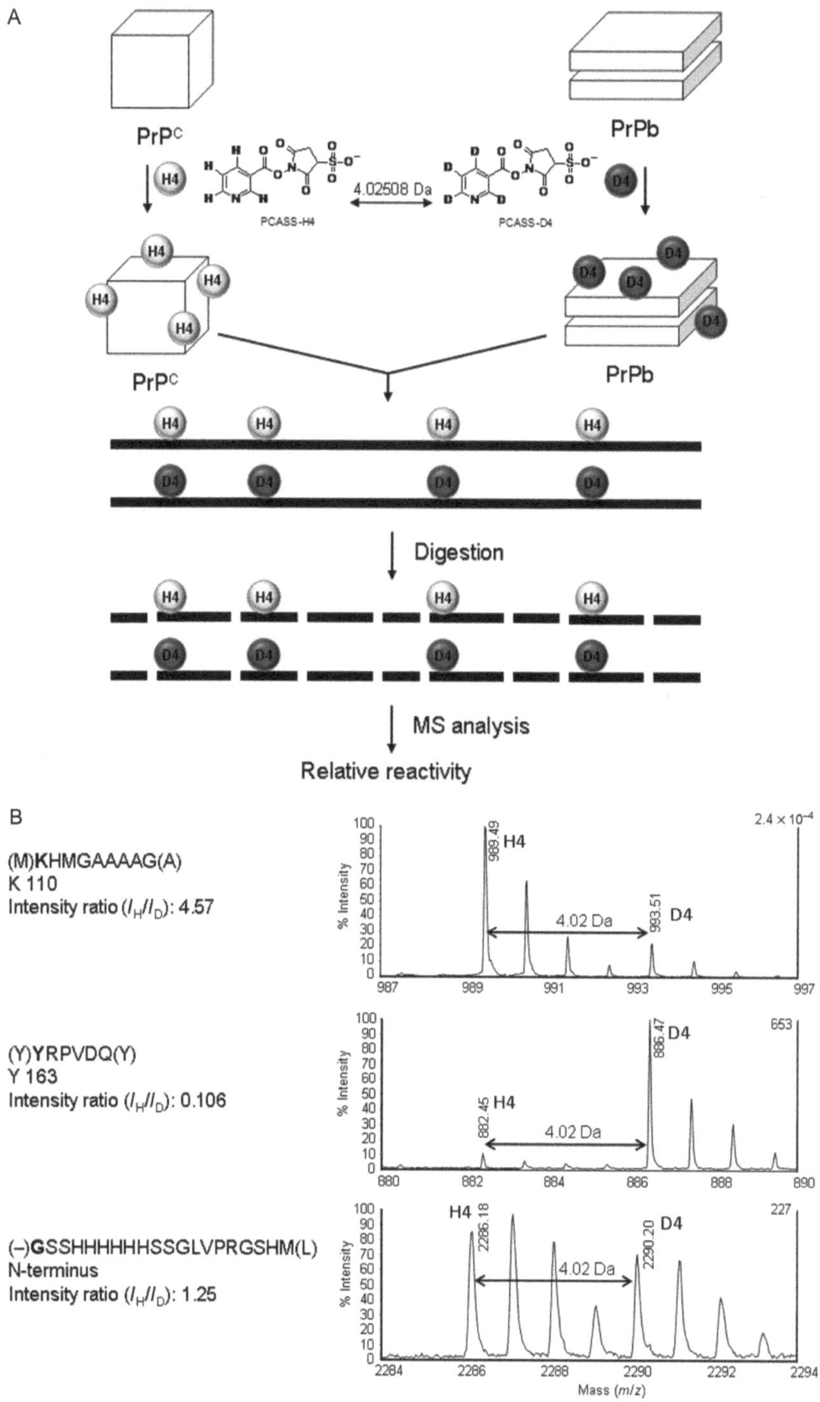
A
PrPC
PrPb
H4
D4
PCASS-H4
PCASS-D4
4.02508 Da
Digestion
MS analysis
Relative reactivity
B
(M)KHMGAAAAG(A)
K 110
Intensity ratio (I_H/I_D): 4.57
(Y)YRPVDQ(Y)
Y 163
Intensity ratio (I_H/I_D): 0.106
(–)GSSHHHHHHSSGLVPRGSHM(L)
N-terminus
Intensity ratio (I_H/I_D): 1.25
% Intensity
989.49
993.51
4.02 Da
2.4 × 10⁻⁴
882.45
886.47
653
2286.18
2290.20
227
Mass (m/z)

(Continued)

Recently, we have expanded this approach for the use of isotopically coded hydrogen peroxide ($H_2{}^{16}O_2$ and $H_2{}^{18}O_2$) as the modification reagent. This allowed us to obtain an additional complementary set of differentially modified methionine and tryptophan residues between PrPC and PrPβ (Serpa, Petrotchenko, Wishart, & Borchers, 2013), which was found to be in good agreement with PCASS modification results.

5. HYDROGEN–DEUTERIUM EXCHANGE

HDX is based on the principle that protein backbone hydrogens can be exchanged with deuterium upon exposure of a protein to a D_2O-based buffer. The exchange rates for individual peptide bond amide hydrogen atoms are dependent on the protein's structure: tightly hydrogen-bonded segments undergo very slow exchange, while disordered regions exchange much more rapidly. The hydrogen bonding of the amide hydrogen in the amide bond of a particular amino acid residue may indicate its involvement in secondary structure elements and/or exposure to the solvent. Short controlled immersion of the protein or protein complex into a D_2O solution will lead to the replacement of the exchangeable hydrogens on the protein surface with deuterium atoms from the solvent. Because deuterium is twice as heavy as hydrogen, the exchange can be readily detected and quantified by mass spectrometry. There are two general strategies to assess the location and extent of the exchange: bottom-up and top-down analysis. In the bottom-up approach, the protein is quickly digested, usually with pepsin under conditions of low pH and low temperature at which the peptide bond amide hydrogen exchange rate is minimal. The peptides produced are then analyzed by mass spectrometry to determine the relative amount of exchange

Figure 6.2—Cont'd Chemical surface modification. (A) Principle of differential surface modification amino acid residue reactivity determination using isotopically coded modification reagents. Proteins in two different states are modified with light and heavy isotopic forms of the reagent, respectively. Following quenching of the reaction, protein samples are combined, digested, and analyzed by mass spectrometry. Differentially modified peptides manifest in the MS spectra as pairs of signals separated by the mass difference between the light and heavy isotopic forms of the reagent used. The ratio of the intensities of the light and heavy forms of the peptides reflects the relative reactivities of the modified sites in the two states of the protein. (B) An example of a differential surface modification study of the prion protein in two conformational states, PrPC and PrPβ. The isotopically coded modification reagent PCASS-H4/D4 was employed. Several residues showed differential reactivities between two forms of the protein. *Reprinted from Serpa, Patterson, et al. (2013), with permission.*

that has occurred. In the top-down approach, the intact protein is exposed to time-controlled incubation in D_2O buffer, and is infused into mass spectrometer and analyzed by MS and MS/MS.

We have developed this top-down method in combination with electron-capture dissociation (ECD)-FTICR MS (Pan, Han, Borchers, & Konermann, 2008, 2009). ECD is a rapid fragmentation technique that produces selective fragmentation of peptide bonds, avoiding hydrogen scrambling (i.e., migration of the amide hydrogens along peptide chain), and produces an extensive series of c- and z-ions covering the protein sequence, with most fragments differing by a single residue. By comparing the masses of the consecutive fragments in both the c- and z-series, the exchange rate at nearly single residue resolution can be determined (Fig. 6.3A).

We have applied this approach to the analysis of both the secondary structures of intact proteins, as well as conformational changes, as in the case of the prion protein conversion from PrPC to PrPβ. The protein solution is continuously mixed in a capillary—first with D_2O, then with an acidic quenching solution, and then the solution is directly infused into the mass spectrometer. Using this approach, we determined that approximately 38 amides were protected from exchange in PrPC, while only 23 are protected in the misfolded PrPβ form. In other words, 15 amides became unprotected when PrP changed from the monomer to the oligomer. The region of deprotection was localized to residues 148–164, which is the stretch of the protein sequence encompassing H1–β2 (Fig. 6.3B). HDX deprotection in this region indicates the loss of the secondary structure (melting of H1, disassembly of the β-sheet involving the β2 strand) and/or disruption of the H1–β2/H2–H3 interface (Serpa, Patterson, et al., 2013).

6. CROSS-LINKING

The idea behind the use of cross-linking to determine a protein's structure is straightforward: to introduce new covalent bonds between pairs of functional groups in the protein in order to identify cross-linked sites and—based on the length of the cross-linking bridges formed—to deduce the distances between these cross-linked sites (Petrotchenko & Borchers, 2010a). These distances, in turn, can be used as constraints in the protein structure model-building process, and/or as characteristic features of the protein's conformational changes.

The workflow in a typical "bottom-up" mass spectrometry-based cross-linking experiment involves cross-linking the protein(s) of interest, optional

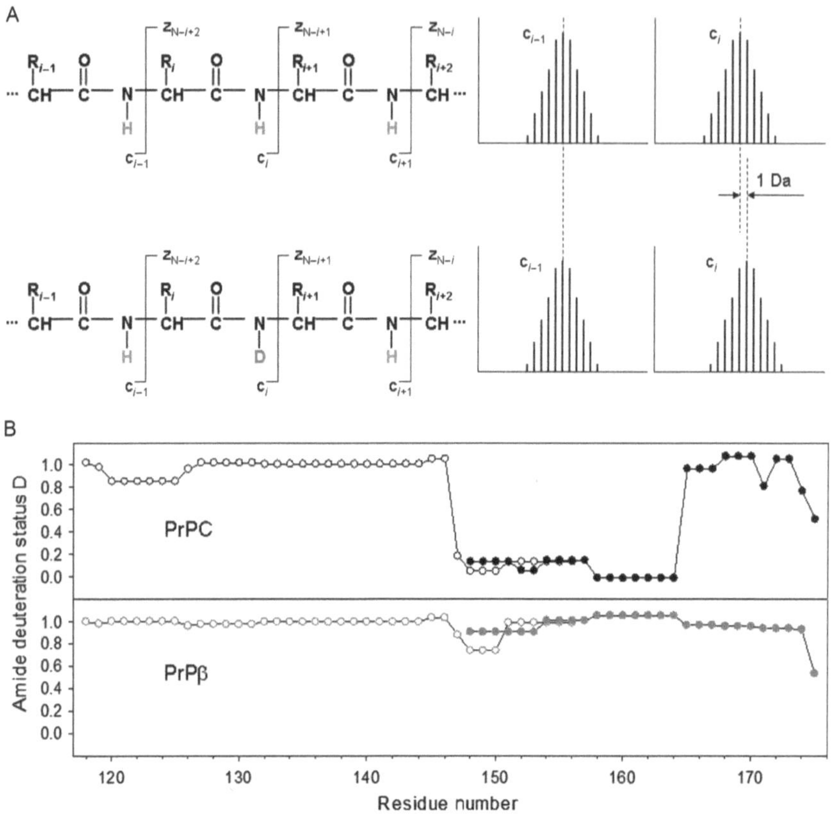

Figure 6.3 Hydrogen–deuterium exchange. (A) Principle of the determination of the amino acid residues deuteration status by top-down ECD–FTICR FTMS. Facile ECD-fragmentation produces scrambling-free series of the c- and z-series of fragments at nearly every peptide bond in the protein. Comparing mass shifts between consecutive fragments in the series allows to us estimate the degree of H/D exchange for every peptide bond amide. Little or no exchange for the amino acid residue would indicate involvement of its particular amide proton in hydrogen bonding. (B) Example of the differential hydrogen–deuterium exchange study of the two states of the prion protein. Hydrogen–deuterium exchange patterns, observed for the PrPC and PrPβ forms, indicates, that there is no significant difference in H-to-D exchange within the N- and C-terminal regions; however, there is a significant difference in exchange for fragments from the 148–164 regions. *Reprinted from Serpa, Patterson, et al. (2013), with permission.*

separation or purification of the cross-linked protein products, digestion of the cross-linked proteins into cross-linked and non-cross-linked peptides, and optional purification or enrichment of the cross-linked peptides (cross-links) (Fig. 6.4A). Finally, mass spectrometric analysis of the inter-peptide cross-links (two peptides connected by the cross-linker bridge) leads

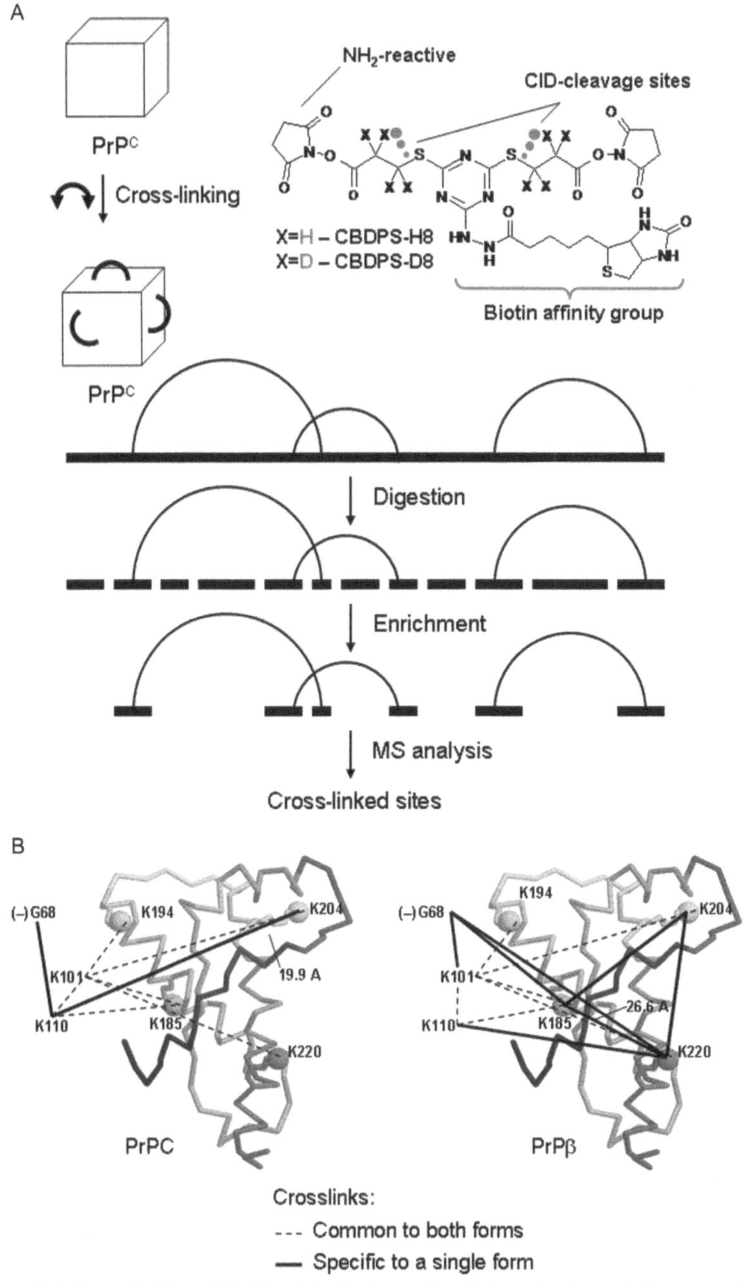

Figure 6.4 Cross-linking. (A) Principle behind the determination of the cross-linked sites. Cross-linked proteins are digested, cross-linked peptides are optionally enriched and analyzed by mass spectrometry. Several MS-oriented features of the cross-linking reagents facilitate detection and identification of the cross-links. Use of isotopically coded affinity-enrichable CID-cleavable reagent CBDPS-H8/D8 allows specific and sensitive detection and unambiguous identification of the cross-links. (B) Example of the differential cross-linking study of the native and aggregated states of prion protein. Several differential CBDPS cross-links are not compatible with the native structure of the protein and point to the nature of the conformational change, which leads to the aggregation. *Reprinted from Serpa, Patterson, et al. (2013), with permission.*

to the identification of the component peptides and the cross-linking sites. Alternatively, the isolation of a small cross-linked protein can be done inside the mass spectrometer, with the cross-linking sites being localized by top-down FTICR MS (Kruppa, Schoeniger, & Young, 2003; Novak & Giannakopulos, 2007; Novak, Young, Schoeniger, & Kruppa, 2003).

Despite the apparently straightforward nature of the chemical cross-linking approach, there are several significant challenges, such as low relative and absolute abundance of the cross-links, the combinatorial nature of the possible combinations of peptides that constitute each cross-link, and the generally higher molecular weight of these interpeptide cross-links. Up to now, these issues have prevented the routine and widespread use of this technique. Fortunately, many of these challenges have been addressed by recent developments in mass spectrometric instrumentation, as well as by the development of new cross-linking reagents (particularly isotopically coded cross-linking reagents), and—not insignificantly—by the development of specialized software for the processing of cross-linking data.

Numerous cross-linking reagents have recently been designed to incorporate special features which facilitate downstream processing and mass spectrometric analysis. These features include affinity tags and charge groups to facilitate selective enrichment of the cross-links, and the incorporation of isotopic coding, mass defect groups, MS/MS reporter groups, and cleavage sites to facilitate mass spectrometric detection and identification (Paramelle, Miralles, Subra, & Martinez, 2013; Petrotchenko & Borchers, 2010a). In order to obtain shorter (and therefore tighter) distance constraints, reagents with broader reactivity such as homo- and hetero-bifunctional photoreactive and zero-length cross-linkers are currently being developed. The complexity of the resulting mixture of cross-linking products is expected to be even higher for nonselective cross-linking reagents, so this approach will require more sophisticated data processing approaches as well.

Digestion with trypsin targets lysines and arginines, while lysine is also the target for amine-reactive cross-linking reagents. Because trypsin does not cleave at modified lysine residues, this combination often results in large cross-linked peptides. To circumvent this problem, double digestion (i.e., the use of an additional enzyme with a different specificity, such as GluC or AspN, has been proposed; Yan et al., 2009). We have recently reported on the successful use of the nonspecific enzyme proteinase K for generating "families" of interpeptide cross-links of an optimal size for mass spectrometric analysis (Petrotchenko et al., 2012).

Enrichment of cross-linked peptides—to separate them from the overwhelming background of non-cross-linked peptides—also facilitates the mass spectrometric detection and assignment of cross-links. Enrichment techniques include gel-filtration chromatography (because a typical interpeptide cross-links is larger than a linear peptide; Leitner et al., 2012), strong cation-exchange chromatography for the tryptic interpeptide cross-links (because tryptic interpeptide cross-links carry twice of positive charges compared to linear non-cross-linked tryptic peptides; Chen et al., 2010), affinity purification using tags which have been incorporated into the structure of a cross-linking reagent (usually a biotin group; Fujii, Jacobsen, Wood, Schoeniger, & Guy, 2004), functional groups for covalent capture (Buncherd et al., 2012; Chowdhury et al., 2009; Sohn et al., 2012; Yan et al., 2009), antigenic groups (Petrotchenko, Doant, & Borchers, 2006), and specific non-covalent interaction groups (Wang & Hakansson, 2008). These recently developed techniques have all been crucial for the successful detection of multiple cross-links.

There are several additional techniques, which have been developed to improve mass spectrometric detection and identification of interpeptide cross-links. These include enzyme-mediated introduction of ^{18}O isotopes during digestion (Back et al., 2002), N-terminal modification of the cross-linked peptides with isotopically coded reagents (Chen, Chen, & Anderson, 1999; Petrotchenko, Serpa, & Borchers, 2010) and the use of metabolically labeled proteins (Taverner, Hall, O'Hair, & Simpson, 2002). All of these techniques can produce "signatures" in the mass spectra that are specific to interpeptide cross-links.

Last but certainly not least, the introduction of high-mass accuracy high-performance high-sensitivity instruments, such as FTICR-based mass spectrometers with multiple new and efficient fragmentation methods, has had a major impact on the progress of the cross-linking approach. Due to the previously mentioned combinatorial nature of the interpeptide cross-links, the accuracy of the mass measurements and facile MS/MS fragmentation are crucial factors for obtaining the correct assignments of the cross-links. The introduction of instruments such as the ThermoFisher Orbitrap with high-mass accuracy and sensitivity—and, importantly, which can be easily interfaced to HPLC—has been a significant breakthrough in detecting and assigning cross-links contained within the complex mixture of cross-linking reaction products.

The large amount of mass spectrometric data produced in a typical cross-linking experiment requires specialized software tools for data analysis.

A number of programs specifically designed for the processing of mass spectrometric cross-linking data have recently been developed (Mayne & Patterton, 2011). Available software packages can produce simple predictions of the masses of cross-linked peptides, or go all the way to proteome-wide MS/MS-based identification of the cross-links. Most often, cross-linking studies are focused on a known protein or protein complex, although cross-linking is starting to be used for the examination of protein interaction networks on the proteome-wide scale (Yang et al., 2012). In our laboratory, we usually employ the ICC-CLASS software package (Petrotchenko & Borchers, 2010b) for interrogation of the LC–MS/MS data from experiments using isotopically coded collision induced dissociation (CID)-cleavable cross-links. This software has components that have been tailored for the analysis of data from cross-linking experiments using specific techniques, such as ^{15}N-labeling, isotopically coded N-terminal modification of the cross-links, etc.

Of all the different types of data that can be determined by mass spectrometry-based structural proteomics, interresidue distance constraints is the most obvious type to be incorporated into protein structure modeling software (see below).

We have used most of these recently developed methods in the cross-linking-combined-with-mass-spectrometry approach for the characterization of the prion protein conversion and for elucidating the structure of the resulting aggregate. Our first line of experiments is usually Lys–Lys cross-linking, using our isotopically coded CID-cleavable affinity-purifiable amine-reactive cross-linker CBDPS-H8/D8 (Petrotchenko, Serpa, & Borchers, 2011). Cross-linking was followed by proteinase K digestion, purification with avidin, and LC–MS/MS analysis on Orbitrap instrument (Fig. 6.4A). The use of proteinase K allowed unambiguous identification of the interpeptide cross-links in the prion protein, which has few tryptic cleavage sites and is resistant to enzymatic digestion by more-specific enzymes when in its aggregated form. We were able to detect and identify 13 cross-links, some of which were preferentially found in either the PrPC or PrPβ forms of the protein. Analysis of this data revealed that some of the cross-links observed in PrPβ (K185–K220 and K204–K220 in the H1–β2/H2–H3 region) were not compatible with the NMR structure of PrPC, suggesting specific sites of conformational change or the formation of new interprotein contacts in PrPβ (Fig. 6.4B).

To further examine not only the conformational changes of the individual PrPβ molecules but also the arrangement of the PrPβ molecules in an aggregate, we utilized an ^{15}N metabolic labeling strategy to discriminate between intra- and interprotein cross-links (Taverner et al., 2002). To obtain tighter distance constraints, we used several zero-length cross-linking reagents. ^{15}N-labeled PrP was produced and mixed 1:1 with ^{14}N-PrP, followed by conversion, cross-linking and enzymatic digestion. While intraprotein cross-links produce only ^{14}N–^{14}N or ^{15}N–^{15}N paired peptides and a doublet signature, interprotein cross-links produce a series of four peaks: ^{14}N–^{14}N, ^{14}N–^{15}N, ^{15}N–^{14}N, or ^{15}N–^{15}N. Cross-links can be further confirmed by a characteristic "signatures" in the fragment ions. An in-house program called $^{14}N^{15}N$ DXMSMS Match was developed to analyze this kind of data. This strategy resulted in the identification of 13 intraprotein cross-links and 11 interprotein cross-links. The intraprotein cross-links confirmed suggested rearrangement of the β1–H1–β2 loop away from the H2–H3 interface, and the introduction of a segment of the N-terminal tail region into the core. Interprotein cross-linking has allowed for the first time, the experimental determination of the stacking of the prion protein monomers within the oligomer.

7. ADDITIONAL MASS SPECTROMETRIC TECHNIQUES FOR THE PROTEIN STRUCTURE ANALYSIS

The collection of the structural proteomics techniques is enhanced by any other complementary mass spectrometry approach that can provide detailed protein structural information. Native ESI-MS, for example, can provide some information on the possible arrangements of the subunits in multicomponent protein assemblies, deduced from the dissociation pattern and the order of the proteins which "fall off" the complex (Marcoux & Robinson, 2013). Ion-mobility MS is able to provide specific conformational characteristics of the protein and protein complexes, derived from their measured cross-sectional areas (Konijnenberg, Butterer, & Sobott, 2013). All of this additional information can be used when models of the final protein structure or the protein's conformational changes are being postulated.

8. COMBINATION OF MULTIPLE STRUCTURAL PROTEOMICS TECHNIQUES

We believe that combining multiple structural proteomics approaches for the characterization of the proteins under study is crucial for solving protein structures. Although each method cannot provide complete structural information on its own, each method provides different and specific structural information on the protein. Thus, a combination of these multiple approaches may provide sufficient complementary information to derive the detailed protein structure. Results from different methods verify and support each other findings and, ultimately, provide confidence in the final result.

For the prion protein study mentioned here, we have used limited proteolysis, chemical surface modification, HDX, and cross-linking as part of our collection of the structural proteomics tools. The data from these multiple approaches are in remarkable agreement and have provided a total of >30 residue-specific constraints, which collectively suggest that the rearrangement of the β1–H1–β2–H2 region is the major conformational difference between PrPC and PrPβ (Fig. 6.5). A conformational change

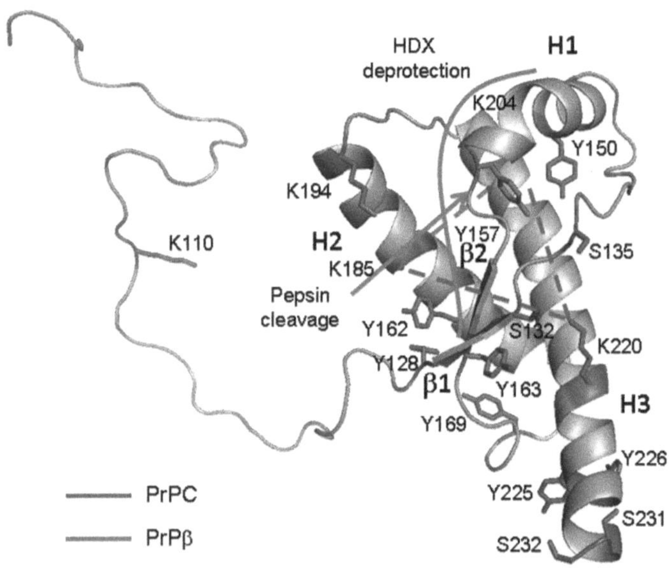

Figure 6.5 Summary of the structural differences between PrPC and PrPβ, as revealed by multiple structural proteomics methods. The residues, which are preferentially modified or cross-linked in the native PrPC and oligomeric PrPβ samples, are highlighted in light grey and dark grey, respectively. The preferential pepsin cleavage site for PrPβ is indicated by a light grey arrow. The region of the structure which loses protection from hydrogen–deuterium exchange in the PrPβ sample is indicated by the light grey arc. The K185–K204, K185–K220, and K204–K220 CBDPS cross-links (light grey dashed lines) are present only in the oligomeric PrPβ sample. The K185–K220 and K204–K220 cross-links are incompatible with the native PrPC structure, which suggests a possible conformational change in the PrPβ aggregated form of the protein. The data from multiple approaches collectively suggest rearrangement of the β1–H1–β2–H2 region in PrPβ. *Reprinted from Serpa, Patterson, et al. (2013), with permission.*

in the H1–β2-rigid loop region and distortion of its contact with helices 2 and 3 would create new hydrophobic patches on the surface of the molecule, which, in turn, could be responsible for driving the aggregation process. Through analysis of the intra- and interprotein constraint data, only one possible dimeric structure was found that satisfied all of the constraints. Thus, this can be considered as the first experimentally determined structure for the early conversion and aggregation events in the prion protein's misfolding process. This study of prion proteins has illustrated—and validated—the utility of applying an entire arsenal of structural proteomics methods to produce a detailed and comprehensive characterization of the conformational changes and the aggregation process. The results obtained thus far have encouraged us to propose a similar approach to the investigation of the structure of multiple protein systems.

9. USE OF EXPERIMENTAL STRUCTURAL PROTEOMICS CONSTRAINTS IN PROTEIN STRUCTURE MODELING

The output of a mass spectrometry-based structural proteomics is a set of characteristics for the amino acid residues of the protein. To make sense of this array of experimental data, these pieces of information need to be translated into a final three-dimensional structure of the protein. Depending on the structural question for a particular study, differing numbers of experimental constraints may be needed to provide the answer. For example, a few long distance cross-linking constraints may indicate a global conformational change, if differentially observed between two conformational states of the protein. A small amount of data on the changes in amino acid residues exposure between a free and bound state upon protein complex formation may designate protein interaction interface. Likewise, a single cross-link can rule out a potential conformational model (Petrotchenko, Pedersen, Borchers, Tomer, & Negishi, 2001).

The ultimate challenge, though, would be to automatically solve a problem in protein structure by simply inputting structural proteomics data. Unfortunately, to our knowledge, to date there is no turn-key protein structure modeling software that is able to incorporate all of the experimental information provided by structural proteomics and independently solve the structure. Human intervention is required in all cases. Again, selection of templates or fold recognition in the threading process can be achieved or influenced by a limited number of structural constraints (Young et al., 2000). However, true *ab initio* protein structure modeling would probably require numerous tight structural constraints. We envision several conceptual ways for accomplishing this. Multiple protein structural models could be generated and subsequently filtered for the selection of the "best" models, based on satisfying structural constraints derived from structural proteomics experiments. Alternatively, the modeling process itself can be somehow guided by incorporating structural constraints into some type of scoring function, influencing pathway of *in silico* folding process. Another possibility would be constraint-guided three-dimensional arrangement of the secondary structural elements to generate an initial fold pattern followed by refinement of the model.

Protein modeling software programs such as Rosetta (Herzog et al., 2012) or NMR-based packages (Schwieters, Kuszewski, Tjandra, & Clore, 2003) can already use distance constraints as input data. Due to the chemical nature of the cross-linking process, however, only cross-links which can form on the surface of the protein, but not those penetrating the protein globule, should be used (Kahraman, Malmström, & Aebersold, 2011).

10. FUTURE DIRECTIONS

Structural proteomics using mass spectrometry for the structural analysis of proteins has a bright and promising future. The widening availability of high-mass accuracy high-performance high-sensitivity instruments, such as Thermo's Orbitrap, which greatly facilitate successful cross-linking applications, will allow the use of mass spectrometry-based structural proteomics by larger numbers of molecular biology researchers. Further development of the mass spectrometric techniques, instrumentation, and methods including top-down analysis, new fragmentation techniques, and gas-phase reactions are certain to continue to have a positive impact on this field. To provide more detailed structural information on proteins will require a new generation of nonselective modification reagents and short-range nonspecific cross-linking reagents. A collection of reagents of varying specificities and characteristics, specifically designed to facilitate downstream mass spectrometric analysis, as well as easy-to-use software for rapid processing of the data are also elements of the structural proteomics toolkit. Interpretation of the resulting structural data is also tightly linked to the development of protein modeling software. Easy-to-use protein modeling software programs which can easily and automatically incorporate and integrate distance and exposure information generated by structural proteomics experiments also needs to be developed. This will lead to rapid progress in this exciting field.

11. CONCLUSIONS

In summary, mass spectrometry-based structural proteomics is already being successfully applied to many aspects of the structural analysis of proteins and protein complexes, including the analysis of protein structures and conformational changes, the determination of protein interaction interfaces, and for elucidating the topology of multisubunit protein complexes. Although not discussed in this review, the first examples of the application of structural proteomics techniques to the identification of proteome-wide protein interactions, have recently been presented (Herzog et al., 2012; Zheng et al., 2011). With the successful integration of multiple types of experimental data into the modeling process, we envision that the "holy grail" of structural proteomics—the autonomous solving of protein structures—will be achieved in the very near future.

ACKNOWLEDGMENT

This work was supported by a Genome Canada, Genome British Columbia, Technology Development Grant.

REFERENCES

Back, J. W., Notenboom, V., de Koning, L. J., Muijsers, A. O., Sixma, T. K., de Koster, C. G., et al. (2002). Identification of cross-linked peptides for protein interaction studies using mass spectrometry and 18O labeling. *Analytical Chemistry*, *74*(17), 4417–4422.

Buncherd, H., Nessen, M. A., Nouse, N., Stelder, S. K., Roseboom, W., Dekker, H. L., et al. (2012). Selective enrichment and identification of cross-linked peptides to study 3-D structures of protein complexes by mass spectrometry. *Journal of Proteomics*, *75*(7), 2205–2215.

Chen, X., Chen, Y. H., & Anderson, V. E. (1999). Protein cross-links: Universal isolation and characterization by isotopic derivatization and electrospray ionization mass spectrometry. *Analytical Biochemistry*, *273*(2), 192–203.

Chen, Z. A., Jawhari, A., Fischer, L., Buchen, C., Tahir, S., Kamenski, T., et al. (2010). Architecture of the RNA polymerase II-TFIIF complex revealed by cross-linking and mass spectrometry. *EMBO Journal*, *29*(4), 717–726.

Chowdhury, S. M., Du, X., Tolić, N., Wu, S., Moore, R. J., Mayer, M. U., et al. (2009). Identification of cross-linked peptides after click-based enrichment using sequential collision-induced dissociation and electron transfer dissociation tandem mass spectrometry. *Analytical Chemistry*, *81*(13), 5524–5532.

Fasold, H., Klappenberger, J., Meyer, C., & Remold, H. (1971). Bifunctional reagents for the crosslinking of proteins. *Angewandte Chemie International Edition in English*, *10*(11), 795–801.

Fujii, N., Jacobsen, R. B., Wood, N. L., Schoeniger, J. S., & Guy, R. K. (2004). A novel protein crosslinking reagent for the determination of moderate resolution protein structures by mass spectrometry (MS3-D). *Bioorganic and Medicinal Chemistry Letters*, *14*(2), 427–429.

Green, N. S., Reisler, E., & Houk, K. N. (2001). Quantitative evaluation of the lengths of homobifunctional protein cross-linking reagents used as molecular rulers. *Protein Science*, *10*(7), 1293–1304.

Herzog, F., Kahraman, A., Boehringer, D., Mak, R., Bracher, A., Walzthoeni, T., et al. (2012). Structural probing of a protein phosphatase 2A network by chemical cross-linking and mass spectrometry. *Science*, *337*(6100), 1348–1352.

Kahraman, A., Malmström, L., & Aebersold, R. (2011). Xwalk: Computing and visualizing distances in cross-linking experiments. *Bioinformatics*, *27*(15), 2163–2164.

Konermann, L., Vahidi, S., & Sowole, M. A. (2014). Mass spectrometry methods for studying structure and dynamics of biological macromolecules. *Analytical Chemistry*, *86*(1), 213–232.

Konijnenberg, A., Butterer, A., & Sobott, F. (2013). Native ion mobility-mass spectrometry and related methods in structural biology. *Biochimica et Biophysica Acta*, *1834*(6), 1239–1256.

Kruppa, G. H., Schoeniger, J., & Young, M. M. (2003). A top down approach to protein structural studies using chemical cross-linking and Fourier transform mass spectrometry. *Rapid Communications in Mass Spectrometry*, *17*, 155–162.

Leitner, A., Reischl, R., Walzthoeni, T., Herzog, F., Bohn, S., Förster, F., et al. (2012). Expanding the chemical cross-linking toolbox by the use of multiple proteases and enrichment by size exclusion chromatography. *Molecular and Cellular Proteomics*, *11*(3), M111.014126. Epub 2012 Jan 27.

Marcoux, J., & Robinson, C. V. (2013). Twenty years of gas phase structural biology. *Structure*, *21*(9), 1541–1550.

Mayne, S. L., & Patterton, H. G. (2011). Bioinformatics tools for the structural elucidation of multi-subunit protein complexes by mass spectrometric analysis of protein-protein cross-links. *Briefings in Bioinformatics*, *12*(6), 660–671.

Novak, P., & Giannakopulos, A. E. (2007). Chemical cross-linking and mass spectrometry as structure determination tools. *European Journal of Mass Spectrometry*, *13*(2), 105–113.

Novak, P., Young, M. M., Schoeniger, J. S., & Kruppa, G. H. (2003). A top-down approach to protein structure studies using chemical cross-linking and Fourier transform mass spectrometry. *European Journal of Mass Spectrometry*, *9*(6), 623–631.

Pan, J., Han, J., C. H. Borchers, C. H., & Konermann, L. (2008). Electron capture dissociation of electrosprayed protein ions for spatially resolved hydrogen exchange measurements. *Journal of the American Chemical Society*, *130*(35), 11574–11575.

Pan, J., Han, J., C. H. Borchers, C. H., & Konermann, L. (2009). Hydrogen/deuterium exchange mass spectrometry with top-down electron capture dissociation for characterizing structural transitions of a 17 kDa protein. *Journal of the American Chemical Society*, *131*(35), 12801–12808.

Paramelle, D., Miralles, G., Subra, G., & Martinez, J. (2013). Chemical cross-linkers for protein structure studies by mass spectrometry. *Proteomics*, *13*(3–4), 438–456.

Peters, K., & Richards, F. M. (1977). Chemical cross-linking: Reagents and problems in studies of membrane structure. *Annual Review of Biochemistry*, *46*, 523–551.

Petrotchenko, E. V., & Borchers, C. H. (2010a). Crosslinking combined with mass spectrometry for structural proteomics. *Mass Spectrometry Reviews*, *29*, 862–876.

Petrotchenko, E. V., & Borchers, C. H. (2010b). ICC-CLASS: Isotopically-coded cleavable crosslinking analysis suite. *BMC Bioinformatics*, *11*(1), 64.

Petrotchenko, E., Doant, T., & Borchers, C. (2006). A novel chromophoric affinity-tagged isotopically-coded crosslinker DGDNBS. Presented at the 54th ASMS Conference on Mass Spectrometry and Allied Topics, Seattle, WA.

Petrotchenko, E. V., Pedersen, L. C., Borchers, C. H., Tomer, K. B., & Negishi, M. (2001). The dimerization motif of cytosolic sulfotransferases. *FEBS Letters*, *490*(1–2), 39–43.

Petrotchenko, E. V., Serpa, J. J., Berjanskii, M., Suriyamongkol, B. P., Wishart, D. S., & Borchers, C. H. (2012). Use of proteinase K non-specific digestion for selective and comprehensive identification of interpeptide crosslinks: Application to prion proteins. *Molecular and Cellular Proteomics*, *11*(7), M111.013524.

Petrotchenko, E. V., Serpa, J. J., & Borchers, C. H. (2010). Use of a combination of isotopically coded cross-linkers and isotopically coded N-terminal modification reagents for selective identification of inter-peptide crosslinks. *Analytical Chemistry*, *82*(3), 817–823.

Petrotchenko, E. V., Serpa, J. J., & Borchers, C. H. (2011). An isotopically-coded CID-cleavable biotinylated crosslinker for structural proteomics. *Molecular and Cellular Proteomics*. *10*(2). http://dx.doi.org/10.1074/mcp.M110.001420.

Schwieters, C. D., Kuszewski, J. J., Tjandra, N., & Clore, G. M. (2003). The Xplor-NIH NMR molecular structure determination package. *Journal of Magnetic Resonance*, *160*(1), 65–73.

Serpa, J. J., Parker, C. E., Petrotchenko, E. V., Han, J., Pan, J., & Borchers, C. H. (2012). Mass spectrometry-based structural proteomics. *European Journal of Mass Spectrometry*, *18*(2), 251–267.

Serpa, J. J., Patterson, A. P., Pan, J., Han, J., Wishart, D. S., Petrotchenko, E. V., et al. (2013). Using multiple structural proteomics approaches for the characterization of prion proteins. *Journal of Proteomics*, *81*, 31–42.

Serpa, J. J., Petrotchenko, E. V., Wishart, D. S., & Borchers, C. H. (2013). Using isotopically-coded hydrogen peroxide as a surface modification reagent for the structural characterization of prion-protein aggregates. Presented at the 61st ASMS Conference on Mass Spectrometry and Allied Topics, Minneapolis, MN.

Sohn, C. H., Agnew, H. D., Lee, J. E., Sweredoski, M. J., Graham, R. L., Smith, G. T., et al. (2012). Designer reagents for mass spectrometry-based proteomics: Clickable cross-linkers for elucidation of protein structures and interactions. *Analytical Chemistry*, *84*(6), 2662–2669.

Taverner, T., Hall, N. E., O'Hair, R. A. J., & Simpson, R. J. (2002). Characterization of an antagonist interleukin-6 dimer by stable isotope labelling, cross-linking and mass spectrometry. *Journal of Biological Chemistry*, *277*(48), 46487–46492.

Wang, B., & Hakansson, K. (2008). Design and evaluation of a novel homobifunctional cross-linker with selective metal dioxide-based enrichment potential. Presented at the 56th ASMS Conference on Mass Spectrometry and Allied Topics, Denver, CO.

Yan, F., Che, F. Y., Rykunov, D., Nieves, E., Fiser, A., Weiss, L. M., et al. (2009). Non-protein based enrichment method to analyze peptide cross-linking in protein complexes. *Analytical Chemistry*, *81*(17), 7149–7159.

Yang, L., Zheng, C., Weisbrod, C. R., Tang, X., Munske, G. R., Hoopmann, M. R., et al. (2012). In vivo application of photocleavable protein interaction reporter technology. *Journal of Proteome Research*, *11*(2), 1027–1041.

Young, M. M., Tang, N., Hempel, J. C., Oshiro, C. M., Taylor, E. W., Kuntz, I. D., et al. (2000). High throughput protein fold identification by using experimental constraints derived from intramolecular cross-links and mass spectrometry. *Proceedings of the National Academy of Science USA*, *97*, 5802–5806.

Zheng, C., Yang, L., Hoopmann, M. R., Eng, J. K., Tang, X., Weisbrod, C. R., et al. (2011). Cross-linking measurements of in vivo protein complex topologies. *Molecular and Cellular Proteomics*, *10*(10), M110.006841.

CHAPTER SEVEN

Organellar Proteomics of Embryonic Stem Cells

Faezeh Shekari[*,†], **Hossein Baharvand**[†,‡,1], **Ghasem Hosseini Salekdeh**[*,§,1]
[*]Department of Molecular Systems Biology at Cell Science Research Center, Royan Institute for Stem Cell Biology and Technology, ACECR, Tehran, Iran
[†]Department of Developmental Biology, University of Science and Culture, ACECR, Tehran, Iran
[‡]Department of Stem Cells and Developmental Biology at Cell Science Research Center, Royan Institute for Stem Cell Biology and Technology, ACECR, Tehran, Iran
[§]Department of Systems Biology, Agricultural Biotechnology Research Institute of Iran, Karaj, Iran
[1]Corrresponding authors: e-mail address: baharvand@royaninstitute.org; salekdeh@royaninstitute.org

Contents

Abstract

Embryonic stem cells (ESCs) are undifferentiated cells with two common remarkable features known as self-renewal and differentiation. Proteomics plays an increasingly important role in understanding molecular mechanisms underlying self-renewal and pluripotency of ESCs and their applications in cell therapy and developmental biology studies. As the function of a protein is strongly associated with its localization in cell, a complete and accurate picture of the proteome of ESCs cannot be achieved without knowing the subcellular locations of proteins. Subcellular fractionation allows enrichment of low abundant proteins and signaling complexes and reduces the complexity of the sample. It also provided insight into tracking proteins that shuttle between different compartments. Despite the substantial interest and efforts in ESC subcellular proteomics area, progress has been relatively limited. In this review, we present an overview on current status of ESCs organelle proteomics research and discuss challenges in subcellular proteomics.

1. INTRODUCTION

Embryonic Stem Cells (ESCs) have commendable attributes including indefinite proliferation potential accompany with preservation of the ability

Advances in Protein Chemistry and Structural Biology, Volume 95
ISSN 1876-1623
http://dx.doi.org/10.1016/B978-0-12-800453-1.00007-5

to differentiate into derivatives of all three germ layers. Their marvelous potentials bring them out as a worthy source for different applications such as cell therapy and developmental biology studies. Moreover, they provide remarkable sources for drug discovery and toxicity testing (Rubin & Haston, 2011; Fig. 7.1). The ultimate goal of stem cell technology is to generate and expand cells that can be used in human diseases and regenerative medicine either directly by transplantation of ESCs or indirectly by transplantation of ESC derivatives. Molecular analysis of the ESCs will uncover and further define molecular mechanisms and signaling pathways involved in the maintenance of the undifferentiated state and regulation of differentiation. A comprehensive understanding of these molecular mechanisms will be essential for the aforementioned ESCs applications. Proteomics may play an increasingly important role in understanding molecular mechanisms underlying self-renewal and pluripotency of ESCs (Baharvand, Fathi, Van Hoof, & Salekdeh, 2007; Reiland, Salekdeh, & Krijgsveld, 2011).

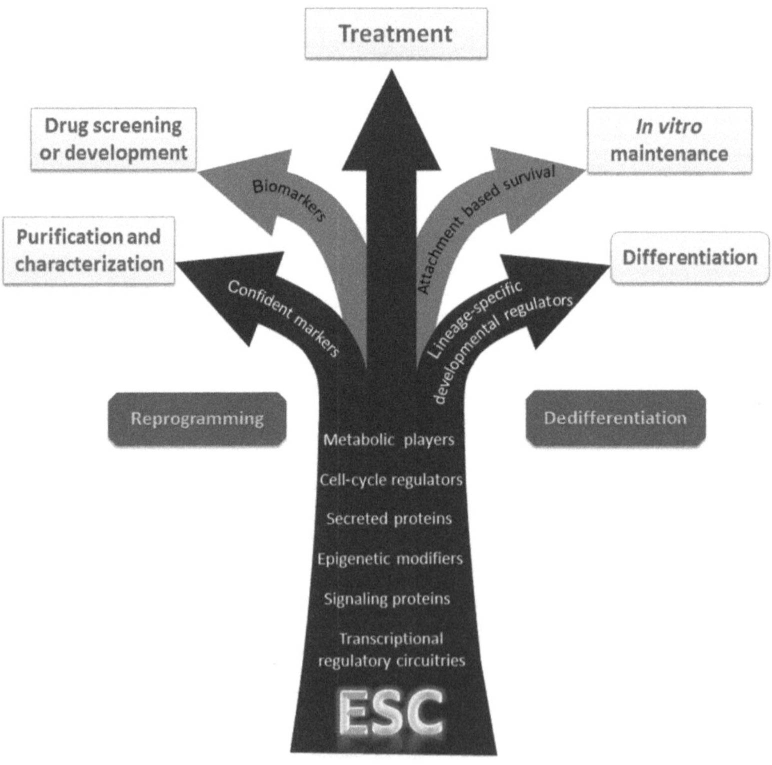

Figure 7.1 Various aspects of ESC technology (branches) involved in the function of plenty of proteins.

Despite advances in proteomics technologies, achieving a combination of high-throughput screening while maintaining high sensitivity for the detection of low copy number proteins still remains a major challenge. This means that the dynamic resolution is restricted and only the most abundant proteins can be detected when whole-cell proteomes are analyzed. Furthermore, many regulatory steps, especially those involved in cell proliferation and differentiation depend on chemical modifications occur after proteins are created that alter their binding ability, enzymatic activity, and stability. Most regulatory proteins such as phosphatases, kinases, or GTPases exist in low copy numbers but very specific subcellular locations. Although the number of genes in the human genome expressed in ESCs is around 8000 (Lee et al., 2006), the alternative splicing of transcribed RNA and post-translational modifications of proteins give rise to a human proteome size that is likely to be significantly larger than the number of estimated genes expressed and result in further complexity of eukaryotic cells (Harrison, Kumar, Lang, Snyder, & Gerstein, 2002).

One of the main layers of increased complexity in eukaryotes is organization of proteins into different cellular compartments in order to localizing various functions into the specific location (Baharvand et al., 2007). For example, cytosolic or nuclear localization of β-catenin or GSK3, two well-known players in self-renewal maintenance or differentiation of ESCs, has been depicted in Fig. 7.2. β-Catenin localization in nucleus following by Wnt signaling leads to activation of differentiation genes transcription (for review see Merrill, 2012; Sokol, 2011). More recently, it has been shown that β-catenin localization in nucleus of human ESC can be prohibited by stabilization of β-catenin in cytoplasm and resulting in self-renewal through a yet unknown mechanism (Kim et al., 2013). The localization of GSK3 in nucleus or cytosol can also result in differentiation or self-renewal, respectively (Bechard & Dalton, 2009; Fig. 7.2).

Obviously, a single-step characterization of an entire proteome seems at least presently rather unfeasible.

Proteomics and subcellular fractionation form an ideal partnership when it comes to enrichment and analysis of low abundant protein and intracellular organelles. Separation of organelles in subcellular fractions termed "subcellular fractionation" (Fleischer & Kervina, 1974) and proteomic analysis of subcellular fractions termed "subcellular proteomics" (Dreger, 2003). Subcellular fractionation is a flexible and adjustable approach resulting in reduced sample complexity and is most efficiently combined with mass spectrometry analysis. This approach allows the separation of organelles based on

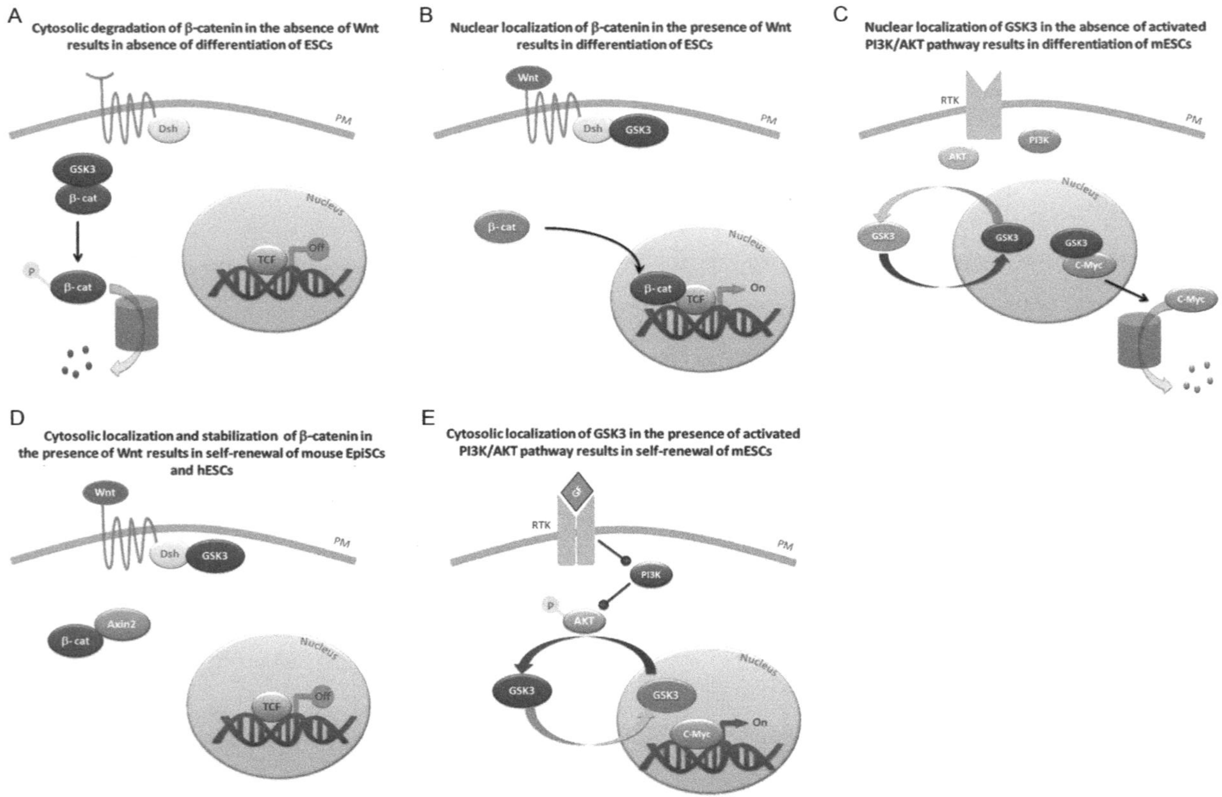

Figure 7.2 Cytosolic or nuclear localization of beta catenin or GSK3 proteins result in different outcome in ESCs.

their physical properties and was initially applied to separate organelles derived from rat liver (Fleischer & Kervina, 1974). In this review, we present an overview on current status of ESCs organelle proteomics research and discuss technical challenges for subcellular proteomics.

2. ORGANELLE PROTEOME ANALYSIS OF ESC

In the early of 1950s, sucrose gradient was used to isolating mitochondria (Holter, Ottesen, & Weber, 1953). In 1974, the Nobel Prize in Physiology or Medicine was awarded jointly to Albert Claude, Christian de Duve, and George E. Palade because of their efforts to develop the basic methodology to exploring organization of the cell; grinding the cells into fragments and then sort them out on a large scale with the aid of the centrifuge (Nobelprize.org, 2013). The main concept is sequential removal of nucleus, mitochondria, and microsomes from disrupted cell. Since then, centrifuge has been used as a powerful tool for traveling to the inside of the cell. Although the original method is still being widely used, there are some emerging methods which are not relying on centrifuge such as fluorescent-assisted organelle sorting (Bock, Steinlein, & Huber, 1997; Gauthier, Sobota, Ferraro, Mains, & Lazure, 2008), antibody based organelle isolation (Hornig-Do et al., 2009; Kausch, Owen, Narayanswami, & Bruce, 1999; Lawson et al., 2006), free-flow electrophoresis (Sengelov & Borregaard, 1999; Weber, Weber, & Eckerskorn, 2004), laser capture microdissection (Pflugradt, Schmidt, Landenberger, Sanger, & Lutz-Bonengel, 2011) and dielectrophoresis (Moschallski et al., 2010; for more review see Lee, Tan, & Chung, 2010; Satori, Kostal, & Arriaga, 2012).

Despite the importance of protein localization based proteomics in ESCs, relatively little research has been done in this field.

Plasma membrane (PM) proteomics of ESCs are among the most desirable proteomic analysis because of the importance of these proteins in cellular signaling and communication. PM proteins have several functional classes including ion channels, receptor tyrosine kinases, G protein-coupled receptors, and integrins. Knowledge about PM proteins is crucial to understand the cellular response to stimuli or changing environmental conditions.

In ESCs, PM proteins recognize growth factors and induce signaling cascades that possibly lead proliferation or differentiation of cells. In addition, PM proteins that are exposed on the outside of the cell may be used as handles that are recognized by antibodies in cell-sorting experiments. The application of antibodies against cell surface marker will allow fractionating heterogeneous populations into distinct classes, which can then be used for further biological studies or cell therapy.

The first large-scale analysis of ESC PM proteome was performed using cell surface biotinylation along with density gradient centrifugation to purify the PM in mouse ESCs (Nunomura et al., 2005; Table 7.1). This approach was also utilized to profile membrane proteome of human ESCs (Gu et al., 2011) and mouse ESCs (Gu et al., 2010; Intoh et al., 2009; Wollscheid et al., 2009). Gel electrophoresis combined with mass spectrometry approaches were also applied for membrane proteomics analysis of mouse (Intoh et al., 2009) and human (Gerwe et al., 2011; Mcquade et al., 2009; Shekari et al., 2011) ESCs. Fourier transform LC–ESI–MS/MS and MS(3) mass spectrometry has been used to evaluate membrane proteome of human ESCs independent of culture conditions (Harkness et al., 2008).Various sample preparation and digestion procedures were also examined to evaluate their efficiency, quality, and compatibility with subsequent mass spectrometry analysis (Dormeyer et al., 2008). Stable isotope labeling by amino acids in cell culture (SILAC) for human ESC lines was succeeded in resolving more than 1000 membrane proteins (Prokhorova et al., 2009; Sarkar et al., 2012). Furthermore, applying chemoproteomic targeting strategy for discovery of cell surface *N*-glycoproteins of mouse ESC and induced pluripotent stem cell (iPSC) lines, resulted in identification of 500 cell surface proteins (Gundry et al., 2012). Despite of these valuable efforts, our knowledge of membrane proteins is still very limited. Approximately 20–30% of all genes in an organism encode integral membrane proteins, which are far beyond current available data. Analysis of membrane proteins requires methods that solve problems such as contamination of intracellular components, protein insolubility, low abundance, and loss of hydrophobic peptides, which prevent protein identification.

Although ER is a runway for ribosomes to translate and/or modify newly synthesized proteins, it has many other important functions. ER is the main site of lipid synthesis and Ca^{2+} storage and membrane contact sites involving the ER provide an important function in both of these exchange reactions to other organelles (Helle et al., 2013). Separation of the ER from the PM results in misregulation of phosphoinositide signaling followed by accumulation of PI4P levels at the PM and constitutively activation of the unfolded protein response (Manford, Stefan, Yuan, Macgurn, & Emr, 2012). The ER and mitochondria contact sites have also many roles including coordination of calcium transfer, regulation of mitochondrial fission,

Table 7.1 Proteomics studies of ESC organelles

	Article	Technical notes	ESC line	Number of identified protein
Membrane part	Harkness et al. (2008)	LTQ-FT	hESC-OD3	3133 identified, 1075 validated
	Dormeyer, Van Hoof, Mummery, Krijgsveld, and Heck (2008)	LTQ-MS	hESC line HUES-7	1077
	Mcquade, Schmidt, Pascovici, Stojanov, and Baker (2009)	IPG/LTQ-linear ion-trap	hESC line SIVF001	2851
	Prokhorova et al. (2009)	SILAC/LTQ-FT Ultra	hESC lines (Odense-3, KMEB1, and HUES9)	1556
	Gerwe et al. (2011)	1D gel electrophoresis/ion trap FT-ICR	hESC lines (BG01, WA09, and abBG02)	775
	Shekari et al. (2011)	BN-PAGE/ MALDI-TOF	hESC line Royan H5	69
	Gu et al. (2011)	Biotin labeling/ LTQ-MS	hESC line HUES3	5405
	Sarkar, Collier, Randall, Muddiman, and Rao (2012)	SILAC/subcellular fractionation/LTQ-Orbitrap XL	hESC line H9	1185
	Intoh et al. (2009)	Biotin labeling/2D DIGE/QTOF MS and iTRAQ/ MALDI-TOF/ TOF	mESC line D3	258 and 659

Continued

Table 7.1 Proteomics studies of ESC organelles—cont'd

	Article	Technical notes	ESC line	Number of identified protein
	Gundry et al. (2012)	Biotin labeling and Cell Surface Capturing/LTQ-Orbitrap	mESC lines R1 and D3	500 *N*-glycoproteins
	Nunomura et al. (2005)	Biotin labeling/2D LC Q-Tof	mESC line D3	324
	Gu et al. (2010)	Biotin labeling/LTQ-MS	mESC line D3	3468
	Wollscheid et al. (2009)	Biotin labeling and cell surface capturing/CHIP-QTOF 6530 (Agilent), LTQ and LTQ-FT	mESC	408 *N*-glycoproteins
Nuclear part	Nasrabadi et al. (2010)	2DE/MALDI-TOF/TOF	monkey ESC	560
	Williamson et al. (2008)	iTRAQ/QSTAR XL	mESC line E14.1	2389
	Barthelery, Jaishankar, Salli, Freeman, et al. (2009)	2DIGE/MALDI-TOF/TOF	hESC line H9 (WA-09)	1521

inflammation, autophagy (Marchi, Patergnani, & Pinton, 2014) and can modulate cell fate (Bravo-Sagua et al., 2013). The implications of ER proteins in ESC self-renewal or pluripotency still remain to be elucidated.

In the case of highly proliferative stem cells, nucleus is the main location for cell division, transcription, and cellular programming besides maintenance of pluripotency. Nucleus isolation following by gel electrophoresis for monkey (Nasrabadi et al., 2010) and human (Barthelery, Jaishankar, Salli, Freeman, & Vrana, 2009) ESCs resolved 560 and 1521 spots, respectively (Table 7.1). Furthermore, nuclear extracts of human ESCs with human mesenchymal stem cells (Jaishankar et al., 2009) or human ESC-derived neural stem cell (Barthelery, Jaishankar, Salli, & Vrana, 2009) were compared using two-dimensional difference gel electrophoresis (2DIGE). Williamson et al. could identify some of important transcription factors through nuclear proteomics (Williamson et al., 2008). Giving the limitation of gel based proteomics, application of quantitative mass spectrometry proteomics approaches will allow a more detailed and comprehensive insight into nuclear proteome.

Despite the importance of mitochondria in both energy processing and life or death decision of cell of stem cells (Chen, Hsu, & Wei, 2012; Liu et al., 2013; Prigione & Adjaye, 2010; Prigione, Fauler, Lurz, Lehrach, & Adjaye, 2010; Rehman, 2010; Son, Jeong, Kwon, & Cho, 2013; Varum et al., 2011), no published reports is available of ESC mitochondrial proteome.

Recently, Sarkar et al. reported proteomics of membrane, nuclear, and cytoplasmic fractions human ESCs for membrane, nuclear, and cytoplasmic fractions (Sarkar et al., 2012; Table 7.1).

3. SUBCELLULAR FRACTIONATION: CURRENT APPROACHES AND CHALLENGES

As shown in Fig. 7.3, the very first step in all subcellular fractionation method is disruption of cellular integrity to releasing intact organelles from membrane enclosure (step 1). Mechanical homogenization by a homogenizer or sonication is the most commonly used approach. The efficiency of cell disruption is often checked out through microscope and, if confirmed, the disrupted cell extract will be subjected to further centrifugal fractionation. There are two main approaches, continuous and discontinuous gradients, for fractionation of organelles on a sucrose gradient (step 2 or 3). Although various solutes (such as Nycodenz; Murayama, Fujimura, Morita, & Shindo, 2001) has been examined in the place of sucrose, it is still routinely being used because it is biologically inert, inexpensive, and dialyzable (Lee et al., 2010).

Centrifugation on discontinuous gradients or differential centrifugation divides cell extracts into two fractions: one containing nucleus and large sheets of membrane and the other containing mitochondria, cytoplasm, and microsomes (step 2). Each of these fractions subjected to further fractionation (steps 4–8). Nucleus is separated from membrane sheets through ultra-centrifugation on a discontinuous gradient (step 4), and mitochondria is isolated by a high-speed centrifugation from microsomes and cytoplasmic fractions (step 5). Further centrifugation of mitochondria on a

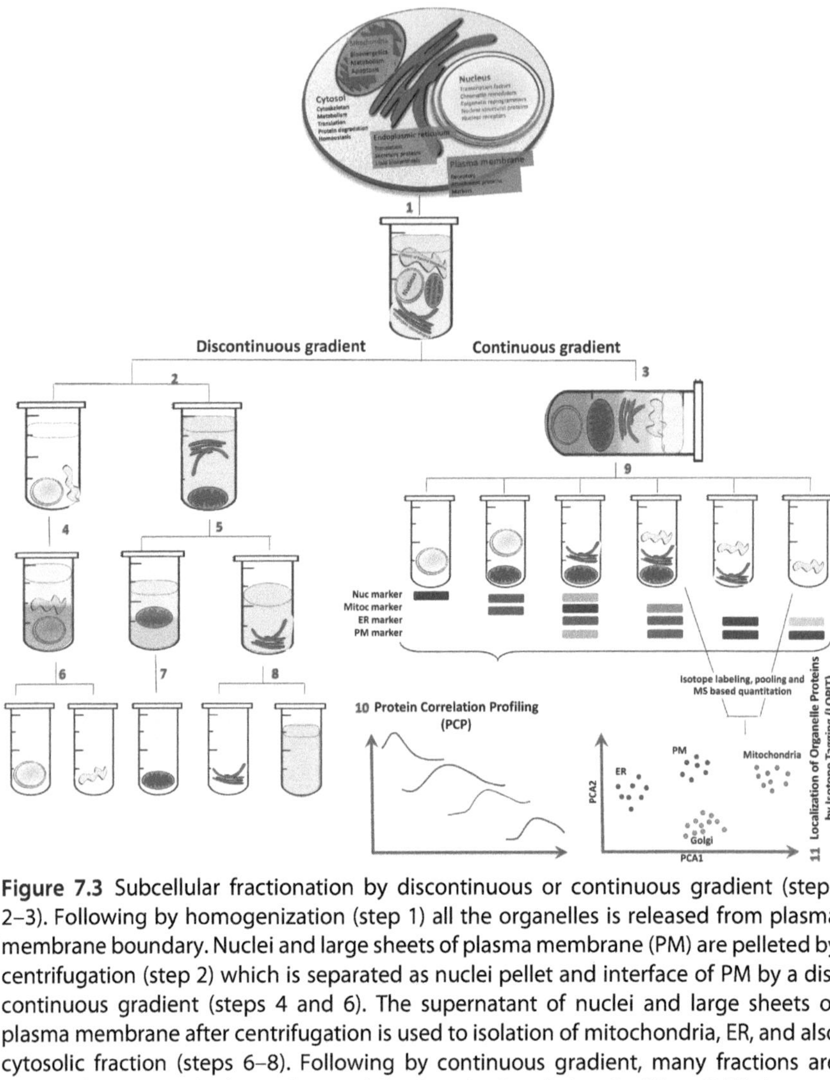

Figure 7.3 Subcellular fractionation by discontinuous or continuous gradient (steps 2–3). Following by homogenization (step 1) all the organelles is released from plasma membrane boundary. Nuclei and large sheets of plasma membrane (PM) are pelleted by centrifugation (step 2) which is separated as nuclei pellet and interface of PM by a discontinuous gradient (steps 4 and 6). The supernatant of nuclei and large sheets of plasma membrane after centrifugation is used to isolation of mitochondria, ER, and also cytosolic fraction (steps 6–8). Following by continuous gradient, many fractions are collected (step 9) which can be used for Protein Correlation Profiling (PCP) (step10) or Localization of Organelle Proteins by Isotope Tagging (LOPIT) (step 11).

discontinuous Nycodenz gradient can improve purity of mitochondria fraction (Okado-Matsumoto & Fridovich, 2001) and another step of ultracentrifugation can separate microsomes and cytoplasmic fraction (step 8). Differential centrifugation is the classic and still a key method of subcellular fractionation since theoretically separates each organelle at different stage. However, it is technically so difficult, if not impossible, to resolve pure organelle without any contamination through differential centrifugation (Yan, Aebersold, & Raines, 2009).

Alternatively, continuous gradients can be followed by protein correlation profiling (PCP; Andersen et al., 2003; Foster et al., 2006) (steps 3 and 9–10) or localization of organelle proteins by isotope tagging (LOPIT; Dunkley, Watson, Griffin, Dupree, & Lilley, 2004) (step 11). Although PCP or LOPIT provide attractive resolution, they need large amount of analysis time and immunoblotting that can be a great bottleneck for these approaches. By combining PCP profile of proteins with χ^2 test, Foster et al. localized 1258 cytosolic proteins from Mouse liver homogenate to eight cytoplasmic organelles (mitochondrion, ER, golgi, ER/golgi vesicles, early endosomes, recycling endosomes, PM, and proteasome; Foster et al., 2006).

The greatest obstacle to subcellular fractionation is purity of fractions which can be undermined by variability in preparation conditions, degree of contamination to cell type and also different organellar behavior during isolation processes (Dreger, 2003). Subcellular fractionation is based on organellar buoyancy which is basically rooted in their composition and structure. Contemplating cellular organelles as dynamically changing component of cells; depend on cellular situation and their function, results in contaminations in subcellular fractionation. These contaminations can be originated from physicochemical, structural, or physiological properties of dynamically changing organelles. For example, organellar fractionation in cells with large mitochondria and small nuclei results in contaminated mitochondria and nuclear fractions in contrast to the cells with small mitochondria and large nuclei (such as ESCs) which suffered from contaminated mitochondria and PM fractions. Moreover, organelles are not isolated individual island in the cells, but they are structurally and physiologically connected.

4. ORGANELLE PROTEOMICS DATABASES AND TOOLS

As discussed above, there are only a few high-throughput experimental proteome datasets on subcellular localization in ESC. Out of about 50 million entries in UniProtKB database (2013_12 release), nearly 6 million proteins have subcellular localization annotation; that means only 12%. This clearly highlights the shortage of information on subcellular localization of proteins and emphasize on the necessity of deep and comprehensive analysis of organellar proteomics (Foster et al., 2006).

Alternatively, many tools and databases have been developed in order to predict subcellular localization of proteins. These include DBSubLoc (Guo, Hua, Ji, & Sun, 2004), LocDB (Rastogi & Rost, 2011), LOCATE (Sprenger et al., 2008) and eSLDB (Pierleoni, Martelli, Fariselli, & Casadio, 2007). Database of Protein Subcellular Localization (DBSubLoc; http://www.bioinfo.tsinghua.edu.cn/~guotao/intro.html) collected the subcellular localization annotation from primary protein database SWISS-PROT and PIR. Protein Localization Database for Human and Arabidopsis (LocDB) is a manually curated database with experimental annotations which contains 13342 human proteins. LOCATE (http://locate.imb.uq.edu.au/) is contains higher number of proteins (about 65,000) and the subcellular locations of selected proteins from RIKEN FANTOM4 (Functional Annotaion of Mammalian genome; http://fantom.gsc.riken.jp/) as determined by a high-throughput, immunofluorescence-based assay, and by manually reviewing over 1700 peer-reviewed publications. Eukaryotic Subcellular Localization DataBase (eSLDB; http://gpcr.biocomp.unibo.it/esldb/index.htm) contains the experimental localizations, when available, the homology-based annotations, when feasible, and predictions performed with machine learning based methods.

Despite the remarkable effort to development prediction tools to localize proteins into the subcellular parts, much progress still needed before specifically for prediction multiple localization of proteins such as beta catenin and GSK3 (Fig. 7.2).

5. CONCLUDING REMARKS

Subcellular proteomics is where cell biology crossed proteomics (Jung, Heller, Sanchez, & Hochstrasser, 2000; Millar & Taylor, 2014). While subcellular fractionation faces many daunting challenges, it is still one of the pillars of proteomics and there are plenty of rooms for further improvement. As protein localization in a cell is strongly associated with its function, a complete and accurate picture of the proteome of cells cannot be achieved without knowing the subcellular location(s) of proteins (Au et al., 2007; Gatto, Vizcaino, Hermjakob, Huber, & Lilley, 2010). Subcellular fractionation allows enrichment of low abundant proteins and signaling complexes and reduces the complexity of the sample. Analyzing organelle proteome allows also tracking proteins that shuttle between different compartments. Recent advances in proteomics and bioinformatics tools are essential for the future of organelle proteomics and we anticipate that continuous advances in proteomics technologies will remarkably enhance our knowledge about organelle proteins and their functions in ESCs.

REFERENCES

Andersen, J. S., Wilkinson, C. J., Mayor, T., Mortensen, P., Nigg, E. A., & Mann, M. (2003). Proteomic characterization of the human centrosome by protein correlation profiling. *Nature, 426*, 570–574.

Au, C. E., Bell, A. W., Gilchrist, A., Hiding, J., Nilsson, T., & Bergeron, J. J. (2007). Organellar proteomics to create the cell map. *Current Opinion in Cell Biology, 19*, 376–385.

Baharvand, H., Fathi, A., Van Hoof, D., & Salekdeh, G. H. (2007). Concise review: Trends in stem cell proteomics. *Stem Cells, 25*, 1888–1903.

Barthelery, M., Jaishankar, A., Salli, U., Freeman, W. M., & Vrana, K. E. (2009). 2-D DIGE identification of differentially expressed heterogeneous nuclear ribonucleoproteins and transcription factors during neural differentiation of human embryonic stem cells. *Proteomics. Clinical Applications, 3*, 505–514.

Barthelery, M., Jaishankar, A., Salli, U., & Vrana, K. E. (2009). Reptin52 expression during in vitro neural differentiation of human embryonic stem cells. *Neuroscience Letters, 452*, 47–51.

Bechard, M., & Dalton, S. (2009). Subcellular localization of glycogen synthase kinase 3beta controls embryonic stem cell self-renewal. *Molecular and Cellular Biology, 29*, 2092–2104.

Bock, G., Steinlein, P., & Huber, L. A. (1997). Cell biologists sort things out: Analysis and purification of intracellular organelles by flow cytometry. *Trends in Cell Biology, 7*, 499–503.

Bravo-Sagua, R., Rodriguez, A. E., Kuzmicic, J., Gutierrez, T., Lopez-Crisosto, C., Quiroga, C., et al. (2013). Cell death and survival through the endoplasmic reticulum-mitochondrial axis. *Current Molecular Medicine, 13*, 317–329.

Chen, C. T., Hsu, S. H., & Wei, Y. H. (2012). Mitochondrial bioenergetic function and metabolic plasticity in stem cell differentiation and cellular reprogramming. *Biochimica et Biophysica Acta, 1820*, 571–576.

Dormeyer, W., Van Hoof, D., Mummery, C. L., Krijgsveld, J., & Heck, A. J. (2008). A practical guide for the identification of membrane and plasma membrane proteins in human embryonic stem cells and human embryonal carcinoma cells. *Proteomics, 8*, 4036–4053.

Dreger, M. (2003). Subcellular proteomics. *Mass Spectrometry Reviews, 22*, 27–56.

Dunkley, T. P., Watson, R., Griffin, J. L., Dupree, P., & Lilley, K. S. (2004). Localization of organelle proteins by isotope tagging (LOPIT). *Molecular & Cellular Proteomics, 3*, 1128–1134.

Fleischer, S., & Kervina, M. (1974). Subcellular fractionation of rat liver. *Methods in Enzymology, 31*, 6–41.

Foster, L. J., De Hoog, C. L., Zhang, Y., Zhang, Y., Xie, X., Mootha, V. K., et al. (2006). A mammalian organelle map by protein correlation profiling. *Cell, 125*, 187–199.

Gatto, L., Vizcaino, J. A., Hermjakob, H., Huber, W., & Lilley, K. S. (2010). Organelle proteomics experimental designs and analysis. *Proteomics, 10*, 3957–3969.

Gauthier, D. J., Sobota, J. A., Ferraro, F., Mains, R. E., & Lazure, C. (2008). Flow cytometry-assisted purification and proteomic analysis of the corticotropes dense-core secretory granules. *Proteomics, 8*, 3848–3861.

Gerwe, B. A., Angel, P. M., West, F. D., Hasneen, K., Young, A., Orlando, R., et al. (2011). Membrane proteomic signatures of karyotypically normal and abnormal human embryonic stem cell lines and derivatives. *Proteomics, 11*, 2515–2527.

Gu, B., Zhang, J., Wang, W., Mo, L., Zhou, Y., Chen, L., et al. (2010). Global expression of cell surface proteins in embryonic stem cells. *PLoS One*, *5*, e15795.

Gu, B., Zhang, J., Wu, Y., Zhang, X., Tan, Z., Lin, Y., et al. (2011). Proteomic analyses reveal common promiscuous patterns of cell surface proteins on human embryonic stem cells and sperms. *PLoS One*, *6*, e19386.

Gundry, R. L., Riordon, D. R., Tarasova, Y., Chuppa, S., Bhattacharya, S., Juhasz, O., et al. (2012). A cell surfaceome map for immunophenotyping and sorting pluripotent stem cells. *Molecular & Cellular Proteomics*, *11*, 303–316.

Guo, T., Hua, S., Ji, X., & Sun, Z. (2004). DBSubLoc: Database of protein subcellular localization. *Nucleic Acids Research*, *32*, D122–D124.

Harkness, L., Christiansen, H., Nehlin, J., Barington, T., Andersen, J. S., & Kassem, M. (2008). Identification of a membrane proteomic signature for human embryonic stem cells independent of culture conditions. *Stem Cell Research*, *1*, 219–227.

Harrison, P. M., Kumar, A., Lang, N., Snyder, M., & Gerstein, M. (2002). A question of size: The eukaryotic proteome and the problems in defining it. *Nucleic Acids Research*, *30*, 1083–1090.

Helle, S. C., Kanfer, G., Kolar, K., Lang, A., Michel, A. H., & Kornmann, B. (2013). Organization and function of membrane contact sites. *Biochimica et Biophysica Acta*, *1833*, 2526–2541.

Holter, H., Ottesen, M., & Weber, R. (1953). Separation of cytoplasmic particles by centrifugation in a density-gradient. *Experientia*, *9*, 346–348.

Hornig-Do, H. T., Gunther, G., Bust, M., Lehnartz, P., Bosio, A., & Wiesner, R. J. (2009). Isolation of functional pure mitochondria by superparamagnetic microbeads. *Analytical Biochemistry*, *389*, 1–5.

Intoh, A., Kurisaki, A., Yamanaka, Y., Hirano, H., Fukuda, H., Sugino, H., et al. (2009). Proteomic analysis of membrane proteins expressed specifically in pluripotent murine embryonic stem cells. *Proteomics*, *9*, 126–137.

Jaishankar, A., Barthelery, M., Freeman, W. M., Salli, U., Ritty, T. M., & Vrana, K. E. (2009). Human embryonic and mesenchymal stem cells express different nuclear proteomes. *Stem Cells and Development*, *18*, 793–802.

Jung, E., Heller, M., Sanchez, J. C., & Hochstrasser, D. F. (2000). Proteomics meets cell biology: The establishment of subcellular proteomes. *Electrophoresis*, *21*, 3369–3377.

Kausch, A. P., Owen, T. P., Jr., Narayanswami, S., & Bruce, B. D. (1999). Organelle isolation by magnetic immunoabsorption. *Biotechniques*, *26*, 336–343.

Kim, H., Wu, J., Ye, S., Tai, C. I., Zhou, X., Yan, H., et al. (2013). Modulation of beta-catenin function maintains mouse epiblast stem cell and human embryonic stem cell self-renewal. *Nature Communications*, *4*, 2403.

Lawson, E. L., Clifton, J. G., Huang, F., Li, X., Hixson, D. C., & Josic, D. (2006). Use of magnetic beads with immobilized monoclonal antibodies for isolation of highly pure plasma membranes. *Electrophoresis*, *27*, 2747–2758.

Lee, T. I., Jenner, R. G., Boyer, L. A., Guenther, M. G., Levine, S. S., Kumar, R. M., et al. (2006). Control of developmental regulators by Polycomb in human embryonic stem cells. *Cell*, *125*, 301–313.

Lee, Y. H., Tan, H. T., & Chung, M. C. (2010). Subcellular fractionation methods and strategies for proteomics. *Proteomics*, *10*, 3935–3956.

Liu, W., Long, Q., Chen, K., Li, S., Xiang, G., Chen, S., et al. (2013). Mitochondrial metabolism transition cooperates with nuclear reprogramming during induced pluripotent stem cell generation. *Biochemical and Biophysical Research Communications*, *431*, 767–771.

Manford, A. G., Stefan, C. J., Yuan, H. L., Macgurn, J. A., & Emr, S. D. (2012). ER-to-plasma membrane tethering proteins regulate cell signaling and ER morphology. *Developmental Cell*, *23*, 1129–1140.

Marchi, S., Patergnani, S., & Pinton, P. (2014). The endoplasmic reticulum-mitochondria connection: one touch, multiple functions. *Biochimica et Biophysica Acta, 1837*(4), 461–469.

Mcquade, L. R., Schmidt, U., Pascovici, D., Stojanov, T., & Baker, M. S. (2009). Improved membrane proteomics coverage of human embryonic stem cells by peptide IPG-IEF. *Journal of Proteome Research, 8*, 5642–5649.

Merrill, B. J. (2012). Wnt pathway regulation of embryonic stem cell self-renewal. *Cold Spring Harbor Perspectives in Biology, 4*, a007971.

Millar, A. H., & Taylor, N. L. (2014). Subcellular proteomics-where cell biology meets protein chemistry. *Frontiers in Plant Science, 5*, 55.

Moschallski, M., Hausmann, M., Posch, A., Paulus, A., Kunz, N., Duong, T. T., et al. (2010). MicroPrep: Chip-based dielectrophoretic purification of mitochondria. *Electrophoresis, 31*, 2655–2663.

Murayama, K., Fujimura, T., Morita, M., & Shindo, N. (2001). One-step subcellular fractionation of rat liver tissue using a Nycodenz density gradient prepared by freezing-thawing and two-dimensional sodium dodecyl sulfate electrophoresis profiles of the main fraction of organelles. *Electrophoresis, 22*, 2872–2880.

Nasrabadi, D., Larijani, M. R., Fathi, A., Gourabi, H., Dizaj, A. V., Baharvand, H., et al. (2010). Nuclear proteome analysis of monkey embryonic stem cells during differentiation. *Stem Cell Reviews, 6*, 50–61.

Nobelprize.org (2013). *Nobelprize.org. Nobel Media AB 2013.* Available from: http://www.nobelprize.org/nobel_prizes/medicine/laureates/1974/presentation-speech.html, Accessed 8 Jan 2014 [Online].

Nunomura, K., Nagano, K., Itagaki, C., Taoka, M., Okamura, N., Yamauchi, Y., et al. (2005). Cell surface labeling and mass spectrometry reveal diversity of cell surface markers and signaling molecules expressed in undifferentiated mouse embryonic stem cells. *Molecular & Cellular Proteomics, 4*, 1968–1976.

Okado-Matsumoto, A., & Fridovich, I. (2001). Subcellular distribution of superoxide dismutases (SOD) in rat liver: Cu, Zn-SOD in mitochondria. *The Journal of Biological Chemistry, 276*, 38388–38393.

Pflugradt, R., Schmidt, U., Landenberger, B., Sanger, T., & Lutz-Bonengel, S. (2011). A novel and effective separation method for single mitochondria analysis. *Mitochondrion, 11*, 308–314.

Pierleoni, A., Martelli, P. L., Fariselli, P., & Casadio, R. (2007). eSLDB: Eukaryotic subcellular localization database. *Nucleic Acids Research, 35*, D208–D212.

Prigione, A., & Adjaye, J. (2010). Modulation of mitochondrial biogenesis and bioenergetic metabolism upon in vitro and in vivo differentiation of human ES and iPS cells. *The International Journal of Developmental Biology, 54*, 1729–1741.

Prigione, A., Fauler, B., Lurz, R., Lehrach, H., & Adjaye, J. (2010). The senescence-related mitochondrial/oxidative stress pathway is repressed in human induced pluripotent stem cells. *Stem Cells, 28*, 721–733.

Prokhorova, T. A., Rigbolt, K. T., Johansen, P. T., Henningsen, J., Kratchmarova, I., Kassem, M., et al. (2009). Stable isotope labeling by amino acids in cell culture (SILAC) and quantitative comparison of the membrane proteomes of self-renewing and differentiating human embryonic stem cells. *Molecular & Cellular Proteomics, 8*, 959–970.

Rastogi, S., & Rost, B. (2011). LocDB: Experimental annotations of localization for Homo sapiens and Arabidopsis thaliana. *Nucleic Acids Research, 39*, D230–D234.

Rehman, J. (2010). Empowering self-renewal and differentiation: The role of mitochondria in stem cells. *Journal of Molecular Medicine, 88*, 981–986.

Reiland, S., Salekdeh, G. H., & Krijgsveld, J. (2011). Defining pluripotent stem cells through quantitative proteomic analysis. *Expert Review of Proteomics, 8*, 29–42.

Rubin, L. L., & Haston, K. M. (2011). Stem cell biology and drug discovery. *BMC Biology, 9*, 42.

Sarkar, P., Collier, T. S., Randall, S. M., Muddiman, D. C., & Rao, B. M. (2012). The subcellular proteome of undifferentiated human embryonic stem cells. *Proteomics, 12*, 421–430.

Satori, C. P., Kostal, V., & Arriaga, E. A. (2012). Review on recent advances in the analysis of isolated organelles. *Analytica Chimica Acta, 753*, 8–18.

Sengelov, H., & Borregaard, N. (1999). Free-flow electrophoresis in subcellular fractionation of human neutrophils. *Journal of Immunological Methods, 232*, 145–152.

Shekari, F., Taei, A., Pan, T. L., Wang, P. W., Baharvand, H., & Salekdeh, G. H. (2011). Identification of cytoplasmic and membrane-associated complexes in human embryonic stem cells using blue native PAGE. *Molecular BioSystems*, 7, 2688–2701.

Sokol, S. Y. (2011). Maintaining embryonic stem cell pluripotency with Wnt signaling. *Development, 138*, 4341–4350.

Son, M. J., Jeong, B. R., Kwon, Y., & Cho, Y. S. (2013). Interference with the mitochondrial bioenergetics fuels reprogramming to pluripotency via facilitation of the glycolytic transition. *The International Journal of Biochemistry & Cell Biology, 45*, 2512–2518.

Sprenger, J., Lynn Fink, J., Karunaratne, S., Hanson, K., Hamilton, N. A., & Teasdale, R. D. (2008). LOCATE: A mammalian protein subcellular localization database. *Nucleic Acids Research, 36*, D230–D233.

Varum, S., Rodrigues, A. S., Moura, M. B., Momcilovic, O., Easley, C. A. T., Ramalho-Santos, J., et al. (2011). Energy metabolism in human pluripotent stem cells and their differentiated counterparts. *PLoS One, 6*, e20914.

Weber, P. J. A., Weber, G., & Eckerskorn, C. (2004). Isolation of organelles and prefractionation of protein extracts using free-flow electrophoresis. *Current Protocols in Protein Science, 32*, 22.5.1–22.5.21.

Williamson, A. J., Smith, D. L., Blinco, D., Unwin, R. D., Pearson, S., Wilson, C., et al. (2008). Quantitative proteomics analysis demonstrates post-transcriptional regulation of embryonic stem cell differentiation to hematopoiesis. *Molecular & Cellular Proteomics*, 7, 459–472.

Wollscheid, B., Bausch-Fluck, D., Henderson, C., O'Brien, R., Bibel, M., Schiess, R., et al. (2009). Mass-spectrometric identification and relative quantification of N-linked cell surface glycoproteins. *Nature Biotechnology, 27*, 378–386.

Yan, W., Aebersold, R., & Raines, E. W. (2009). Evolution of organelle-associated protein profiling. *Journal of Proteomics, 72*, 4–11.

CHAPTER EIGHT

Screening of Protein–Protein and Protein–DNA Interactions Using Microarrays: Applications in Biomedicine

Juan Casado-Vela*[,1,2], Manuel Fuentes†, José Manuel Franco-Zorrilla*
*Centro Nacional de Biotecnología, Spanish National Research Council (CSIC), Madrid, Spain
†Centro de Investigación del Cáncer/IBMCC (USAL/CSIC), IBSAL, Departamento de Medicina, Unidad de Proteomics & Servicio General de Citometría, University of Salamanca, Salamanca, Spain
[1]Corresponding author: e-mail address: jcasado@cnb.csic.es; jcasado@atomm.es
[2]Current address: Atomm -R&D consulting services-. www.atomm.com. José Abascal 57, 7D. 28003 Madrid

Contents

Abstract

In this report, we focus on two different array-based technologies that enable large-scale screening of protein interactions. First, protein arrays focus on the identification of protein–protein interactions (PPIs). Second, DNA arrays have also evolved to explore the identification of protein–DNA interactions (PDIs), offering novel tools to control key biological processes. Such a tool is termed protein-binding DNA arrays (also protein–DNA arrays or protein-binding microarrays). These two array-based technologies share unrivaled screening capabilities and constitute valid approaches to address biological questions at the molecular level and, eventually, may be used in biomedical applications. Outstanding achievements of these technologies and their eventual application in biomedicine are discussed here, including the identification and characterization of

Advances in Protein Chemistry and Structural Biology, Volume 95
ISSN 1876-1623
http://dx.doi.org/10.1016/B978-0-12-800453-1.00008-7

biomarkers, screening of PPIs, detection of protein posttranslational modifications and biofluid profiling. Advantages and limitations of protein arrays, protein-binding arrays, and other proteomic technologies are also discussed here. Finally, we built a list of dedicated databases and on-line resources comprising updated information on human PPIs and PDIs that can serve as a toolbox for researchers in the field.

1. INTRODUCTION

The pool of molecules concomitantly contributing to regulate and keep homeostasis in living organisms includes: nucleic acids, proteins (including enzymes and peptides), and metabolites. All of them play essential roles and, thus, understanding their interaction is a fundamental step that may foster biomedical applications. Currently, there is no single technique able to cope with this complex mixture of molecules at the same time. For that reason, different analytical techniques combined with enrichment or purification protocols is required to address their identification, modulation, and to measure their dynamic changes (Fig. 8.1).

The analytical techniques used for the characterization of each molecule are very different and the interpretation of data typically requires specialization and expertise. Thus, four areas of science emerged, termed genomics, transcriptomics, metabolomics and proteomics. These areas share a common basic aim, which is the large-scale identification and characterization of pools of biological molecules. From a general and simplistic perspective, genomics and transcriptomics address the study of the DNA and RNA molecules, metabolomics focuses on the study of "small molecules", differing in chemical composition and properties. Finally, proteomics addresses the study of proteins, enzymes and peptides. As displayed in Fig. 8.1, the complexity of each -omic approach is significantly higher (genomics $<$metabolomics$<$proteomics), due to the increasing repertoire of different molecules with high complexity to be analyzed.

From all the possible interactions among biomolecules within cells, in this report we focus on protein interactions (specifically, protein–protein interactions (PPIs), and protein–DNA interactions (PDIs)) and relevant techniques enabling their characterization in high-throughput format. Proteins carry out the majority the biochemical reactions within cells and may also function as signal messengers or as gene transcription factors. Thus, proteins constitute frequently targets for drug design and might also serve as biomarkers in biomedical applications. Importantly, living organisms include a

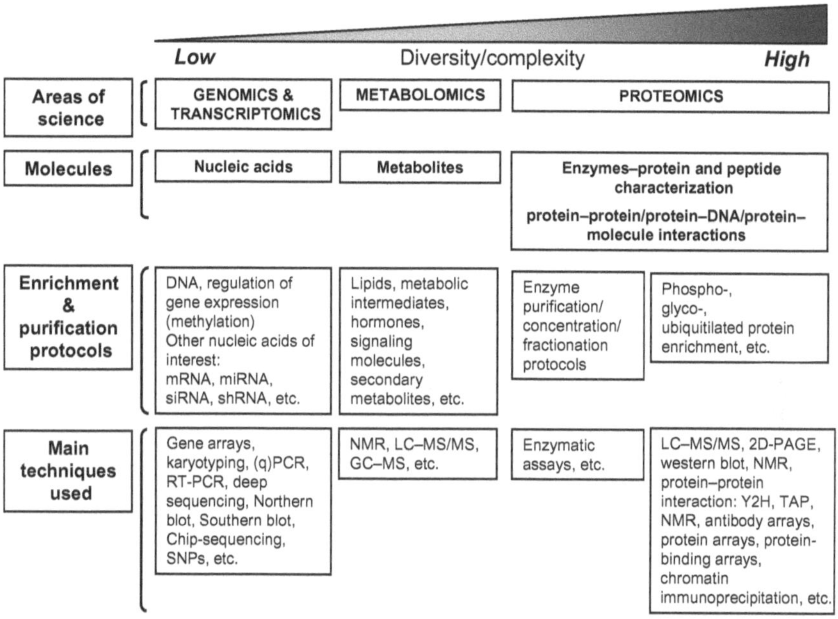

Figure 8.1 (A) Overview of the main types of compounds typically studied, well-established enrichment protocols, and main techniques used. (B) A comprehensive view of living organisms requires the functional integration of data comprising three closely related branches of study, genomics/transcriptomics, proteomics/enzymology/peptidomics, and metabolomics. Abbreviations: (q)PCR, quantitative polymerase chain reaction; RT-PCR, real-time polymerase chain reaction; SNPs, single nucleotide polymorphism; Chip, chromatin immunoprecipitation; NMR, nuclear magnetic resonance; LC–MS/MS, liquid chromatography coupled to tandem mass spectrometry; GC–MS, gas chromatography coupled to tandem mass spectrometry, 2D-PAGE, two dimensional polyacrylamide gel electrophoresis; Y2H, yeast two hybrid; TAP, tandem affinity purification.

number of mechanisms leading to the generation of an ample repertoire of proteins, which relies on three main processes: first, at deoxyribonucleic acid (DNA) level (i.e., gene polymorphisms), second, at precursor messenger ribonucleic acid (pre-mRNA) or messenger ribonucleic acid (mRNA) level (i.e., alternative splicing or differential splicing) and, finally, at the protein level (i.e., posttranslational modification). On top of this, other sources of complexity include the occurrence of chimeric proteins (Casado-Vela, Lacal, & Elortza, 2013). It has to be also noted that proteins rarely exist as isolated entities inside living organisms and typically exert their biological, biochemical or signaling activity either through binary interactions with other molecules or by forming complexes and aggregates. Therefore, unraveling the identity and the dynamic changes of such protein interactions becomes crucial since they may provide the basis for protein regulation and, thus, tools for controlling physiological alterations and signaling networks where proteins are involved in.

Nowadays, there are several technologies available to address the study PPIs. Yeast two-hybrid assay (Y2H) is a well-established methodology that permits the rapid identification of binary interactions between a chosen test protein, termed "bait" and an interacting protein(s), termed prey(s) (Fields & Song, 1989). Y2H methods were reviewed by Bruckner, Polge, Lentze, Auerbach, and Schlattner (2009). Basically, the identification of interacting proteins by Y2H is achieved by expressing the bait protein as a hybrid, fused to the DNA-binding domain of a transcription factor, and screening it against a library of prey candidates that are fused to the corresponding transcriptional activation domain. Both fusions are expressed in yeast cells that carry a reporter gene whose expression is under the control of the transcription factor, such that the interaction of the "two hybrids" leads to expression of the reporter. Other methods used to decipher PPIs include affinity purification methods that are based on antibodies mediated enrichment and tandem affinity purifications (TAP; see, Xu et al., 2010 for review) which is also being widely used for protein complex purification. These techniques, in combination with mass spectrometry have led to valuable information about protein complexes partners and PPI. Lately, array based methods as nucleic acid programmable protein array (NAPPA; Sibani & LaBaer, 2011) or protein *in situ* arrays (PISA; He, Stoevesandt, & Taussig, 2008) became popular approaches to test interactions in high-throughput format. Despite of robustness and development of these approaches, it has to be noted that none of the methods used to the study of PPIs are comparably better than other and typically lead to the identification of different subsets of interactions (Braun et al., 2009; Chen et al., 2004; Chen, Rajagopala, Stellberger, & Uetz, 2010; Chen, Zhou, Sanders, Nolan, & Cai, 2009). Analysis of protein complexes using affinity purification followed by mass spectrometry typically identifies directly and indirectly associated proteins, whereas Y2H analyses identify direct, binary PPIs (Braun et al., 2009). Moreover, different Y2H systems can be used, which also leads to detection of markedly different subsets of interacting proteins, reviewed in (Chen et al., 2010). Therefore, the results derived from different PPI analyses are complementary.

Previous publications inferred different estimations of the human interactome (Table 8.1) with figures widely differing from ~13,000 to

Table 8.1 Previous estimations of the human interactome, brief description, and references

Estimation of the human interactome (publication year)	Description and references
375,000 (2005)	These authors used literature-mining algorithms and then estimated the number of protein interactions assuming 25,000 human genes (Ramani, Bunescu, Mooney, & Marcotte, 2005).
154,000–369,000 (2006)	The authors quoted that their estimation includes protein complexes (Hart, Ramani, & Marcotte, 2006).
650,000 (2008)	This estimation relies on data retrieved from Y2H[a] experiments and database searches (Stumpf et al., 2008)
130,000 (2009)	This number of protein interactions exclusively considered binary interactions and considered data from four repeat Y2H screens (Venkatesan et al., 2009)
13,217[b] (2012)	This estimation considered the longest protein isoform[c] of 20,846 human protein sequences. The size of the interactome was estimated using computational methods based on structural inference. The authors quoted that the figure proposed includes self-interactions (Tyagi, Hashimoto, Shoemaker, Wuchty, & Panchenko, 2012).

[a]Y2H: yeast-two-hybrid.
[b]Including self-interactions and based on structural inference.
[c]See Casado-Vela, Cebrian, del Pulgar, et al. (2011) for definitions on protein isoform and protein species.

~370,000. Even the definition of the term "protein interaction" significantly varies among references (binary interactions or protein complexes, stable interactions versus weak or transient, predicted or computationally inferred interactions, etc.).

Different experimental approaches may also be used to achieve information on PDIs as follows: nitrocellulose-binding assays (Woodbury & von Hippel, 1983), gel shift analysis (Jansen, Gronenborn, & Clore, 1987), southwestern blotting (Bowen, Steinberg, Laemmli, & Weintraub, 1980; Miskimins, Roberts, McClelland, & Ruddle, 1985), yeast reporter constructs (Hanes & Brent, 1991), electrophoretic mobility shift assays (EMSA; Lane, Prentki, & Chandler, 1992) and nuclease foot printing (Hampshire, Rusling, Broughton-Head, & Fox, 2007). All those techniques are low throughput and are usually considered too laborious if the aim is the analysis of a wide range of DNA sequence variants able to bind proteins. More importantly, the previous mentioned protein–DNA techniques are limited to a small number of DNA sequences tested. More recently, the combination of classical methodologies and high-throughput techniques is contributing to the characterization of PDIs. These methodologies include bacterial one-hybrid (Meng, Brodsky, & Wolfe, 2005) and those based on the systematic evolution of ligands by exponential enrichment (SELEX) coupled to large-scale DNA sequencing (Jolma et al., 2010; Roulet et al., 2002). Protein-binding DNA arrays (also found as protein–DNA arrays or protein-binding microarrays in the literature) offer a high-throughput strategy for the analysis of specific PDIs. Briefly, protein-binding DNA arrays consist of a matrix of double stranded DNA probes arrayed on a solid surface that can be probed with the test protein. DNA–protein complexes are detected by fluorescence emission, either directly from the protein labeled with a fluorochrome, or through an immunochemical reaction (Berger et al., 2006; Warren et al., 2006). Protein–DNA arrays (PDI-arrays) allow interrogating thousands of DNA molecules in a single experiment and identifying the DNA sequences specifically recognized by a particular protein. With a particular interest on transcription factors, the use of PDI-arrays is contributing in the elucidation of the cis-elements responsible for gene expression and thus, in the deciphering of the transcriptional code (Badis et al., 2009).

Regarding proteins with DNA-binding properties, transcription factors constitute a subset of proteins intensively studied due to their eventual biomedical role. These proteins regulate gene transcription, mainly through binding to particular DNA sequences, thus, activating or inhibiting their transcription. In 2004 (Babu, Luscombe, Aravind, Gerstein, & Teichmann, 2004), it was proposed that more than 2600 transcription factor can be found in human cells. Nevertheless, that figure is just an assumption based on the occurrence of DNA-binding domains in the sequence of proteins. The number of proteins with DNA-binding properties still remains unknown.

From the literature, it becomes evident that our understanding of the human interactome is in its early stages. Nevertheless, there is ample agreement on the idea that deciphering PPIs and PDIs may be crucial to understand protein function. In this sense, array technologies open the possibility to address this task and allow the identification of both PPIs and PDIs, which constitute an "always-pursued" objective in science. Therefore, array technology opens the door for addressing a plethora of biological questions through sensitive and high-throughput (i.e., several thousand proteins/molecules can be screened in a single experiment) approaches. The rapid development and miniaturization of arrays led to remarkable achievements in biomedicine. Miniaturization seems a must, in our race towards the identification of novel protein biomarkers.

2. PROTEIN ARRAYS

Protein arrays may be concisely defined as collections of proteins attached to known positions on solid surfaces. Albeit the apparent simplicity of the concept "protein array," different terms and definitions may be found in the literature such as forward-phase or reverse-phase protein arrays, tissue or antibody protein arrays and cell-free or cell-based protein arrays (LaBaer & Ramachandran, 2005; Lee, Magee, Gaster, LaBaer, & Wang, 2013; Liotta et al., 2003; Matarraz, Gonzalez-Gonzalez, Jara, Orfao, & Fuentes, 2011). Importantly, different criteria may be used to classify protein arrays, such as the type and source of the proteins printed on the arrays, their applications or the technologies used to build and detect interactions (Casado-Vela, Gonzalez-Gonzalez, et al., 2013). A list of types of protein arrays described in the literature, concise description, main applications, and references is included in Table 8.2.

Since the introduction of the first arrays published in 2000–2001 (Haab, Dunham, & Brown, 2001; MacBeath & Schreiber, 2000; Miller, Butler, Teh, & Haab, 2001) both printing and detection technologies evolved rapidly, mainly because of their potential applications. Originally, the idea of building protein arrays was based on the immobilization of purified target proteins (obtained by heterologous cell-based systems or using recombinant protein technology) directly onto defined positions on the array surface. This idea may be extended to the attachment of known sets of antibodies to build antibody arrays (Fig. 8.2).

Figure 8.2A shows an overview of the general workflow and key steps involved in protein array and antibody array analysis. In protein arrays, the target proteins are attached on defined positions of the array surfaces and they are covered with a solution of the query protein of interest. Conversely, in antibody arrays a set of known antibodies are printed on defined positions

Table 8.2 Type, description, and main applications of protein arrays

Type	Description	Main applications
Forward phase protein arrays (Espina et al., 2003; Gulmann, Sheehan, Kay, Liotta, & Petricoin, 2006; Liotta et al., 2003; Paweletz et al., 2001; Wang et al., 2006)	Purified protein preparations in solution are used to cover the surface of arrays where target proteins are immobilized on defined positions of the array.	Typically, these arrays aim to identify molecules (immobilized on defined positions the surface of arrays) that bind to a protein of interest (layered on the array surface as a solution).
Reverse phase protein arrays (Chandra, Reddy, & Srivastava, 2011; Espina et al., 2003; Gulmann et al., 2006; Liotta et al., 2003; Paweletz et al., 2001; Wang et al., 2006)	Complex protein mixtures extracted or purified from any biological sample (e.g., cellular lysate, tissue lysate, proteins extracted or purified from tumor specimens, core needle biopsies or biofluids such as blood, serum, plasma, cerebrospinal fluid, synovial fluid, etc.,) are directly attached on the surface of the arrays. Tissue protein arrays may be considered a subtype of reverse phase protein arrays. The proteins of interest are present in the sample(s) under study. Tissue protein arrays are defined as the ordered attachment of tissue core slices (ranging from few and up to thousands cores) extracted from donor blocks (fresh, frozen, or paraffinated) on a two-dimensional planar surface—definition adapted from Gulmann et al. (2006).	Reverse phase arrays typically aim the identification of proteins present in complex mixtures directly immobilized on the surface of the array. Especially valuable for screening purposes in clinical research. High-throughput immunohistochemical studies, identification of auto-antibodies, and allo-antibodies present in biological fluids.

Table 8.2 Type, description, and main applications of protein arrays—cont'd

Type	Description	Main applications
Differential profiling and screening protein arrays (Schweitzer, Predki, & Snyder, 2003)	This type of protein arrays consists on the detection of proteins differentially expressed among samples, enabling differential protein profiling of normal/control versus diseased/treated.	Typically used to identify and quantify the occurrence and differential expression profiling of multiple proteins simultaneously. The aim of differential profiling is the eventual identification of biomarkers, representative of an alteration/disease/pathology with potential diagnostic or prognostic value.
Functional protein arrays (Schweitzer et al., 2003)	Focused on the identification and characterization of the specific function of proteins, their regulation, and their interaction with other molecules (including protein–protein, protein–peptide, protein–lipid, protein–nucleic acid, or protein–small molecule/drug interactions).	The objective is the detection of multiple interactions with low reagent consumption in a cost-effective manner and in a reduced time scale. Functional protein arrays also aim the detection of posttranslational modifications, which may modulate protein function.
Arrays from biological samples (Paweletz et al., 2001)	Protein pools extracted from biological samples. In this case, the pool of proteins present in biological samples includes a wide diversity of protein species (Casado-Vela, Cebrian, del Pulgar, et al., 2011, for review) derived after alternative splicing and posttranslational modification processes. The occurrence of single nucleotide polymorphisms could also increase protein diversity.	These arrays are of special interest in clinical applications since the sample (mainly protein extracts from biopsies) represent the *statu quo* of the individuals.

Continued

Table 8.2 Type, description, and main applications of protein arrays—cont'd

Type	Description	Main applications
Cell-based protein arrays (proteins are produced using heterologous cells) (Ptacek et al., 2005)	Cellular systems are used as protein factories. Complex mixtures of proteins (cells or tissue lysates) are attached on two-dimensional planar surfaces. In some cases, another dimension is included when lysate is based on subcellular fractionation.	Thousands of different proteins individually produced and purified are spotted on the surface of arrays. Low density (up to hundreds of proteins) or high-density (thousands of proteins) arrays may be printed in a "tailor-made" manner.
Cell-free protein arrays (proteins are produced using cell extracts directly on the surface of arrays) (He & Taussig, 2001; Ramachandran et al., 2004, 2008; Tao & Zhu, 2006)	*Protein in situ arrays* (*PISA*) (He & Taussig, 2001). DNA molecules are used as templates to produce proteins that become attached on the surface immediately after their synthesis through recognition of a tag sequence. *Puromycin capture protein arrays* (*PuCa*) (Tao & Zhu, 2006). DNA molecules are transcribed into mRNA *in vitro* and the 3′-end of the mRNA is hybridized with single stranded DNA (ssDNA) oligonucleotides modified with biotin and puromycin. The mRNAs molecules are, then, arrayed on a streptavidin-coated slide and translated after immersion with a cell-free lysate. The puromycin molecules attached to each DNA are able to capture and immobilize the nascent proteins	Cell-free protein arrays can be used in basic research, but the aim is to build protein arrays including subsets of proteins that may serve for diagnostic or prognostic purposes in clinical routine.

Table 8.2 Type, description, and main applications of protein arrays—cont'd

Type	Description	Main applications
	Nucleic acid programmable protein arrays (*NAPPA*) (Ramachandran et al., 2004, 2008). cDNAs encoding for the proteins of interest are printed on the surface nearby positions where antibodies selectively bind proteins *de novo* synthesized directly on the array surface using commercially available *in vitro* expression kits.	
Antibody arrays (Wingren & Borrebaeck, 2009)	Polyclonal and/or monoclonal antibodies, preferably well characterized, are attached on defined positions on the surface of arrays enabling the detection of cognate-binding molecules, tissue lysates, biofluids, and cell surface epitopes.	Antibodies specifically binding proteins (or specifically binding lipids, sugars, or other small molecules) are produced, purified, and printed on the surface of protein arrays. Depending on the specificity of the antibodies, many different molecules may be screened. The main limitation is the purity, specificity, and affinity of the antibodies printed on the surface of arrays.

of the array, enabling the binding of cognate protein partners. Although protein arrays and antibody arrays are conceptually different, they both rely on the detection and quantification of fluorescent signals on defined specific areas of arrays. Thus, protein arrays and antibody arrays share the technologies used for digital image acquisition and image analysis, but differ in the algorithms used for data interpretation. In the case of PPIs, the identification of protein-binding partners is typically calculated using the *z*-score value—see Díez et al. (2012) for detailed description on the calculation of this and other statistical strategies used in microarray data analysis. In antibody arrays, two different conditions (such as control versus treatment) are typically

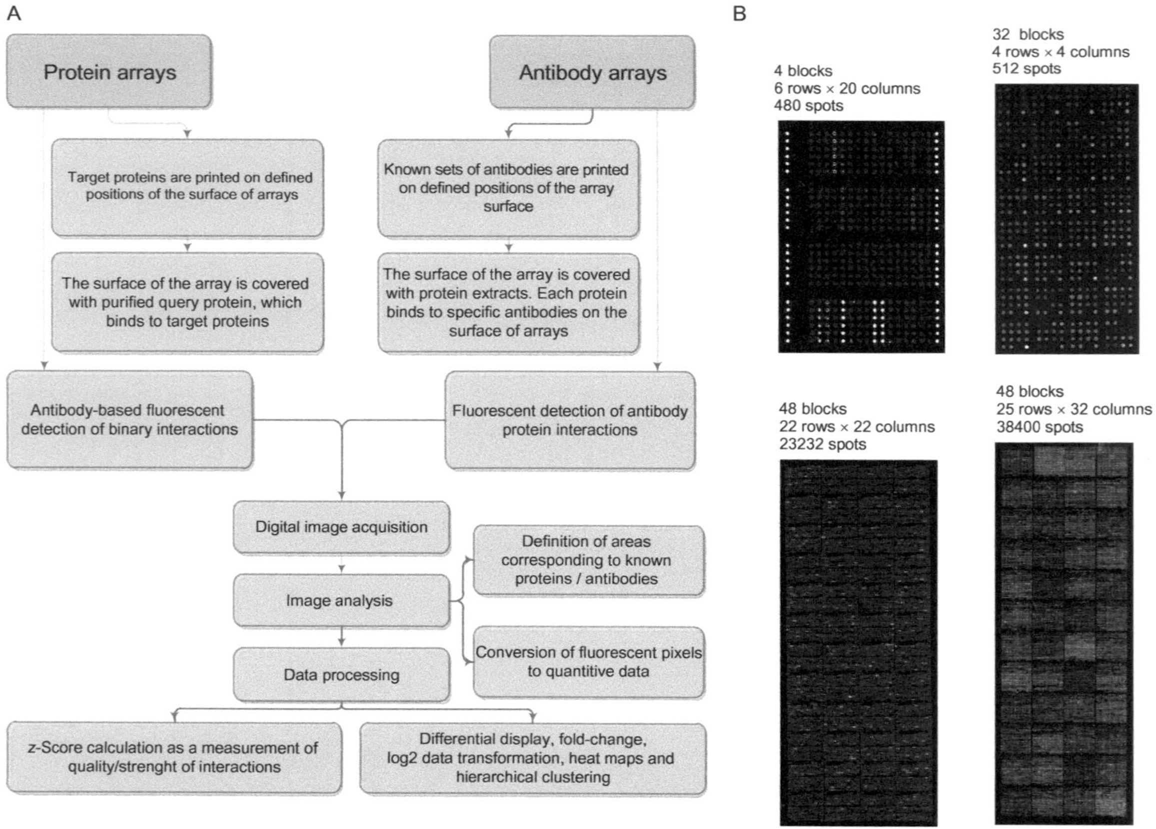

Figure 8.2 (A) Outline of the general experimental workflow for protein array and antibody arrays analysis. (B) Overview of different array configurations showing increasing number of features/spots printed and detected as round fluorescent spot signals. Low-density arrays (less than 1000 spots) and high-density arrays (typically over 1000 spots) are shown. The highest density array (bottom-right corner) corresponds to HuProt™ arrays, bearing more than 17,000 full length human proteins distributed in 38,400 spots printed on a single array surface the image was provided by Dr. Ronny Schmidt from Cambridge Protein Arrays Ltd.

compared, facilitating data comparison using differential display image analysis, fold-change measurements, heat map, and hierarchical clustering analysis.

Protein arrays also constitute flexible tool since, regardless of the nature of the molecules of interest (proteins, antibodies, or even peptides), they all may be attached onto solid surfaces in different physical configurations and in a wide range of spot densities to adapt to experimental needs Fig. 8.2B. This task requires both automation and miniaturization, typically achieved using automated spotting platforms. To exemplify this, low-density protein arrays (480 spots and 512 spots, respectively, printed with different block, row and column disposition) and high-density arrays (printed arrays including 23232 and 38400 spots) are shown in Fig. 8.2B. In all cases, protein spots (differing in signal intensities) were detected after digital image acquisition.

Regarding the source of the proteins printed on arrays, different techniques can be used. In antibody arrays, a wide variety of antibodies is currently available in the market and the technologies to produce polyclonal and monoclonal antibodies are well known. On the other hand, a wide range of protein purification strategies have been described, but the use of bacterial systems as protein factories is still frequent. Despite the proven utility of cellular systems to produce proteins, several limitations hampered their use in protein arrays (Angenendt, Kreutzberger, Glokler, & Hoheisel, 2006): (a) different cellular systems typically differ in protein production yields, which may also differ from batch-to-batch, (b) purification of the protein(s) products of interest is required and the purification yield could be difficult to predict, (c) after purification, the proteins could be improperly folded, not functional or insoluble, (d) the posttranslational modification status of the proteins produced may differ depending on the cellular system used, and (e) the production and purification of recombinant proteins are tedious and slow. As a mean to circumvent some of the inherent limitations of cell-based protein arrays, cell-free approaches emerged as a valid alternative. Three different cell-free protein array technologies have been reported in the literature: PISA (He & Taussig, 2001), puromycin capture protein arrays (PuCa; Tao & Zhu, 2006), and NAPPA (Ramachandran et al., 2004, 2008). Briefly, in the three cell-free protein array strategies the proteins are synthesized from their corresponding messenger ribonucleic acid (mRNA) or complementary deoxyribonucleic acid (cDNA) templates directly on the surface of arrays and using *in vitro* cell-free coupled transcription and/or translation expression systems. As a result, PISA, PuCa, and

NAPPA render protein arrays including a matrix of different proteins attached on defined positions of the arrays. Out of the three cell-free approaches, NAPPA evolved at a rapid pace due to several advantages. With the NAPPA strategy, protein synthesis may be easily achieved in a few hours just by immersion of the mRNA or cDNA molecules of interest within appropriate cellular lysates. Such lysates contain the necessary machinery (transcription, translation factors, chaperones, etc.,) to trigger transcription and/or translation to produce the corresponding protein products in a cell-independent manner (reviewed in Katzen, Chang, and Kudlicki, 2005). It is important to mention that cell-free coupled transcription and translation is preferred because it usually achieves higher protein yields and eliminates mRNA handling at the same time (Arduengo, Schenborn, & Hurst, 2007). From the three cell-free strategies and bearing in mind the increasing number of publications in the field, it seems that the NAPPA strategy is gaining ground in the scientific community for the screening of PPIs.

As depicted in Fig. 8.3A, in NAPPA arrays, cDNA encoding for the protein(s) of interest are configured to append a common protein tag on their N- or C-terminus. Commonly, the tag is in the C-terminus, which provides a triple advantage: first, it ensures that the proteins bound onto the array correspond to full-length proteins. Second, the tag facilitates protein detection using anti-tag antibodies. Finally, the surface of the array is coated with a single type of antibody, as a capture reagent, selectively binding a common protein tag that serves as anchor for all the different protein products. Proteins are *de novo* synthesized directly on the array using commercially available *in vitro* expression systems. The novel proteins are further captured by anti-tag antibodies coprinted close to each specific cDNA molecule. As a result, a layer of proteins (termed "target" proteins) is immobilized forming a network where the position of every protein is well established. Using this strategy, PPIs can be detected as follows: soluble cDNA encoding for a protein of interest (query protein) is simultaneously coexpressed with a set of target proteins directly on the surface of arrays. This step is carried out using the same *in vitro* expression system but the plasmid vector coding for the query and the target proteins typically harbor different tags. Target proteins bind to each specific defined positions on the surface of arrays and remain exposed for interaction with the query protein (Figure 8.3B). The surface of the array is then covered with a solution of an antibody specifically binding to the tag of the query protein. Eventually, the antibody reveals those positions of the array where a query–target protein interaction is taking place, thus, revealing the occurrence of PPIs.

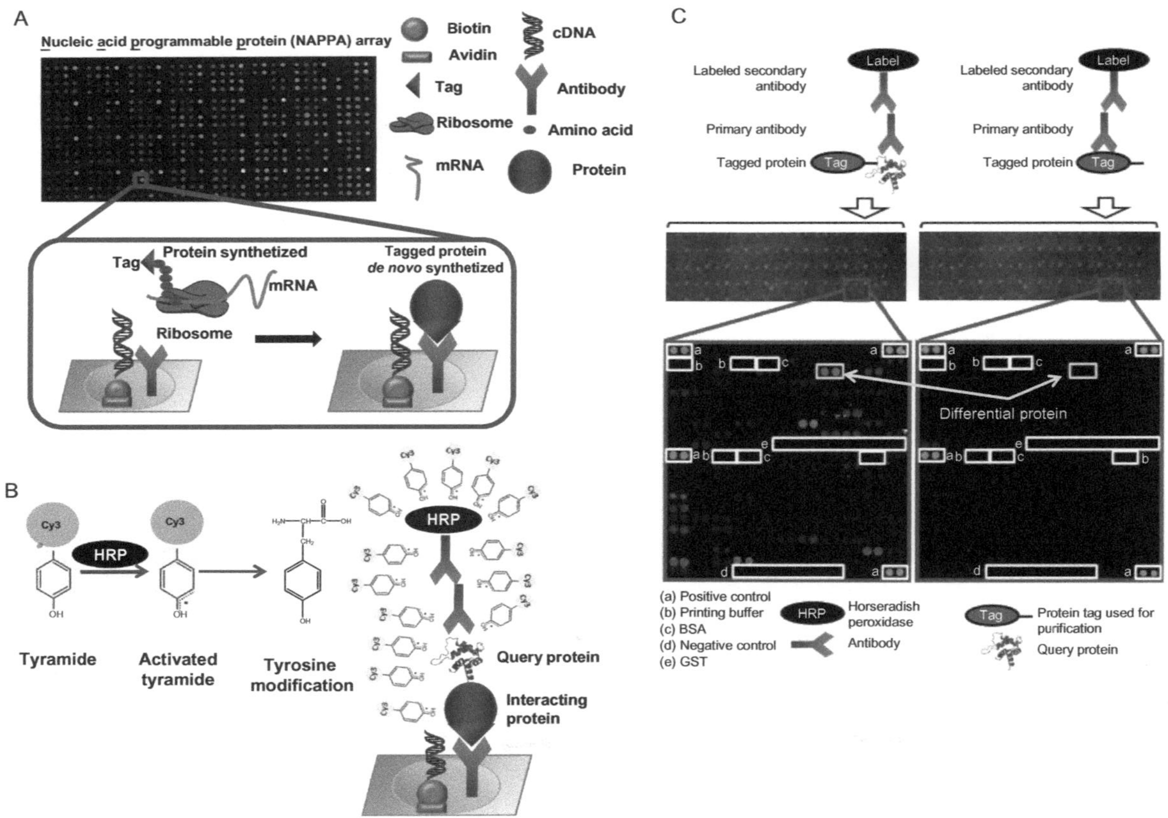
A
Nucleic acid programmable protein (NAPPA) array
Biotin
Avidin
Tag
Ribosome
mRNA
cDNA
Antibody
Amino acid
Protein
Tag
Protein synthetized
mRNA
Ribosome
Tagged protein
de novo synthetized
B
Tyramide
HRP
Activated tyramide
Tyrosine modification
HRP
Query protein
Interacting protein
C
Labeled secondary antibody
Primary antibody
Tagged protein
Label
Tag
Differential protein
(a) Positive control
(b) Printing buffer
(c) BSA
(d) Negative control
(e) GST
Horseradish peroxidase
Antibody
Protein tag used for purification
Query protein

Figure 8.3C shows a differential display analysis corresponding to the identification of PPIs detected after the comparison of two fluorescent images. The image on the left corresponds to NAPPA arrays incubated with a tagged protein. The image on the right corresponds to a control experiment using only the tag. Close-up views of the digital images clearly show a number of protein partners (printed in duplicates).

2.1. Recent achievements of the protein arrays and their application to address the study of the human proteome

Protein arrays showed to be successfully applied for screening of PPIs and derived applications such as protein expression profiling, biomarker discovery, screening of PPIs, identification of posttranslational modifications and, eventually, for clinical diagnostic purposes.

Focusing on PPIs, we would like to emphasize the capabilities of protein arrays as a tool enabling the screening of PPIs in a large-scale manner. Several hundred binary PPIs may be detected in a single experiment, as it was reported for amyloid-β—a protein involved in Alzheimer's disease (Virok et al., 2011)—or NF-κB—an essential modulator involved in cellular signal transduction (Fenner, Scannell, & Prehn, 2010). Indeed, high-density protein arrays enabling the screening of PPIs are commercially available, which include several thousands of human proteins (or fragments of the proteins) printed on standard glass slide (7.5 × 2.5 cm). The main problem now resides

Figure 8.3 (A) Fluorescence image protein array. Those spots displaying higher luminescence correspond to binary protein–protein interactions. A zoomed view on a single spot displays a schematic view of the main processes taking place during *de novo* protein synthesis before incubation of the protein of interest (*de novo* protein synthesis, antibody-based protein binding to known positions on the array). (B) Schematic view of fluorescent (Cy3) tyramide activation by horseradish peroxidase (right panel displays fluorescent signal amplification in those array positions where the query protein binds to interacting proteins *de novo* synthesized on the array surface). Fluorescent signal amplification only takes place if binary protein–protein interactions take place. (C) Differential display analysis corresponding to the identification of protein–protein interactions detected after the comparison of two fluorescent images. The image on the left corresponds to NAPPA arrays incubated with a tagged protein. The image on the right corresponds to a control experiment using only the tag. Close-up views of the digital images clearly show a number of protein partners (printed in duplicates). Quality controls including fluorescent dyes, printing buffer, BSA, GST, and negative controls are also displayed.

in discerning those interactions with biological meaning and their modulation.

Regarding differential profiles of protein expression, a number of studies support the idea that protein arrays are among the best technologies enabling reliable detection of low abundant proteins, even in complex mixtures of proteins such as blood/serum. As an example, protein arrays allow the identification of cytokines in serum (Huang, Huang, Fan, & Lin, 2001; Kader et al., 2005; Mor et al., 2005). Cytokines comprises a family of more than one hundred small proteins that are involved in different cellular processes (e.g., proliferation, inflammation, immunity, migration, fibrosis, repair, and angiogenesis; Feldmann, 2008; McInnes & Schett, 2007) and seem to be also involved in inflammatory disorders and cancer (Huang et al., 2001, 2005). Therefore, cytokines hold tremendous potential as therapeutic targets (Feldmann, 2008) and the detection of the whole repertoire of cytokines is pursued. In this regard, it has been demonstrated that the applicability of antibody protein arrays to detect the presence of 169 different proteins, including cytokines. Although the list of antibodies arrayed and their cognate-binding proteins was not pointed out (Mor et al., 2005), the authors described significant increased cytokine levels in sera from patients with ovarian cancer. Noteworthy this report represents a relevant approach demonstrating the applicability of protein arrays to detect low abundant proteins in serum.

Currently, new proteomics tools have been progressively introduced to unravel changes under pathologic conditions. Conceivably, biomarker discovery might be described as the identification and quantification of protein(s) characteristic of defined biological responses, such as pathogenic processes or treatment responses (Aronson, 2005). Many examples in the literature aimed the identification of protein biomarkers in cancer (Hudson, Pozdnyakova, Haines, Mor, & Snyder, 2007; Matarraz et al., 2011). In a report by Hudson et al. (2007) the authors built high-density protein arrays containing >5000 human proteins and probed them with sera from ovarian cancer patients. After comparison with control sera, 94 tumoral antigens were identified acting as potential ovarian cancer biomarkers. As an extrapolation of these findings, the applicability of protein arrays for biomarker discovery can be extended to other disorders including liver disease, immunological disorders, and bacterial infection as previously reviewed (Wilson, Liotta, & Petricoin, 2010).

In the field of posttranslational modifications, it is well known that phosphorylation is one of the most important and common ways of regulating protein function and biological processes. For that reason, this type of protein modification deserves special attention. Remarkable applications of protein arrays include the identification of phosphorylated proteins and protein kinase-mediated signaling networks in humans (Feilner et al., 2005; Ptacek et al., 2005; Tarrant & Cole, 2009). A brilliant application of protein arrays to address the study of protein phosphorylation was published by Nielsen (Nielsen, Cardone, Sinskey, MacBeath, & Sorger, 2003). These authors built an array of monoclonal antibodies specific for phosphorylated +/− forms of EGFR, ErbB2, and TfR, three proteins involved in the ErBb signal transduction pathway (Yarden & Sliwkowski, 2001). In that report, the authors combined detection of proteins of interest, detection of posttranslational modifications, measurement of protein abundance, and protein kinetics after treatment with EGF (an upstream regulator of the ErBb pathway) at different time points. The use of protein arrays to detect other posttranslational modifications may also be found in the literature, such as ADP-ribosylation a protein modification still poorly characterized (Feijs et al., 2013) but in all cases, the identification of protein posttranslational modifications rely on the availability of well characterized antibodies specifically binding the modifications of interest.

The applicability of protein arrays for clinical diagnostic purposes resides in the possibility of performing large-scale (i.e., up to several thousand proteins may be monitored in a single experiment) comparisons of protein expression patterns occurring in two or more biological samples, allowing the detection of those altered protein profiles that harbor biomedical implications. Belov and coworkers reported a remarkable application of protein arrays that serves to diagnose leukemia and even discern among three major subtypes of leukemia (acute myeloid leukemia, AML; acute lymphocytic leukemia, ALL; and chronic lymphocytic leukemia, CLL) by incubating the arrays directly with leukocyte sample preparations from donor samples (Belov, de la Vega, dos Remedios, Mulligan, & Christopherson, 2001; Belov, Huang, Barber, Mulligan, & Christopherson, 2003). For this purpose, the authors built antibody arrays containing a panel of antibodies specifically binding cluster of differentiation (CD) antigens. With this approach, focusing on the screening of proteins leukocyte surface antigens, the authors demonstrated that antibody arrays in combination with purified leukocyte blood preparations constitute a valid alternative to other conventional techniques (i.e., flow cytometry) currently in use to detect chronic lymphocytic leukemia. Interestingly, a similar approach also served to detect the occurrence of tumor-infiltrating lymphocytes found in colorectal cancer specimens surgically resected (Ellmark et al., 2006).

The construction of customized antibody arrays specifically targeting modified proteins is also feasible. As an example, Chen et al. (2009) described the constructions of arrays bearing a small subset of three antibodies specifically binding glycoproteins (alpha-1B-glycoprotein, serum amyloid P-component, and antithrombin-III). The authors claimed that the different expression levels in these proteins, proposed as biomarkers, could serve for early detection or pancreatic cancer.

Other examples of the applicability of protein arrays that hold potential for diagnostic purposes include their use for the detection of antibodies present in blood, serum, or biofluids raised against intrinsic (auto-antibodies) or extrinsic antigens (alo-antibodies). An outstanding report showing the applicability of protein arrays for detection of auto-antibodies was published by Madoz-Gurpide in 2008 (Madoz-Gurpide, Kuick, Wang, Misek, & Hanash, 2008). In this report, reverse phase protein arrays were constructed by immobilizing lung adenocarcinomas cancer cell protein extracts. These arrays were used to detect auto-antibodies present in serum samples from lung cancer patients. The occurrence of increased levels of auto-antibodies in serum was evidenced through their binding to specific antigens present in lung cancer cell models attached on protein arrays. Consequently, this technique represents a promising tool for early diagnosis of cancer (Anderson & LaBaer, 2005; Anderson et al., 2008, 2011; Locker et al., 1999; Madoz-Gurpide et al., 2008), but could also be extended to other diseases or alterations leading increased levels of auto-antibodies in blood or biological fluids.

Finally, protein arrays may also be used for identification of human pathogens. A brilliant example of the last application was reported by Beare et al. (2008), who virtually expressed the whole proteome (i.e., 1988 open reading frames, representing 97.2% of the coding sequences) of a human pathogen termed *Coxiella burnetii* using cell-free protein expression systems. The resulting proteins were purified and attached on arrays to identify the occurrence of antibodies in sera from individuals infected with this bacterium. As a result, the authors identified 44 immunoreactive proteins. The advantages of such protein array including the proteome of this pathogenic bacterium were double: first, protein arrays allowed discerning infected from noninfected individuals. Second, the results served as a solid basis for the design and development of vaccines against immunogenic proteins. Recent reports conducted using the same pathogenic bacteria as a model validated those observations (Vigil et al., 2010, 2011). The examples above demonstrate that protein arrays represent a promising tool in biomedical research and open new avenues for the development of protein arrays that can serve as reliable diagnostic tests in clinical practice.

2.2. Advantages and limitations of protein arrays

Both, the technologies used for building protein arrays and the increasing number of applications contribute to provide advantageous characteristics to this technology (Table 8.3). Remarkable characteristics of protein array technology include reduced reagent consumption and the ability to keep a tight control of the experimental conditions. Roger Ekins and coworkers described these binding events based on miniaturization as the key parameter. They predicted that a system that uses small amounts of capture molecules and a small amount of sample can be more sensitive than a system using one hundred times more material. This is true if $K < 0.1$ where K is the affinity constant between ligand and target. The capture ligand is presented in a confined area of the array, reducing its diffusion. The binding event with its specific target takes place with the highest possible capture molecule concentration and therefore, the highest signal intensities and optimal signal-to-noise ratios can be achieved in these small spots (Ekins & Chu, 1992; Ekins, Chu, & Biggart, 1990). An immunoassay in an array format displays sensitivities in the pM to fM range, enabling testing low-abundant (pg/mL) analytes in crude proteomes with a small volume of sample. In many cases, the sample to test is minimal so protein microarrays show a relevant advantage in clinical applications.

Notably, protein array technology offers improved reproducibility compared to techniques such as Y2H. In this regard (Ito et al., 2001) compared their results with two previous reports also using Y2H technology, and pinpointed a surprisingly low overlap among three datasets (<20% in every case). Lack of reproducibility could jeopardize the comparability and reliability of the results obtained in high-throughput screening experiments and evidenced the need of carrying out technical replicates in order to identify genuine interactions, which is also applicable to other protein interaction screening techniques. In terms of sensitivity, LC–MS/MS and gel-based analysis of complex protein mixtures frequently require protein enrichment, fractionation, or high-abundant protein depletion steps in order to detect low abundant proteins that could be biologically relevant—see Ly and Wasinger (2011) and Millioni et al. (2011) for comparative description on enrichment versus depletion strategies. Protein fractionation, enrichment, and depletion strategies may be tedious and frequently require higher sample

Table 8.3 Advantages and limitations of protein arrays to address the study of protein–protein interactions

Advantages

☑ Large-scale monitoring experiments (several thousand proteins can be monitored in a single experiment).
☑ Wide range of applications: screening, differential profiling, detection of biomarkers, functional characterization of proteins, protein–protein/peptide/molecule interactions.
☑ Low sample consumption (especially relevant for clinical applications).
☑ Fast experiments, allowing detection/quantitation/characterization of up to thousands of proteins in a single experiment.
☑ Experimental conditions can be easily controlled.
☑ Increased protein detection levels compared to other high-throughput proteomic technologies (e.g., LC–MS/MS).
☑ A range of different chemistries may be used to modify solid surface (typically glass) to achieve protein binding.
☑ Customizable, they can be constructed "ad hoc" for specific purposes.
☑ Proprietary antibodies raised in-house may be printed.

Disadvantages

☒ Typically focused on binary protein–protein interactions (not optimized for the identification of protein complexes).
☒ Increased possibility of detecting false positives as potential interactors.
☒ Validation experiments (coimmunoprecipitation, Western-blot, colocalization, etc.,) are required.
☒ The posttranslational modification status of the proteins attached on arrays is difficult to control.
☒ The largest "whole proteome" currently reported corresponds to a relatively simple model organism (yeast), complete eukaryotic proteomes remain unknown.
☒ Lack of ability to measure binding/dissociation constants.
☒ The concentration of the target protein/molecule in the solution cannot be measured.
☒ The highest density arrays reported to date include ~23,000 different positions on a single array.
☒ Several variables may alter the functionality of the proteins during the production and/or storage of the protein arrays.
☒ Complexity of experimental design, lack of standardized protocols.
☒ The proteins arrayed may not be properly folded, fully functional, or optimally oriented after binding to solid surfaces.
☒ Deep characterization of the specificity and stability of the antibodies used in antibody arrays is required.

volumes, which could be limiting in the case of clinical samples. Comparatively, the probability of detecting low abundant proteins present in small sample volumes containing complex protein mixtures seems to be significantly higher in protein array analyses. Indeed, protein detection limits as low as 3–5 amol (Kim, Daniel, & Mirkin, 2009; Thaxton et al., 2009) of protein were reported, requiring only minute amounts of sample material and without the need for enrichment protocols (Brase et al., 2010).

Some of the outstanding achievements of protein arrays reported in previous sections underline the applicability of protein arrays to detect low abundant proteins (such as cytokines and biomarkers) in complex protein mixtures (e.g., serum samples) without the need of sample fractionation or protein enrichment protocols. Therefore, the sensitivity of protein arrays is comparable to liquid chromatography coupled to selected reaction monitoring (SRM) technology (Lange, Picotti, Domon, & Aebersold, 2008). Indeed, reports using SRM approaches accurately quantified some low abundant proteins such as cancer biomarkers in human body fluids (Huttenhain et al., 2012) or quantified proteins at concentration below fifty copies per cell (Picotti, Bodenmiller, Mueller, Domon, & Aebersold, 2009). Despite the similar sensitivities achieved by protein arrays and SRM, protein arrays offer unrivaled and versatile capabilities to characterize protein–protein/molecule interactions (Hurst et al., 2009; Katz et al., 2011; Kersten et al., 2004). However, it is important to highlight the limitations of protein arrays to characterize protein partners (Table 8.3), including lack of ability to measure binding/dissociation constants or to measure the concentration of the target protein/molecule in the solution.

Other inherent limitation of protein arrays is that an ideal protein array should include the whole repertoire of proteins of the organism of interest. In the case of humans, this may be considered as a utopian thought. The number of different protein species in humans is vast (i.e., including alternative-spliced variants, posttranslational modifications, single nucleotide polymorphisms, and chimeric proteins; Casado-Vela, Cebrian, del Pulgar, et al., 2011) and greatly exceeds the current ability to attach proteins on a single array. The printed densities of arrays reported in the literature greatly vary from few hundreds to up to 38,400 different spots (see examples of high-density arrays in Fig. 8.2). The largest protein array in terms of number of different proteins that we found in the literature included 16,368 unique full-length human open reading frames (Jeong et al., 2012). In fact, protein arrays bearing more than 17,000 human recombinant proteins printed on arrays bearing 38,400 spots are already available (www.cdi-lab.com;

www.cambridgeproteinarrays.com). A major limitation of protein arrays is that the biochemical diversity of the proteins found in humans precludes the possibility of producing and purifying every single protein, regardless of the strategy used to produce those proteins. To exemplify this, it is widely accepted that transcription factors and membrane proteins are often difficult to express and/or purify. In this sense, protein arrays based on cell-free protein expression systems represent a promising alternative, overcoming some limitations of protein synthesis using cellular systems (i.e., in cellular systems improperly folded proteins tend to aggregate and are frequently proteolyzed and degraded). High protein yields were reported in a number of membrane proteins using cell-free protein expression systems (Schwarz, Dotsch, & Bernhard, 2008). Whereas the advantages of cell-free protein expression systems, different cell extracts used for protein production show different tolerance to additives. Protein arrays based on cell-free protein expression allow rapid and customized production of proteins from different templates. Protein expression using cell-free models is affordable and can be accomplished within a short period of time (typically 2–3 h is enough to translate all the proteins on a given array). More importantly, proteins are produced just prior to use precluding potential deleterious storage effects and provide high stable arrays.

Even if the whole human proteome could be attached on the surface of arrays, a number of variables may alter the functionality of the proteins during the production and/or storage of the protein arrays. The list of main factors significantly affecting the production of protein and antibody arrays was already reported more than one decade ago (Kusnezow, Jacob, Walijew, Diehl, & Hoheisel, 2003), but they are still frequently underestimated. A key variable is the need to produce properly folded and functionally active proteins, which may be also modulated by posttranslational modification. Different cellular lysates may influence the posttranslational modification status of the proteins produced, which may cause an effect on the function of the proteins. As an example, insect cell extracts allow phosphorylation (Casado-Vela, Martinez-Torrecuadrada, & Casal, 2009) and N-glycosylation of proteins (Harrison & Jarvis, 2006). Recent reports support that posttranslational modification and serine/threonine/tyrosine phosphorylation is possible in bacteria. Indeed, bacteria possess serine/threonine kinases (structurally similar to those found in eukaryotic organisms), tyrosine kinases (termed BY-kinases), and other kinases not yet categorized. The occurrence of bacterial kinases and the increasing number of phosphorylated proteins in bacterial systems contrast with previous reports stating that bacterial systems lacks most of the eukaryotic posttranslational machinery (see Sahdev, Khattar, and Saini 2008, for review). Hence, it is important to underline that the phosphorylation efficiency of bacteria is significantly lower (10–100 ×; Mijakovic & Macek, 2012) compared to eukaryotic systems and the phosphorylation patterns may significantly differ—reviewed in Mijakovic and Macek (2012). Therefore, the selection of appropriate cell lysates able to mimic the posttranslational modification status of the proteins could be advantageous.

More importantly, the plethora of potential interactions retrieved using protein arrays raises questions about the possibility of including false positives in the list of potential interactors and introduces the necessity of including further validation experiments such as coimmunoprecipitation followed by Western-blot, ELISA, or colocalization approaches. Therefore, any strategy precluding the potential occurrence of false positives represents a valuable advantage. In this sense, validation of the specificity and suitability of the antibodies used as reagents in protein arrays may significantly contribute to that task (Spurrier, Ramalingam, & Nishizuka, 2008). A recent report (Gujral et al., 2012) used protein arrays to screen for proteins and phospho-proteins acting as signaling molecules in breast cancer. In their study, the authors emphasized that out of several thousand antibodies tested (most of them commercially available) only 5% exhibited sufficient specificity, supporting the idea that checking antibody specificity is of paramount importance. In addition, detection methods for protein microarrays are mainly based on dyes or labels, which do not allow real-time detection; then, it is not easy to obtain info regarding binding parameters by using protein arrays for studying PPIs.

Finally, protein arrays, as other -omic technologies, provide large amounts of data in a high-throughput manner. Thus, the major current challenge is the standardization, analysis, integration, and eventual interpretation of the huge amount of data generated with protein array technology (Casado-Vela, Cebrian, del Pulgar, et al., 2011; Díez et al., 2012).

3. PROTEIN-BINDING DNA ARRAYS AND THEIR APPLICATION TO ADDRESS THE STUDY OF DNA-BINDING PROTEINS

DNA-binding proteins are involved in a number of key cellular processes such as DNA repair and modification, transcriptional regulation, recombination, replication, and restriction. Thus, the identification of the list of proteins able to bind DNA and the elucidation of the specific sequences within DNA that are recognized constitute two main objectives. Of particular interest, is the case of transcription factors (TFs), responsible for the transcriptional regulation through recognition and binding to specific DNA sequences located inside regulatory domains. The elucidation of DNA-motifs recognized by TFs is a main task for elucidating the so called "transcriptional regulatory code" (Harbison et al., 2004). Chromatin immunoprecipitation (ChIP) of TF–DNA complexes and analysis of DNA fragments with oligonucleotide arrays (ChIP-chip) or deep-sequencing (ChIP-seq) provides an *in vivo* landscape of the target genes of TFs—reviewed in Park (2009). However, ChIP-based methodologies make the study of TF-binding sites at large scale an unaffordable task, given the necessity of specific antibodies against the TFs under study. Alternatively, PDI-arrays offer a valuable alternative to analyze and detect specific DNA sequences-binding proteins.

Figure 8.4 shows a schematic overview of protein-binding DNA arrays and their application to unravel PDIs. With this approach, single-stranded DNA molecules (ssDNA) are converted into double-stranded DNAs (dsDNA) in a primer extension reaction using universal primers complementary to all the DNA sequences. In this manner, known dsDNA sequences remain attached on known positions of the array. Such array of molecules can be directly incubated with soluble preparations of the protein of interest (purified or bacterial protein extracts may be used). Finally, PDIs are immunologically probed with Cy5 fluorescent-labeled specific antibodies raised against protein epitopes (or against epitope tags). Fluorescent-labeled primary or secondary antibodies may be used to detect the interactions. Subsequently, specific sequences within DNA that bind query proteins are identified after image processing using dedicated software.

Several fluorochrome-based detection strategies may be used. For instance, Warren et al. (2006) conjugated Cy3 at a particular cysteine residue on an unstructured portion of the protein Exd that did not affect its functionality. More versatile and easier are those strategies based on immunological detection of DNA-transcription factor complexes with fluorochrome-conjugated antibodies. In this case, PDI-array is incubated with the transcription factor of interest fused to an epitope tag (e.g., glutathione S-transferase, GST), and bound PDI-array incubated with a labeled antibody (Mukherjee et al., 2004). Alternatively, the protein may be fused to a fluorescent protein (e.g., DsRed), and fluorescence directly detected without need for multistep incubations (Kim, Lee, et al., 2009).

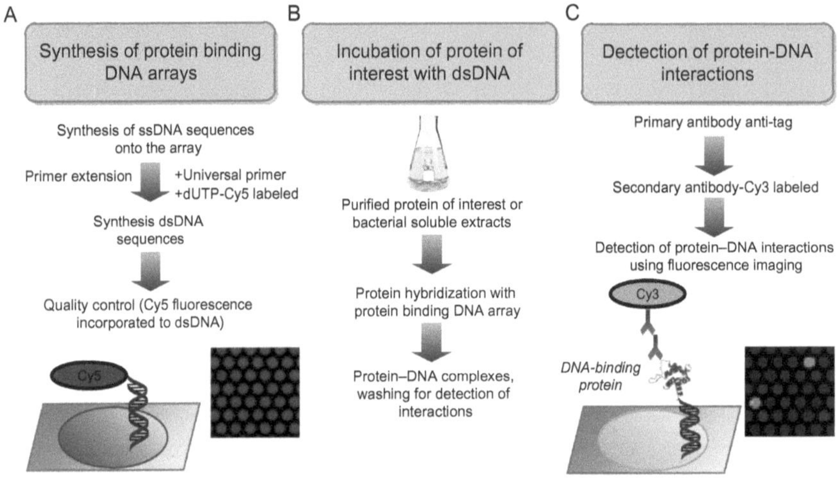

Figure 8.4 Schematic overview of the detection protein–DNA interactions using protein-binding DNA arrays. (A) Single-stranded DNA molecules (ssDNA) are converted into double-stranded DNAs (dsDNA) in a primer extension reaction using universal primers complementary to all the DNA sequences. Cy5-dUTP is incorporated to monitor the synthesis of the second strand and to normalize the signal intensity. A detail of the image after scanning at the red channel where the Cy5 signal is uniformly distributed is shown. (B) dsDNA arrays are then incubated with the protein of interest, purified from bacterial cultures. Whole bacterial cell lysates may be used for incubation of array. Alternatively, purified protein preparations may be also used. (C) DNA–protein complexes are probed with specific antibodies raised against the epitope tag and subsequently with a Cy5-conjugated secondary antibody. After scanning the slide, DNA–protein complexes are revealed as bright spots, displayed as light-grey spots.

The current number of different DNA sequences represented in the PDI-arrays depends on three main factors: the design of dsDNAs; the length of the oligonucleotide spotted onto the array; and the resolution of the array (i.e., number of different spots). Accordingly, optimum design choice is critical for obtaining as much information as possible in a single experiment. In their hairpin-based design, Warren et al. (2006) included four copies for all the possible 8-mer sequences synthesized in a high-density microarray (131,584 features per array; NimbleGen systems). The group of Martha Bulyk followed a different strategy to compact all the possible 10-mer in a 44 k microarray based on the use of a universal linker sequence for primer extension. The optimization of the design of DNA molecules with pseudo-random properties in a maximally compact manner is also of paramount importance (Berger & Bulyk, 2009; Berger et al., 2006). The possibility of multiplexing this design (e.g., four 44 k microarrays synthesized in a single slide, as manufactured by Agilent Technologies), allows analyzing DNA-binding properties of several different proteins in a single experiment. Using a similar approach for optimizing the number of different DNA molecules, we increased the resolution of the interrogated patterns up to 11 bp. This design not only allows detecting longer motifs than other previous designs, but also increases the number of DNA probes containing lower order sequences, enhancing the statistical power for identification of specific DNA-motifs (Franco-Zorrilla & Solano, 2014; Godoy et al., 2011).

More importantly, the design of PDI-arrays is critical for determination of specific sequences recognized by DNA-binding proteins. With this respect, two types of molecules have been used in PDI-arrays: short dsDNAs designed as self-complementary palindromes (Warren et al., 2006); and short dsDNAs originated by primer extension using a universal primer complementary to a common DNA sequence present in all the DNA oligonucleotides (Mukherjee et al., 2004). Other issue for consideration in the design refers to the number of different DNA sequences represented in the PDI-array. Because most metazoan DNA-binding proteins recognize less than 10 bp (Berger et al., 2006; Wingender et al., 2001), PDI-arrays should be capable of identifying sequences preferred by almost any DNA-binding protein. In addition, the number of different DNA sequences increases exponentially with the length of the interrogated fragment. For example, we can obtain all possible 8 bp-long dsDNA sequences (8-mer) in 32,768 different molecules ($=4^8/2$), but all possible 9-mers, 10-mer, and 11-mer should represent 131,072, 524,288, and 2,097,152 dsDNA probes, respectively.

3.1. Recent achievements of protein-binding DNA arrays and their application to address the study of the human proteome

The identification of regulatory networks governing animal development, core biological processes, and responses to environmental stimuli is critical for understanding the central role of transcriptional programs and TFs in these processes. This becomes particularly interesting in the case of human diseases, as about 40% of cancer genes (genes which, when mutated, are responsible to the development of cancer) correspond to nucleic acid-binding proteins (Furney, Higgins, Ouzounis, & Lopez-Bigas, 2006), and a third of human developmental disorders have been attributed to dysfunctional TFs (Vaquerizas, Kummerfeld, Teichmann, & Luscombe, 2009). This understanding requires cataloging target genes of TFs, as effectors of the cellular responses to intrinsic and environmental stimuli. Comprehensive characterization of DNA-binding specificities of TFs then emerges as a critical bottleneck in the identification of target genes.

PDI-arrays offer an extremely useful tool for the identification of TF-binding DNA motifs *in vitro*. Berger et al. (2008) used PDI-arrays for the identification of DNA-binding patterns of the majority (168) of mouse homeodomain proteins, defining 65 distinct DNA-motifs. Interestingly, distinct DNA-motifs correlated well with amino acid sequence similarity among the homeodomain proteins, revealing striking structure–function relationships. Moreover, this large-scale analysis of mouse homedomains allowed predicting DNA-binding specificities among animal homeodomains (Berger et al., 2008).

Similar approaches were followed in the characterization of DNA-binding profiles of yeast and mouse TFs belonging to different structural classes (Badis et al., 2009; Zhu et al., 2009). In the case of yeast, Zhu et al. (2009) characterized the DNA-binding profiles for 89 TFs, some of them previously undiscovered as DNA-binding proteins. In addition, data obtained using PDI-arrays correlated well with available data on DNA-binding *in vivo* obtained from ChIP experiments (Zhu et al., 2009). Thus, given the good agreement between *in vitro* and *in vivo* experiments, opens novel possibilities for the identification of targets genes of TFs. Moreover, PDI-array data can be used to further refine ChIP-based data in order to distinguish between direct or indirect targets of TFs. Similarly to that described for yeast TFs, Badis et al. (2009) analyzed 104 mouse TFs belonging to 22 structural classes. The main achievement of this study relates the complexity in DNA recognition of TFs, where approximately half of the proteins tested recognized multiple sequence motifs. When PDI-array data were used to analyze ChIP-based results, both primary and secondary DNA motifs where enriched among bound DNA fragments by two different TFs, indicating that data obtained *in vitro* had a good agreement with TF-targets observed *in vivo*.

In summary, PDI-arrays have been extremely useful for the identification of the DNA sequences recognized by TFs in some model organisms, and these data are likely to be similar for conserved homolog proteins in other organisms. Moreover, PDI-array data can be used to refine ChIP data and, when integrated with additional strategies, such as phylogenomic shadowing or nucleosome occupancy data, will serve to predict *cis*-regulatory elements and target genes.

Early studies on the regulation of gene expression have demonstrated that TFs regulate the expression of their target genes in a combinatorial manner (McKenna & O'Malley, 2002; Remenyi, Scholer, & Wilmanns, 2004). That means that the expression of a given gene depends on the combinatorial action of several TFs, each recognizing a specific DNA-sequence motif. Recent studies highlighted the relevance of the establishment of particular PPIs for transcriptional regulation in different tissues, both in mouse and human (Ravasi et al., 2010). In spite that physical interactions among TFs do not necessarily involve a combined interaction of the protein complex with DNA, this opens the possibility that heterodimeric TFs recognize specific DNA sequences. With this respect, PDI-arrays may also offer a tool for determining DNA-binding specificity of heterodimeric TFs, beyond their uses indicated above. Actually, Grove et al. (2009) performed an integrated analysis of several paralog basic helix–loop–helix (bHLH) TFs, and demonstrated that bHLH proteins that do not form heterodimers bind to DNA as homodimers. By contrast, bHLH proteins that participate in heterodimeric interactions do not bind to DNA on their own, but rather they need the formation of the corresponding heterodimers to recognize and bind DNA in a sequence-specific manner (Grove et al., 2009).

3.2. Advantages and limitations of protein-binding DNA arrays

Major advantages and disadvantages have been pointed out in previous paragraphs and are summarized in Table 8.4. As stated above, protein-binding DNA arrays offer a valuable alternative for the identification of DNA-sequences specifically recognized by DNA-binding proteins. In principle, the advantages of PDI-arrays are basically the same as in protein arrays.

PDI array data are very useful on their own thanks to the identification of DNA-sequences recognized by the protein, helping in the design of artificial promoters and/or TFs involved in a particular biological process. However, these data are extremely useful when a combination of strategies is followed for the identification of targets genes. With this respect, the analysis of the promoter regions of genes presumably regulated by a particular TF will help to define actual target genes of TFs. In this sense, the presence of the specific DNA-sequence recognized by the TF in the proximal promoter of a gene presumably regulated by the TF, will be an indicative of its regulatory potential. Transcriptomic assays specifically designed for the TF under study (e.g., knocked-out cells or organisms) will yield a transcriptional landscape of the gene network regulated by the TF, and the presence of specific *cis*-elements will discriminate between direct and indirect targets (Godoy et al., 2011). This strategy may be even more accessible when analyzing publicly available

Table 8.4 Advantages and limitations of protein-binding DNA arrays to address the study of protein–DNA interactions

Advantages
☑ Identification of specific DNA sequences recongized by DNA-binding proteins *in vitro*. ☑ Relatively simple methodology. ☑ Fast: the entire process takes two working days. ☑ No need of purification and/or labeling of the protein. ☑ Suitable for any commercially available antibody. ☑ Pararallelizable when using multiplexed arrays: several different proteins can be assayed at once. ☑ Relatively cheap with multiplexed arrays. ☑ Data obtained in PDI-arrays correlate well with data obtained *in vivo*. ☑ Aplicable to known heterodimers.
Disadvantages
☒ *In vitro* system that do not yield actual targets *in vivo*. ☒ Restricted to short sequences, between 8 and 11 bp. Bipartite motifs are rarely identified. ☒ Depends on the solubility and correct folding of the protein expressed in *E. coli*. ☒ Not suitable for proteins requiring post-translational modifications for DNA-binding. ☒ Not suitable for proteins requiring unknown specific heterodimerizations.

transcriptomic assays. Public repositories such as GEO (http://www.ncbi.nlm.nih.gov/geo/) or ArrayExpress (http://www.ebi.ac.uk/arrayexpress/) contain more than 40,000 complete transcriptomic assays involving above one million different samples (as in December 2013). Thus, simple searches in databases of transcriptomic assays involving the TF of interest, or a structurally similar TF, or a biological process in wich are involved may be an easy alternative for identifying target genes (Godoy et al., 2011).

4. DATABASES AND WEB RESOURCES FOR PPIs AND FOR PDIs

An increasing range of public databases currently allow the retrieval of information on protein interactions, including predictions of interactions and even modeling the pathways involved, reviewed in (Fernandez-Suarez & Galperin, 2013). The information compiled in such databases constitutes a valuable resource for information regarding interactions based on previous experiments and to compare with novel experimental data. For that reason, we also include here a directory of valuable resources (Table 8.5) including the list, description, and links gathering information of PPIs and PDIs.

An issue for consideration is that many databases include data derived from prediction algorithms and computational methods. In this regard, a number of algorithms and computational methods currently coexist and can be used to infer the occurrence of PPIs (Gomez, Choi, & Wu, 2008; Gong et al., 2008; Jessulat et al., 2011; Mishra, 2012; Pitre et al., 2008; Skrabanek, Saini, Bader, & Enright, 2008). These algorithms rely on one or more features—such as genomic sequence, topological genomic clustering, protein sequence, protein structure, protein functional/structural domains, or evolutionary relationship—and may also take advantage of known PPI datasets to test, train, and improve the quality of their predictions. A comparative overview of prediction algorithms is beyond the scope of this report, but it is important to underline that computational algorithms frequently take advantage of reliable training datasets (i.e., *bona-fide* list of protein interactions) to test and to improve their predictions.

More importantly, the list of candidate protein partners retrieved may significantly differ among databases, due to the fact that the amount and quality of the information deposited in each database are not really comparable (Klingstrom & Plewczynski, 2011, reviewed PPI databases and their sources of information). The overall information overlap among databases is limited and, thus, gathering information from as many databases as possible may represent an advantage if thorough information on the interactome of a specific protein is the objective. This task currently constitutes an obstacle that may be prohibitive in terms of time.

The reasons above justify the current trend towards the development of web-based search engines able to gather protein interactors from multiple databases at the same time and yielding updated information. Examples of such engines include PSICQUIC (Aranda et al., 2011), DASMI (Blankenburg et al., 2009), and BIPS (Garcia-Garcia, Schleker, Klein-Seetharaman, & Oliva, 2012), recently developed and publicly available for the scientific community. These web tools significantly simplify the screening of information.

Nevertheless, relevant information affecting protein interactions (such as the specific physical and biochemical parameters affecting to those interactions) is frequently overlooked (Schleker et al., 2012) and, more importantly, still suffers from high rates of false positives and errors

Table 8.5 Name, references link, and brief description of resources on human protein–protein interactions (PPIs) and protein–DNA interactions (PDIs)

Name, acronym (reference)	Web link	Full name and description
2P2Idb (Bourgeas, Basse, Morelli, & Roche, 2010)	http://www.hupo.org/research/hpp/	Hand-curated database dedicated to the structure of protein–protein complexes with known small molecule inhibitors
3did (Mosca, Ceol, Stein, Olivella, & Aloy, 2013)	http://www.3did.irbbarcelona.org/	3D interacting domains. Domain–domain interactions in proteins with known 3D structures.
3D-Interologs (Lo, Chen, & Yang, 2010)	http://gemdock.life.nctu.edu.tw/3d-interologs/	Protein–protein interactions in various evolutionary lineages
AANT (Hoffman et al., 2004)	http://aant.icmb.utexas.edu/	Amino acid–nucleotide interaction database. Categorizes all amino acid–nucleotide interactions from experimentally determined protein–nucleic acid structures, and provides users with a graphic interface for visualizing these interactions in aggregate.
BioGRID (Chatr-Aryamontri et al., 2013)	http://thebiogrid.org/	Genetic and physical interactions in yeast, worm, fly, and human
BioLiP (Yang, Roy, & Zhang, 2013)	http://zhanglab.ccmb.med.umich.edu/BioLiP/	Semi-manually curated database for high-quality, biologically relevant ligand–protein-binding interactions
CancerResource (Ahmed et al., 2011)	http://bioinf-data.charite.de/cancerresource/	Cancer-relevant proteins and compound interactions

CCSB Interactome database (reference not available to cite this database)	http://interactome.dfci.harvard.edu/index.php?page=home	Database including data from different organisms, including humans
CellCircuits (Mak, Daly, Gruebel, & Ideker, 2007)	http://www.cellcircuits.org/search/index.html	Molecular network models: from pairwise molecular interactions to whole pathways
ConsensusPathDB (Kamburov, Stelzl, Lehrach, & Herwig, 2013)	http://cpdb.molgen.mpg.de/	Integrates interaction networks in Homo sapiens including binary and complex protein–protein interactions, genetic, metabolic, signaling, gene regulatory, and drug-target interactions, as well as biochemical pathways
HPRD (Muthusamy, Thomas, Prasad, & Pandey, 2013)	http://www.hprd.org/ http://www.humanproteinpedia.org/	Human protein reference database. Includes abundant information on protein characterization, mass spectrometry, and protein–protein interaction as follows: Coimmunoprecipitation and mass spectrometry-based protein–protein interaction, coimmunoprecipitation, and Western blotting based protein–protein interaction, fluorescence-based experiments, immunohistochemistry, mass spectrometric analysis, protein, and peptide microarray, Western blotting, yeast two-hybrid based protein–protein interaction.

Continued

Table 8.5 Name, references link, and brief description of resources on human protein–protein interactions (PPIs) and protein–DNA interactions (PDIs)—cont'd

Name, acronym (reference)	Web link	Full name and description
IntAct (Kerrien et al., 2012)	http://www.ebi.ac.uk/intact/main.xhtml	Database and analysis tools for protein interaction data. All interactions are derived from literature curation or direct user submissions and are freely available.
JASPAR (Portales-Casamar et al., 2010)	http://jaspar.genereg.net	Open-access database for eukaryotic transcription factor-binding profiles
KDBI (Ji et al., 2003)	http://bidd.nus.edu.sg/group/kdbi/kdbi.asp	Kinetic data of bio-molecular interactions database. Kinetic data of bio-molecular interaction (KDBI) is a collection of experimentally determined kinetic data of protein–protein, protein–RNA, protein–DNA, protein–ligand, RNA–ligand, DNA–ligand binding, or reaction events described in the literature.
KUPS (Chen, Jeong, & Dermyer, 2011)	http://www.ittc.ku.edu/chenlab/kups/	The University of Kansas Proteomics Service (KUPS) provides high-quality protein–protein interaction (PPI) datasets for researchers who are interested in elucidating PPIs with *in silico* methods.

MatrixDB (Chautard, Fatoux-Ardore, Ballut, Thierry-Mieg, & Ricard-Blum, 2011)	http://matrixdb.ibcp.fr/	Extracellular matrix interaction database.
MiMI (Tarcea et al., 2009)	http://mimi.ncibi.org/MimiWeb/main-page.jsp	The Michigan molecular interactions database provides access to the knowledge and data merged and integrated from numerous protein interactions databases.
MINT (Licata et al., 2012)	http://mint.bio.uniroma2.it/mint/Welcome.do	Focuses on experimentally verified protein–protein interactions mined from the scientific literature by expert curators. The full MINT dataset can be freely downloaded.
Negatome (Smialowski et al., 2010)	http://mips.helmholtz-muenchen.de/proj/ppi/negatome/	Proteins and domain pairs which are unlikely engaged in direct physical interactions. The database currently contains experimentally supported noninteracting protein pairs derived from two distinct sources: by manual curation of literature and by analyzing protein complexes from the PDB.
NPIDB (Spirin, Titov, Karyagina, & Alexeevski, 2007)	http://npidb.belozersky.msu.ru/	Nucleic acid–protein interaction database. Includes a collection of files in the PDB format containing structural information on DNA–protein and RNA–protein complexes, and a number of online tools for analysis of the complexes.

Continued

Table 8.5 Name, references link, and brief description of resources on human protein–protein interactions (PPIs) and protein–DNA interactions (PDIs)—cont'd

Name, acronym (reference)	Web link	Full name and description
Pathguide (Cary, Bader, & Sander, 2005)	http://pathguide.org/	Provides an overview of more than 190 web-accessible biological pathway and network databases. These include databases on metabolic pathways, signaling pathways, transcription factor targets, gene regulatory networks, genetic interactions, protein–compound interactions, and protein–protein interactions.
PepCyber: P~Pep (Gong et al., 2008)	http://pepcyber.umn.edu/PPEP/	Human protein interactions mediated by phosphoprotein-binding domains.
PID (Schaefer et al., 2009)	http://pid.nci.nih.gov/	Pathway interaction database. Interactions and cellular processes assembled into human signaling pathways.
PINA (Cowley et al., 2012)	http://cbg.garvan.unsw.edu.au/pina/	Protein interaction network analysis. Integrated platform for protein interaction network construction, filtering, analysis, visualization, and management. It integrates protein–protein interaction data from several public databases and builds a complete, nonredundant protein interaction datasets.

PIN database (Luc & Tempst, 2004)	http://pin.mskcc.org/	Proteins interacting in the nucleus. Proteins interacting in nucleus including information from human (and yeast).
PINT Protein–Protein Interactions Thermodynamic Database (Kumar & Gromiha, 2006)	http://www.bioinfodatabase.com/pint/	The protein–protein interactions thermodynamic database is a collection of experimental data of several thermodynamic parameters along with literature, sequence and structural information, and experimental conditions.
PIPs (McDowall, Scott, & Barton, 2009)	http://www.compbio.dundee.ac.uk/www-pips/	Database of predicted human protein–protein interactions
piSite (Higurashi, Ishida, & Kinoshita, 2009)	http://pisite.hgc.jp/	Database of protein interaction sites using multiple binding states in PDB
ProtChemSI (Kalinina, Wichmann, Apic, & Russell, 2011)	http://pcidb.russelllab.org/	Protein–chemical structural interactions database includes all existing 3D structures of complexes of proteins with low molecular weight ligands.
POINT (Huang et al., 2004)	http://point.bioinformatics.tw/Welcome.do; jsessionid=82EDC3103003E47353C764517501AAFE	Prediction of human protein–protein interactome
PrePPI (Zhang, Petrey, Garzon, Deng, & Honig, 2013)	http://genomequebec.mcgill.ca/PReMod/	Predicted transcriptional regulatory modules in the human genome

Continued

Table 8.5 Name, references link, and brief description of resources on human protein–protein interactions (PPIs) and protein–DNA interactions (PDIs)—cont'd

Name, acronym (reference)	Web link	Full name and description
PROMISCUOUS (von Eichborn et al., 2011)	http://bioinformatics.charite.de/promiscuous/	Protein–protein and drug–protein interactions resource
PSIbase (Gong et al., 2005)	http://psibase.kobic.re.kr/	Molecular interaction database focusing on structural interaction of proteins and their domains
SNAPPI (Jefferson, Walsh, Roberts, & Barton, 2007)	http://www.compbio.dundee.ac.uk/SNAPPI/downloads.jsp	Structures, interfaces, and alignments for protein–protein interactions
TissueNet (Barshir et al., 2013)	http://netbio.bgu.ac.il/tissuenet/	Tissue distribution of protein–protein interactions
TRANSFAC (Wingender, 2008)	http://www.biobase-international.com/product/transcription-factor-binding-sites	Private database for eukaryotic transcription factor-binding profiles
TRIP (Shin et al., 2012)	http://www.trpchannel.org/	Protein–protein interactions for mammalian TRP channels
UniHI (Chaurasia & Futschik, 2012)	http://www.unihi.org/	Integrated tool for the exploration of the human interactome
UniPROBE (Robasky & Bulyk, 2011)	http://the_brain.bwh.harvard.edu/uniprobe/	Online database of protein-binding microarray data on protein–DNA interactions

(Tyagi et al., 2012). For that reason, despite the increasing usability of databases, literature searches on peer-reviewed journals still constitutes the main source of information on PPIs in two ways: first, literature searches may retrieve information that may not be included in databases. Second, the comparison of the list of protein interactors retrieved upon database and literature searches enable manual correction of involuntary mistakes due to, for example, gene acronym redundancies (Casado-Vela, Matthiesen, Selles, & Naranjo, 2013).

5. CONCLUSIONS AND FUTURE PERSPECTIVES

The two array based technologies described here (protein arrays and protein-binding DNA arrays) offer the possibility of detecting protein interactions through large-scale screenings. Protein array technology opens the possibility of building customized arrays using specific subsets of printed molecules, including well characterized antibodies found in many labs and not yet commercialized.

Two main characteristics seem to contribute to the rapid evolution of arrays for protein interaction screening. First, the optimization of the technologies used in the construction of arrays including: miniaturization, automation, protein-binding detection protocols, and low sample consumption (Casado-Vela, Gonzalez-Gonzalez, et al., 2013; Corthals, Wasinger, Hochstrasser, & Sanchez, 2000). Second, protein array technology provides unrivaled possibilities to answer different biological questions. This assumption relies on the ability to build array ad hoc, including custom-made or specific sets of and in a cost-effective manner. Thus, from a practical point of view, arrays constitute a promising tool for patient-tailored or "personalized" medicine, defined as the selection of specific therapeutics to treat disease in a particular individual.

Many valuable examples showing the applicability of arrays to biomedicine are included here, which support the validity of arrays as a tool for functional proteomics focusing on the screening of interactions. Nevertheless, the number and refinement of applications of array technologies seem to be continuously expanding.

It is obvious that the array technologies typically provide large sets of information. As outlined in this report, both the information gathered using these approaches and detailed description of the conditions used should be available, since it constitutes a valuable resource of information and may constitute the basis for future validation experiments.

An important idea to bear in mind is that the current number of protein-coding genes is ~21,500 and each gene encodes for, at least, one full length protein, updated from Casado-Vela, Cebrian, Gomez del Pulgar, and Lacal (2011). This means that current technologies already allow the attachment of one single copy of each protein (or the full-length DNA sequence encoding for each protein) of the whole human proteome. Although this objective seems achievable from the theoretical point of view, several hurdles need to be surpassed to reach this aim. One of the main constraints is that the complexity of the human proteome seems massive, especially if we consider the idea of including protein species (i.e., proteins bearing any chemical modification, including posttranslational modifications; Jungblut et al., 1996), protein isoforms (derived after alternative splicing of a gene; Casado-Vela, Cebrian, del Pulgar, et al., 2011) and even chimeric proteins naturally found in humans (Casado-Vela, Lacal, et al., 2013; Frenkel-Morgenstern et al., 2012; Li, Wang, Mor, & Sklar, 2008). All those variables deserve extra consideration if the objective is the holistic and comprehensive understanding of the human proteome. Therefore, the construction of protein arrays including defined sets of proteins, antibodies, or DNA sequences specifically selected to answer specific questions seems to be both, more practical and affordable to an increasing number of researchers.

Although arrays constitute a versatile platform enabling the selection and printing of the subsets of molecules of interest (either proteins or DNA sequences), it is important to highlight some limitations. In this sense, despite the rapid advances, not even the complete interactome for eukaryotic model organisms such as *Saccharomyces cerevisiae* or *Caenorhabditis elegans*, are not currently known. (Tran, Choi, & Zhang, 2013). This limitation also applies to humans, which evidences that we are still in early stages towards unraveling the human interactome and its modulation.

Finally, it has to be noted that protein arrays, antibody arrays, and protein-binding DNA arrays should be used in combination with alternative proteomics technologies (Casado-Vela, Rodriguez-Suarez, et al., 2009), since their combination typically lead to complementary results and contribute to enrich the interpretation of data.

ACKNOWLEDGMENTS

J. C.-V. is a JAE-DOC (CSIC) holder supported by Ministerio de Economía y Competitividad, Spain, co-funded by the European Social Fund.

REFERENCES

Ahmed, J., Meinel, T., Dunkel, M., Murgueitio, M. S., Adams, R., Blasse, C., et al. (2011). CancerResource: A comprehensive database of cancer-relevant proteins and compound interactions supported by experimental knowledge. *Nucleic Acids Research, 39*, D960–D967. http://dx.doi.org/10.1093/nar/gkq910.

Anderson, K. S., & LaBaer, J. (2005). The sentinel within: Exploiting the immune system for cancer biomarkers. *Journal of Proteome Research, 4*(4), 1123–1133. http://dx.doi.org/10.1021/pr0500814.

Anderson, K. S., Ramachandran, N., Wong, J., Raphael, J. V., Hainsworth, E., Demirkan, G., et al. (2008). Application of protein microarrays for multiplexed detection of antibodies to tumor antigens in breast cancer. *Journal of Proteome Research*, 7(4), 1490–1499. http://dx.doi.org/10.1021/pr700804c.

Anderson, K. S., Sibani, S., Wallstrom, G., Qiu, J., Mendoza, E. A., Raphael, J., et al. (2011). Protein microarray signature of autoantibody biomarkers for the early detection of breast cancer. *Journal of Proteome Research, 10*(1), 85–96. http://dx.doi.org/10.1021/pr100686b.

Angenendt, P., Kreutzberger, J., Glokler, J., & Hoheisel, J. D. (2006). Generation of high density protein microarrays by cell-free in situ expression of unpurified PCR products. *Molecular & Cellular Proteomics, 5*(9), 1658–1666. http://dx.doi.org/10.1074/mcp.T600024-MCP200.

Aranda, B., Blankenburg, H., Kerrien, S., Brinkman, F. S., Ceol, A., Chautard, E., et al. (2011). PSICQUIC and PSISCORE: Accessing and scoring molecular interactions. *Nature Methods, 8*(7), 528–529. http://dx.doi.org/10.1038/nmeth.1637.

Arduengo, M., Schenborn, E., & Hurst, R. (2007). The role of cell-free rabbit reticulocyte expression systems in functional proteomics. In W. A. Kudlicki, F. Katzen, & R. P. Bennett (Eds.), *Cell-free protein expression* (pp. 1–18). Carlsbad, CA: Landes Bioscience.

Aronson, J. K. (2005). Biomarkers and surrogate endpoints. *British Journal of Clinical Pharmacology, 59*(5), 491–494. http://dx.doi.org/10.1111/j.1365-2125.2005.02435.x.

Babu, M. M., Luscombe, N. M., Aravind, L., Gerstein, M., & Teichmann, S. A. (2004). Structure and evolution of transcriptional regulatory networks. *Current Opinion in Structural Biology, 14*(3), 283–291. http://dx.doi.org/10.1016/j.sbi.2004.05.004.

Badis, G., Berger, M. F., Philippakis, A. A., Talukder, S., Gehrke, A. R., Jaeger, S. A., et al. (2009). Diversity and complexity in DNA recognition by transcription factors. *Science, 324*(5935), 1720–1723. http://dx.doi.org/10.1126/science.1162327.

Barshir, R., Basha, O., Eluk, A., Smoly, I. Y., Lan, A., & Yeger-Lotem, E. (2013). The TissueNet database of human tissue protein-protein interactions. *Nucleic Acids Research, 41*, D841–D844. http://dx.doi.org/10.1093/nar/gks1198.

Beare, P. A., Chen, C., Bouman, T., Pablo, J., Unal, B., Cockrell, D. C., et al. (2008). Candidate antigens for Q fever serodiagnosis revealed by immunoscreening of a Coxiella burnetii protein microarray. *Clinical and Vaccine Immunology, 15*(12), 1771–1779. http://dx.doi.org/10.1128/CVI.00300-08.

Belov, L., de la Vega, O., dos Remedios, C. G., Mulligan, S. P., & Christopherson, R. I. (2001). Immunophenotyping of leukemias using a cluster of differentiation antibody microarray. *Cancer Research, 61*(11), 4483–4489.

Belov, L., Huang, P., Barber, N., Mulligan, S. P., & Christopherson, R. I. (2003). Identification of repertoires of surface antigens on leukemias using an antibody microarray. *Proteomics, 3*(11), 2147–2154. http://dx.doi.org/10.1002/pmic.200300599.

Berger, M. F., Badis, G., Gehrke, A. R., Talukder, S., Philippakis, A. A., Pena-Castillo, L., et al. (2008). Variation in homeodomain DNA binding revealed by high-resolution analysis of sequence preferences. *Cell, 133*(7), 1266–1276. http://dx.doi.org/10.1016/j.cell.2008.05.024.

Berger, M. F., & Bulyk, M. L. (2009). Universal protein-binding microarrays for the comprehensive characterization of the DNA-binding specificities of transcription factors. *Nature Protocols*, *4*(3), 393–411. http://dx.doi.org/10.1038/nprot.2008.195.

Berger, M. F., Philippakis, A. A., Qureshi, A. M., He, F. S., Estep, P. W., 3rd., & Bulyk, M. L. (2006). Compact, universal DNA microarrays to comprehensively determine transcription-factor binding site specificities. *Nature Biotechnology*, *24*(11), 1429–1435. http://dx.doi.org/10.1038/nbt1246.

Blankenburg, H., Finn, R. D., Prlic, A., Jenkinson, A. M., Ramirez, F., Emig, D., et al. (2009). DASMI: Exchanging, annotating and assessing molecular interaction data. *Bioinformatics*, *25*(10), 1321–1328. http://dx.doi.org/10.1093/bioinformatics/btp142.

Bourgeas, R., Basse, M. J., Morelli, X., & Roche, P. (2010). Atomic analysis of protein-protein interfaces with known inhibitors: The 2P2I database. *PLoS One*, *5*(3), e9598. http://dx.doi.org/10.1371/journal.pone.0009598.

Bowen, B., Steinberg, J., Laemmli, U. K., & Weintraub, H. (1980). The detection of DNA-binding proteins by protein blotting. *Nucleic Acids Research*, *8*(1), 1–20. http://dx.doi.org/10.1093/nar/8.1.1.

Brase, J. C., Mannsperger, H., Frohlich, H., Gade, S., Schmidt, C., Wiemann, S., et al. (2010). Increasing the sensitivity of reverse phase protein arrays by antibody-mediated signal amplification. *Proteome Science*, *8*, 36. http://dx.doi.org/10.1186/1477-5956-8-36.

Braun, P., Tasan, M., Dreze, M., Barrios-Rodiles, M., Lemmens, I., Yu, H., et al. (2009). An experimentally derived confidence score for binary protein-protein interactions. *Nature Methods*, *6*(1), 91–97. http://dx.doi.org/10.1038/nmeth.1281.

Bruckner, A., Polge, C., Lentze, N., Auerbach, D., & Schlattner, U. (2009). Yeast two-hybrid, a powerful tool for systems biology. *International Journal of Molecular Sciences*, *10*(6), 2763–2788. http://dx.doi.org/10.3390/ijms10062763.

Cary, M. P., Bader, G. D., & Sander, C. (2005). Pathway information for systems biology. *FEBS Letters*, *579*(8), 1815–1820. http://dx.doi.org/10.1016/j.febslet.2005.02.005.

Casado-Vela, J., Cebrian, A., del Pulgar, M. T., Sanchez-Lopez, E., Vilaseca, M., Menchen, L., et al. (2011). Lights and shadows of proteomic technologies for the study of protein species including isoforms, splicing variants and protein post-translational modifications. *Proteomics*, *11*(4), 590–603. http://dx.doi.org/10.1002/pmic.201000287.

Casado-Vela, J., Cebrian, A., Gomez del Pulgar, M. T., & Lacal, J. C. (2011). Approaches for the study of cancer: Towards the integration of genomics, proteomics and metabolomics. *Clinical & Translational Oncology*, *13*(9), 617–628. http://dx.doi.org/10.1007/s12094-011-0707-9.

Casado-Vela, J., Gonzalez-Gonzalez, M., Matarraz, S., Martinez-Esteso, M., Vilella, M., Sayagues, J., et al. (2013). Protein arrays: Recent achievements and their application to study the human proteome. *Current Proteomics*, *10*(2), 83–87. http://dx.doi.org/10.2174/1570164611310020003.

Casado-Vela, J., Lacal, J. C., & Elortza, F. (2013). Protein chimerism: Novel source of protein diversity in humans adds complexity to bottom-up proteomics. *Proteomics*, *13*(1), 5–11. http://dx.doi.org/10.1002/pmic.201200371.

Casado-Vela, J., Martinez-Torrecuadrada, J. L., & Casal, J. I. (2009). Differential phosphorylation patterns between the Cyclin-A2/CDK2 complex and their monomers. *Protein Expression and Purification*, *66*(1), 15–21. http://dx.doi.org/10.1016/j.pep.2009.02.007.

Casado-Vela, J., Matthiesen, R., Selles, S., & Naranjo, J. R. (2013). Protein-protein interactions: Gene acronym redundancies and current limitations precluding automated data integration. *Proteomes*, *1*(1), 3–24. http://dx.doi.org/10.3390/proteomes1010003.

Casado-Vela, J., Rodriguez-Suarez, E., Iloro, I., Ametzazurra, A., Alkorta, N., Garcia-Velasco, J. A., et al. (2009). Comprehensive proteomic analysis of human endometrial fluid aspirate. *Journal of Proteome Research*, *8*(10), 4622–4632. http://dx.doi.org/10.1021/pr9004426.

Chandra, H., Reddy, P. J., & Srivastava, S. (2011). Protein microarrays and novel detection platforms. *Expert Review of Proteomics*, *8*(1), 61–79. http://dx.doi.org/10.1586/epr.10.99.

Chatr-Aryamontri, A., Breitkreutz, B. J., Heinicke, S., Boucher, L., Winter, A., Stark, C., et al. (2013). The BioGRID interaction database: 2013 update. *Nucleic Acids Research*, *41*, D816–D823. http://dx.doi.org/10.1093/nar/gks1158.

Chaurasia, G., & Futschik, M. (2012). The integration and annotation of the human interactome in the UniHI Database. *Methods in Molecular Biology*, *812*, 175–188. http://dx.doi.org/10.1007/978-1-61779-455-1_10.

Chautard, E., Fatoux-Ardore, M., Ballut, L., Thierry-Mieg, N., & Ricard-Blum, S. (2011). MatrixDB, the extracellular matrix interaction database. *Nucleic Acids Research*, *39*, D235–D240. http://dx.doi.org/10.1093/nar/gkq830.

Chen, Q., Chen, J., Sun, T., Shen, J., Shen, X., & Jiang, H. (2004). A yeast two-hybrid technology-based system for the discovery of PPARgamma agonist and antagonist. *Analytical Biochemistry*, *335*(2), 253–259. http://dx.doi.org/10.1016/j.ab.2004.09.004.

Chen, X. W., Jeong, J. C., & Dermyer, P. (2011). KUPS: Constructing datasets of interacting and non-interacting protein pairs with associated attributions. *Nucleic Acids Research*, *39*, D750–D754. http://dx.doi.org/10.1093/nar/gkq943.

Chen, Y. C., Rajagopala, S. V., Stellberger, T., & Uetz, P. (2010). Exhaustive benchmarking of the yeast two-hybrid system. *Nature Methods*, 7(9), 667–668. http://dx.doi.org/10.1038/nmeth0910-667.

Chen, J., Zhou, J., Sanders, C. K., Nolan, J. P., & Cai, H. (2009). A surface display yeast two-hybrid screening system for high-throughput protein interactome mapping. *Analytical Biochemistry*, *390*(1), 29–37. http://dx.doi.org/10.1016/j.ab.2009.03.013.

Corthals, G. L., Wasinger, V. C., Hochstrasser, D. F., & Sanchez, J. C. (2000). The dynamic range of protein expression: A challenge for proteomic research. *Electrophoresis*, *21*(6), 1104–1115. http://dx.doi.org/10.1002/(SICI)1522-2683(20000401)21:6<1104::AID-ELPS1104>3.0.CO;2-C.

Cowley, M. J., Pinese, M., Kassahn, K. S., Waddell, N., Pearson, J. V., Grimmond, S. M., et al. (2012). PINA v2.0: Mining interactome modules. *Nucleic Acids Research*, *40*, D862–D865. http://dx.doi.org/10.1093/nar/gkr967.

Díez, P., Dasilva, N., Gonzalez-Gonzalez, M., Matarraz, S., Casado-Vela, J., Orfao, A., et al. (2012). Data analysis strategies for protein microarrays. *Microarrays*, *1*, 64–83. http://dx.doi.org/10.3390/microarrays1020064.

Ekins, R., & Chu, F. (1992). Multianalyte microspot immunoassay. The microanalytical 'compact disk' of the future. *Annales de Biologie Clinique*, *50*(5), 337–353.

Ekins, R., Chu, F., & Biggart, E. (1990). Multispot, multianalyte, immunoassay. *Annales de Biologie Clinique*, *48*(9), 655–666.

Ellmark, P., Belov, L., Huang, P., Lee, C. S., Solomon, M. J., Morgan, D. K., et al. (2006). Multiplex detection of surface molecules on colorectal cancers. *Proteomics*, *6*(6), 1791–1802. http://dx.doi.org/10.1002/pmic.200500468.

Espina, V., Mehta, A. I., Winters, M. E., Calvert, V., Wulfkuhle, J., Petricoin, E. F., et al. (2003). Protein microarrays: Molecular profiling technologies for clinical specimens. *Proteomics*, *3*(11), 2091–2100. http://dx.doi.org/10.1002/pmic.200300592.

Feijs, K. L., Kleine, H., Braczynski, A., Forst, A. H., Herzog, N., Verheugd, P., et al. (2013). ARTD10 substrate identification on protein microarrays: Regulation of GSK3beta by mono-ADP-ribosylation. *Cell Communication and Signaling*, *11*(1), 5. http://dx.doi.org/10.1186/1478-811X-11-5.

Feilner, T., Hultschig, C., Lee, J., Meyer, S., Immink, R. G., Koenig, A., et al. (2005). High throughput identification of potential Arabidopsis mitogen-activated protein kinases substrates. *Molecular & Cellular Proteomics*, *4*(10), 1558–1568. http://dx.doi.org/10.1074/mcp.M500007-MCP200.

Feldmann, M. (2008). Many cytokines are very useful therapeutic targets in disease. *The Journal of Clinical Investigation*, *118*(11), 3533–3536. http://dx.doi.org/10.1172/JCI37346.

Fenner, B. J., Scannell, M., & Prehn, J. H. (2010). Expanding the substantial interactome of NEMO using protein microarrays. *PLoS One*, *5*(1), e8799. http://dx.doi.org/10.1371/journal.pone.0008799.

Fernandez-Suarez, X. M., & Galperin, M. Y. (2013). The 2013 Nucleic Acids Research Database Issue and the online molecular biology database collection. *Nucleic Acids Research*, *41*, D1–D7. http://dx.doi.org/10.1093/nar/gks1297.

Fields, S., & Song, O. (1989). A novel genetic system to detect protein-protein interactions. *Nature*, *340*(6230), 245–246. http://dx.doi.org/10.1038/340245a0.

Franco-Zorrilla, J. M., & Solano, R. (2014). High-throughput analysis of protein-DNA binding affinity. *Methods in Molecular Biology*, *1062*, 697–709. http://dx.doi.org/10.1007/978-1-62703-580-4_36.

Frenkel-Morgenstern, M., Lacroix, V., Ezkurdia, I., Levin, Y., Gabashvili, A., Prilusky, J., et al. (2012). Chimeras taking shape: Potential functions of proteins encoded by chimeric RNA transcripts. *Genome Research*, *22*(7), 1231–1242. http://dx.doi.org/10.1101/gr.130062.111.

Furney, S. J., Higgins, D. G., Ouzounis, C. A., & Lopez-Bigas, N. (2006). Structural and functional properties of genes involved in human cancer. *BMC Genomics*, 7, 3. http://dx.doi.org/10.1186/1471-2164-7-3.

Garcia-Garcia, J., Schleker, S., Klein-Seetharaman, J., & Oliva, B. (2012). BIPS: BIANA Interolog Prediction Server. A tool for protein-protein interaction inference. *Nucleic Acids Research*, *40*, W147–W151. http://dx.doi.org/10.1093/nar/gks553.

Godoy, M., Franco-Zorrilla, J. M., Perez-Perez, J., Oliveros, J. C., Lorenzo, O., & Solano, R. (2011). Improved protein-binding microarrays for the identification of DNA-binding specificities of transcription factors. *The Plant Journal*, *66*(4), 700–711. http://dx.doi.org/10.1111/j.1365-313X.2011.04519.x.

Gomez, S. M., Choi, K., & Wu, Y. (2008). Prediction of protein-protein interaction networks. *Current Protocols in Bioinformatics*. *8*, 8.2.1–8.2.14. http://dx.doi.org/10.1002/0471250953.bi0802s22, Chapter 8, Unit 8.2.

Gong, S., Yoon, G., Jang, I., Bolser, D., Dafas, P., Schroeder, M., et al. (2005). PSIbase: A database of Protein Structural Interactome map (PSIMAP). *Bioinformatics*, *21*(10), 2541–2543. http://dx.doi.org/10.1093/bioinformatics/bti366.

Gong, W., Zhou, D., Ren, Y., Wang, Y., Zuo, Z., Shen, Y., et al. (2008). PepCyber:P PEP: A database of human protein protein interactions mediated by phosphoprotein-binding domains. *Nucleic Acids Research*, *36*, D679–D683. http://dx.doi.org/10.1093/nar/gkm854.

Grove, C. A., De Masi, F., Barrasa, M. I., Newburger, D. E., Alkema, M. J., Bulyk, M. L., et al. (2009). A multiparameter network reveals extensive divergence between *C. elegans* bHLH transcription factors. *Cell*, *138*(2), 314–327. http://dx.doi.org/10.1016/j.cell.2009.04.058.

Gujral, T. S., Karp, R. L., Finski, A., Chan, M., Schwartz, P. E., Macbeath, G., et al. (2012). Profiling phospho-signaling networks in breast cancer using reverse-phase protein arrays. *Oncogene*, *32*, 3470–3476. http://dx.doi.org/10.1038/onc.2012.378.

Gulmann, C., Sheehan, K. M., Kay, E. W., Liotta, L. A., & Petricoin, E. F. (2006). Array-based proteomics: Mapping of protein circuitries for diagnostics, prognostics, and therapy guidance in cancer. *The Journal of Pathology*, *208*(5), 595–606. http://dx.doi.org/10.1002/path.1958.

Haab, B. B., Dunham, M. J., & Brown, P. O. (2001). Protein microarrays for highly parallel detection and quantitation of specific proteins and antibodies in complex solutions. *Genome Biology*, *2*(2), 1–13.

Hampshire, A. J., Rusling, D. A., Broughton-Head, V. J., & Fox, K. R. (2007). Footprinting: A method for determining the sequence selectivity, affinity and kinetics of DNA-binding ligands. *Methods*, *42*(2), 128–140. http://dx.doi.org/10.1016/j.ymeth.2007.01.002.

Hanes, S. D., & Brent, R. (1991). A genetic model for interaction of the homeodomain recognition helix with DNA. *Science*, *251*(4992), 426–430. http://dx.doi.org/10.1126/science.1671176.

Harbison, C. T., Gordon, D. B., Lee, T. I., Rinaldi, N. J., Macisaac, K. D., Danford, T. W., et al. (2004). Transcriptional regulatory code of a eukaryotic genome. *Nature*, *431*(7004), 99–104. http://dx.doi.org/10.1038/nature02800.

Harrison, R. L., & Jarvis, D. L. (2006). Protein N-glycosylation in the baculovirus-insect cell expression system and engineering of insect cells to produce "mammalianized" recombinant glycoproteins. *Advances in Virus Research*, *68*, 159–191. http://dx.doi.org/10.1016/S0065-3527(06)68005-6.

Hart, G. T., Ramani, A. K., & Marcotte, E. M. (2006). How complete are current yeast and human protein-interaction networks? *Genome Biology*, 7(11), 120. http://dx.doi.org/10.1186/gb-2006-7-11-120.

He, M., Stoevesandt, O., & Taussig, M. J. (2008). In situ synthesis of protein arrays. *Current Opinion in Biotechnology*, *19*(1), 4–9. http://dx.doi.org/10.1016/j.copbio.2007.11.009.

He, M., & Taussig, M. J. (2001). Single step generation of protein arrays from DNA by cell-free expression and in situ immobilisation (PISA method). *Nucleic Acids Research*, *29*(15), e73.

Higurashi, M., Ishida, T., & Kinoshita, K. (2009). PiSite: A database of protein interaction sites using multiple binding states in the PDB. *Nucleic Acids Research*, *37*, D360–D364. http://dx.doi.org/10.1093/nar/gkn659.

Hoffman, M. M., Khrapov, M. A., Cox, J. C., Yao, J., Tong, L., & Ellington, A. D. (2004). AANT: The Amino Acid-Nucleotide Interaction Database. *Nucleic Acids Research*, *32*, D174–D181. http://dx.doi.org/10.1093/nar/gkh128.

Huang, R. P., Huang, R., Fan, Y., & Lin, Y. (2001). Simultaneous detection of multiple cytokines from conditioned media and patient's sera by an antibody-based protein array system. *Analytical Biochemistry*, *294*(1), 55–62. http://dx.doi.org/10.1006/abio.2001.5156.

Huang, T. W., Tien, A. C., Huang, W. S., Lee, Y. C., Peng, C. L., Tseng, H. H., et al. (2004). POINT: A database for the prediction of protein-protein interactions based on the orthologous interactome. *Bioinformatics*, *20*(17), 3273–3276. http://dx.doi.org/10.1093/bioinformatics/bth366.

Huang, R. P., Yang, W., Yang, D., Flowers, L., Horowitz, I. R., Cao, X., et al. (2005). The promise of cytokine antibody arrays in the drug discovery process. *Expert Opinion on Therapeutic Targets*, *9*(3), 601–615. http://dx.doi.org/10.1517/14728222.9.3.601.

Hudson, M. E., Pozdnyakova, I., Haines, K., Mor, G., & Snyder, M. (2007). Identification of differentially expressed proteins in ovarian cancer using high-density protein microarrays. *Proceedings of the National Academy of Sciences of the United States of America*, *104*(44), 17494–17499. http://dx.doi.org/10.1073/pnas.0708572104.

Hurst, R., Hook, B., Slater, M. R., Hartnett, J., Storts, D. R., & Nath, N. (2009). Protein-protein interaction studies on protein arrays: Effect of detection strategies on signal-to-background ratios. *Analytical Biochemistry*, *392*(1), 45–53. http://dx.doi.org/10.1016/j.ab.2009.05.028.

Huttenhain, R., Soste, M., Selevsek, N., Rost, H., Sethi, A., Carapito, C., et al. (2012). Reproducible quantification of cancer-associated proteins in body fluids using targeted proteomics. *Science Translational Medicine*, *4*(142), 142ra194. http://dx.doi.org/10.1126/scitranslmed.3003989.

Ito, T., Chiba, T., Ozawa, R., Yoshida, M., Hattori, M., & Sakaki, Y. (2001). A comprehensive two-hybrid analysis to explore the yeast protein interactome. *Proceedings of the National Academy of Sciences of the United States of America, 98*(8), 4569–4574. http://dx.doi.org/10.1073/pnas.061034498.

Jansen, C., Gronenborn, A. M., & Clore, G. M. (1987). The binding of the cyclic AMP receptor protein to synthetic DNA sites containing permutations in the consensus sequence TGTGA. *The Biochemical Journal, 246*(1), 227–232.

Jefferson, E. R., Walsh, T. P., Roberts, T. J., & Barton, G. J. (2007). SNAPPI-DB: A database and API of Structures, interfaces and alignments for protein-protein interactions. *Nucleic Acids Research, 35*, D580–D589. http://dx.doi.org/10.1093/nar/gkl836.

Jeong, J. S., Jiang, L., Albino, E., Marrero, J., Rho, H. S., Hu, J., et al. (2012). Rapid identification of monospecific monoclonal antibodies using a human proteome microarray. *Molecular & Cellular Proteomics. 11*(6), O111.016253. http://dx.doi.org/10.1074/mcp.O111.016253.

Jessulat, M., Pitre, S., Gui, Y., Hooshyar, M., Omidi, K., Samanfar, B., et al. (2011). Recent advances in protein-protein interaction prediction: Experimental and computational methods. *Expert Opinion on Drug Discovery, 6*(9), 921–935. http://dx.doi.org/10.1517/17460441.2011.603722.

Ji, Z. L., Chen, X., Zhen, C. J., Yao, L. X., Han, L. Y., Yeo, W. K., et al. (2003). KDBI: Kinetic Data of Bio-molecular interactions database. *Nucleic Acids Research, 31*(1), 255–257. http://dx.doi.org/10.1093/nar/gkg067.

Jolma, A., Kivioja, T., Toivonen, J., Cheng, L., Wei, G., Enge, M., et al. (2010). Multiplexed massively parallel SELEX for characterization of human transcription factor binding specificities. *Genome Research, 20*(6), 861–873. http://dx.doi.org/10.1101/gr.100552.109.

Jungblut, P., Thiede, B., Zimny-Arndt, U., Muller, E. C., Scheler, C., Wittmann-Liebold, B., et al. (1996). Resolution power of two-dimensional electrophoresis and identification of proteins from gels. *Electrophoresis, 17*(5), 839–847. http://dx.doi.org/10.1002/elps.1150170505.

Kader, H. A., Tchernev, V. T., Satyaraj, E., Lejnine, S., Kotler, G., Kingsmore, S. F., et al. (2005). Protein microarray analysis of disease activity in pediatric inflammatory bowel disease demonstrates elevated serum PLGF, IL-7, TGF-beta1, and IL-12p40 levels in Crohn's disease and ulcerative colitis patients in remission versus active disease. *The American Journal of Gastroenterology, 100*(2), 414–423. http://dx.doi.org/10.1111/j.1572-0241.2005.40819.x.

Kalinina, O. V., Wichmann, O., Apic, G., & Russell, R. B. (2011). Combinations of protein-chemical complex structures reveal new targets for established drugs. *PLoS Computational Biology*, 7(5), e1002043. http://dx.doi.org/10.1371/journal.pcbi.1002043.

Kamburov, A., Stelzl, U., Lehrach, H., & Herwig, R. (2013). The ConsensusPathDB interaction database: 2013 update. *Nucleic Acids Research, 41*, D793–D800. http://dx.doi.org/10.1093/nar/gks1055.

Katz, C., Levy-Beladev, L., Rotem-Bamberger, S., Rito, T., Rudiger, S. G., & Friedler, A. (2011). Studying protein-protein interactions using peptide arrays. *Chemical Society Reviews, 40*(5), 2131–2145. http://dx.doi.org/10.1039/c0cs00029a.

Katzen, F., Chang, G., & Kudlicki, W. (2005). The past, present and future of cell-free protein synthesis. *Trends in Biotechnology, 23*(3), 150–156. http://dx.doi.org/10.1016/j.tibtech.2005.01.003.

Kerrien, S., Aranda, B., Breuza, L., Bridge, A., Broackes-Carter, F., Chen, C., et al. (2012). The IntAct molecular interaction database in 2012. *Nucleic Acids Research, 40*, D841–D846. http://dx.doi.org/10.1093/nar/gkr1088.

Kersten, B., Possling, A., Blaesing, F., Mirgorodskaya, E., Gobom, J., & Seitz, H. (2004). Protein microarray technology and ultraviolet crosslinking combined with mass spectrometry for the analysis of protein-DNA interactions. *Analytical Biochemistry, 331*(2), 303–313. http://dx.doi.org/10.1016/j.ab.2004.05.008.

Kim, D., Daniel, W. L., & Mirkin, C. A. (2009). Microarray-based multiplexed scanometric immunoassay for protein cancer markers using gold nanoparticle probes. *Analytical Chemistry*, *81*(21), 9183–9187. http://dx.doi.org/10.1021/ac9018389.

Kim, M. J., Lee, T. H., Pahk, Y. M., Kim, Y. H., Park, H. M., Choi, Y. D., et al. (2009). Quadruple 9-mer-based protein binding microarray with DsRed fusion protein. *BMC Molecular Biology*, *10*, 91. http://dx.doi.org/10.1186/1471-2199-10-91.

Klingstrom, T., & Plewczynski, D. (2011). Protein-protein interaction and pathway databases, a graphical review. *Briefings in Bioinformatics*, *12*(6), 702–713. http://dx.doi.org/10.1093/bib/bbq064.

Kumar, M. D., & Gromiha, M. M. (2006). PINT: Protein-protein interactions thermodynamic database. *Nucleic Acids Research*, *34*, D195–D198. http://dx.doi.org/10.1093/nar/gkj017.

Kusnezow, W., Jacob, A., Walijew, A., Diehl, F., & Hoheisel, J. D. (2003). Antibody microarrays: An evaluation of production parameters. *Proteomics*, *3*(3), 254–264. http://dx.doi.org/10.1002/pmic.200390038.

LaBaer, J., & Ramachandran, N. (2005). Protein microarrays as tools for functional proteomics. *Current Opinion in Chemical Biology*, *9*(1), 14–19. http://dx.doi.org/10.1016/j.cbpa.2004.12.006.

Lane, D., Prentki, P., & Chandler, M. (1992). Use of gel retardation to analyze protein-nucleic acid interactions. *Microbiological Reviews*, *56*(4), 509–528.

Lange, V., Picotti, P., Domon, B., & Aebersold, R. (2008). Selected reaction monitoring for quantitative proteomics: A tutorial. *Molecular Systems Biology*, *4*, 222. http://dx.doi.org/10.1038/msb.2008.61.

Lee, J. R., Magee, D. M., Gaster, R. S., LaBaer, J., & Wang, S. X. (2013). Emerging protein array technologies for proteomics. *Expert Review of Proteomics*, *10*(1), 65–75. http://dx.doi.org/10.1586/epr.12.67.

Li, H., Wang, J., Mor, G., & Sklar, J. (2008). A neoplastic gene fusion mimics trans-splicing of RNAs in normal human cells. *Science*, *321*(5894), 1357–1361. http://dx.doi.org/10.1126/science.1156725.

Licata, L., Briganti, L., Peluso, D., Perfetto, L., Iannuccelli, M., Galeota, E., et al. (2012). MINT, the molecular interaction database: 2012 update. *Nucleic Acids Research*, *40*, D857–D861. http://dx.doi.org/10.1093/nar/gkr930.

Liotta, L. A., Espina, V., Mehta, A. I., Calvert, V., Rosenblatt, K., Geho, D., et al. (2003). Protein microarrays: Meeting analytical challenges for clinical applications. *Cancer Cell*, *3*(4), 317–325. http://dx.doi.org/10.1016/S1535-6108(03)00086-2.

Lo, Y. S., Chen, Y. C., & Yang, J. M. (2010). 3D-interologs: An evolution database of physical protein- protein interactions across multiple genomes. *BMC Genomics*, *11*(Suppl 3), S7. http://dx.doi.org/10.1186/1471-2164-11-S3-S7.

Locker, G. J., Steger, G. G., Gnant, M. F., Steiner, B., Simonitsch, I., Krainer, M., et al. (1999). Induction of immunomediated diseases by recombinant human granulocyte-macrophage colony-stimulating factor during cancer treatment? *Journal of Immunotherapy*, *22*(1), 85–89.

Luc, P. V., & Tempst, P. (2004). PINdb: A database of nuclear protein complexes from human and yeast. *Bioinformatics*, *20*(9), 1413–1415. http://dx.doi.org/10.1093/bioinformatics/bth114.

Ly, L., & Wasinger, V. C. (2011). Protein and peptide fractionation, enrichment and depletion: Tools for the complex proteome. *Proteomics*, *11*(4), 513–534. http://dx.doi.org/10.1002/pmic.201000394.

MacBeath, G., & Schreiber, S. L. (2000). Printing proteins as microarrays for high-throughput function determination. *Science*, *289*(5485), 1760–1763. http://dx.doi.org/10.1126/science.289.5485.1760.

Madoz-Gurpide, J., Kuick, R., Wang, H., Misek, D. E., & Hanash, S. M. (2008). Integral protein microarrays for the identification of lung cancer antigens in sera that induce a humoral immune response. *Molecular & Cellular Proteomics*, 7(2), 268–281. http://dx.doi.org/10.1074/mcp.M700366-MCP200.

Mak, H. C., Daly, M., Gruebel, B., & Ideker, T. (2007). Cell Circuits: A database of protein network models. *Nucleic Acids Research*, *35*, D538–D545. http://dx.doi.org/10.1093/nar/gkl937.

Matarraz, S., Gonzalez-Gonzalez, M., Jara, M., Orfao, A., & Fuentes, M. (2011). New technologies in cancer. Protein microarrays for biomarker discovery. *Clinical & Translational Oncology*, *13*(3), 156–161. http://dx.doi.org/10.1007/s12094-011-0635-8.

McDowall, M. D., Scott, M. S., & Barton, G. J. (2009). PIPs: Human protein-protein interaction prediction database. *Nucleic Acids Research*, *37*, D651–D656. http://dx.doi.org/10.1093/nar/gkn870.

McInnes, I. B., & Schett, G. (2007). Cytokines in the pathogenesis of rheumatoid arthritis. *Nature Reviews. Immunology*, 7(6), 429–442. http://dx.doi.org/10.1038/nri2094.

McKenna, N. J., & O'Malley, B. W. (2002). Combinatorial control of gene expression by nuclear receptors and coregulators. *Cell*, *108*(4), 465–474.

Meng, X., Brodsky, M. H., & Wolfe, S. A. (2005). A bacterial one-hybrid system for determining the DNA-binding specificity of transcription factors. *Nature Biotechnology*, *23*(8), 988–994. http://dx.doi.org/10.1038/nbt1120.

Mijakovic, I., & Macek, B. (2012). Impact of phosphoproteomics on studies of bacterial physiology. *FEMS Microbiology Reviews*, *36*(4), 877–892. http://dx.doi.org/10.1111/j.1574-6976.2011.00314.x.

Miller, J. C., Butler, E. B., Teh, B. S., & Haab, B. B. (2001). The application of protein microarrays to serum diagnostics: Prostate cancer as a test case. *Disease Markers*, *17*(4), 225–234.

Millioni, R., Tolin, S., Puricelli, L., Sbrignadello, S., Fadini, G. P., Tessari, P., et al. (2011). High abundance proteins depletion vs low abundance proteins enrichment: Comparison of methods to reduce the plasma proteome complexity. *PLoS One*, *6*(5), e19603. http://dx.doi.org/10.1371/journal.pone.0019603.

Mishra, S. (2012). Computational prediction of protein-protein complexes. *BMC Research Notes*, *5*(1), 495. http://dx.doi.org/10.1186/1756-0500-5-495.

Miskimins, W. K., Roberts, M. P., McClelland, A., & Ruddle, F. H. (1985). Use of a protein-blotting procedure and a specific DNA probe to identify nuclear proteins that recognize the promoter region of the transferrin receptor gene. *Proceedings of the National Academy of Sciences of the United States of America*, *82*(20), 6741–6744.

Mor, G., Visintin, I., Lai, Y., Zhao, H., Schwartz, P., Rutherford, T., et al. (2005). Serum protein markers for early detection of ovarian cancer. *Proceedings of the National Academy of Sciences of the United States of America*, *102*(21), 7677–7682. http://dx.doi.org/10.1073/pnas.0502178102.

Mosca, R., Ceol, A., Stein, A., Olivella, R., & Aloy, P. (2013). 3did: A catalog of domain-based interactions of known three-dimensional structure. *Nucleic Acids Research*, *42*, D374–D379. http://dx.doi.org/10.1093/nar/gkt887.

Mukherjee, S., Berger, M. F., Jona, G., Wang, X. S., Muzzey, D., Snyder, M., et al. (2004). Rapid analysis of the DNA-binding specificities of transcription factors with DNA microarrays. *Nature Genetics*, *36*(12), 1331–1339. http://dx.doi.org/10.1038/ng1473.

Muthusamy, B., Thomas, J. K., Prasad, T. S., & Pandey, A. (2013). Access guide to human proteinpedia. *Current Protocols in Bioinformatics*, *41*, 1–15. http://dx.doi.org/10.1002/0471250953.bi0121s41, Chapter 1, Unit 1 21.

Nielsen, U. B., Cardone, M. H., Sinskey, A. J., MacBeath, G., & Sorger, P. K. (2003). Profiling receptor tyrosine kinase activation by using Ab microarrays. *Proceedings of the National Academy of Sciences of the United States of America*, *100*(16), 9330–9335. http://dx.doi.org/10.1073/pnas.1633513100.

Park, P. J. (2009). ChIP-seq: Advantages and challenges of a maturing technology. *Nature Reviews. Genetics*, *10*(10), 669–680. http://dx.doi.org/10.1038/nrg2641.

Paweletz, C. P., Charboneau, L., Bichsel, V. E., Simone, N. L., Chen, T., Gillespie, J. W., et al. (2001). Reverse phase protein microarrays which capture disease progression show activation of pro-survival pathways at the cancer invasion front. *Oncogene*, *20*(16), 1981–1989. http://dx.doi.org/10.1038/sj.onc.1204265.

Picotti, P., Bodenmiller, B., Mueller, L. N., Domon, B., & Aebersold, R. (2009). Full dynamic range proteome analysis of *S. cerevisiae* by targeted proteomics. *Cell*, *138*(4), 795–806. http://dx.doi.org/10.1016/j.cell.2009.05.051.

Pitre, S., Alamgir, M., Green, J. R., Dumontier, M., Dehne, F., & Golshani, A. (2008). Computational methods for predicting protein-protein interactions. *Advances in Biochemical Engineering/Biotechnology*, *110*, 247–267. http://dx.doi.org/10.1007/10_2007_089.

Portales-Casamar, E., Thongjuea, S., Kwon, A. T., Arenillas, D., Zhao, X., Valen, E., et al. (2010). JASPAR 2010: The greatly expanded open-access database of transcription factor binding profiles. *Nucleic Acids Research*, *38*, D105–D110. http://dx.doi.org/10.1093/nar/gkp950.

Ptacek, J., Devgan, G., Michaud, G., Zhu, H., Zhu, X., Fasolo, J., et al. (2005). Global analysis of protein phosphorylation in yeast. *Nature*, *438*(7068), 679–684. http://dx.doi.org/10.1038/nature04187.

Ramachandran, N., Hainsworth, E., Bhullar, B., Eisenstein, S., Rosen, B., Lau, A. Y., et al. (2004). Self-assembling protein microarrays. *Science*, *305*(5680), 86–90. http://dx.doi.org/10.1126/science.1097639.

Ramachandran, N., Raphael, J. V., Hainsworth, E., Demirkan, G., Fuentes, M. G., Rolfs, A., et al. (2008). Next-generation high-density self-assembling functional protein arrays. *Nature Methods*, *5*(6), 535–538. http://dx.doi.org/10.1038/nmeth.1210.

Ramani, A. K., Bunescu, R. C., Mooney, R. J., & Marcotte, E. M. (2005). Consolidating the set of known human protein-protein interactions in preparation for large-scale mapping of the human interactome. *Genome Biology*, *6*(5), R40. http://dx.doi.org/10.1186/gb-2005-6-5-r40.

Ravasi, T., Suzuki, H., Cannistraci, C. V., Katayama, S., Bajic, V. B., Tan, K., et al. (2010). An atlas of combinatorial transcriptional regulation in mouse and man. *Cell*, *140*(5), 744–752. http://dx.doi.org/10.1016/j.cell.2010.01.044.

Remenyi, A., Scholer, H. R., & Wilmanns, M. (2004). Combinatorial control of gene expression. *Nature Structural & Molecular Biology*, *11*(9), 812–815. http://dx.doi.org/10.1038/nsmb820.

Robasky, K., & Bulyk, M. L. (2011). UniPROBE, update 2011: Expanded content and search tools in the online database of protein-binding microarray data on protein-DNA interactions. *Nucleic Acids Research*, *39*, D124–D128. http://dx.doi.org/10.1093/nar/gkq992.

Roulet, E., Busso, S., Camargo, A. A., Simpson, A. J., Mermod, N., & Bucher, P. (2002). High-throughput SELEX SAGE method for quantitative modeling of transcription-factor binding sites. *Nature Biotechnology*, *20*(8), 831–835. http://dx.doi.org/10.1038/nbt718.

Sahdev, S., Khattar, S. K., & Saini, K. S. (2008). Production of active eukaryotic proteins through bacterial expression systems: A review of the existing biotechnology strategies. *Molecular and Cellular Biochemistry*, *307*(1–2), 249–264. http://dx.doi.org/10.1007/s11010-007-9603-6.

Schaefer, C. F., Anthony, K., Krupa, S., Buchoff, J., Day, M., Hannay, T., et al. (2009). PID: The Pathway Interaction Database. *Nucleic Acids Research*, *37*, D674–D679. http://dx.doi.org/10.1093/nar/gkn653.

Schleker, S., Sun, J., Raghavan, B., Srnec, M., Muller, N., Koepfinger, M., et al. (2012). The current Salmonella-host interactome. *Proteomics. Clinical Applications*, *6*(1–2), 117–133. http://dx.doi.org/10.1002/prca.201100083.

Schwarz, D., Dotsch, V., & Bernhard, F. (2008). Production of membrane proteins using cell-free expression systems. *Proteomics*, *8*(19), 3933–3946. http://dx.doi.org/10.1002/pmic.200800171.

Schweitzer, B., Predki, P., & Snyder, M. (2003). Microarrays to characterize protein interactions on a whole-proteome scale. *Proteomics*, *3*(11), 2190–2199. http://dx.doi.org/10.1002/pmic.200300610.

Shin, Y. C., Shin, S. Y., Chun, J. N., Cho, H. S., Lim, J. M., Kim, H. G., et al. (2012). TRIP database 2.0: A manually curated information hub for accessing TRP channel interaction network. *PLoS One*, 7(10), e47165. http://dx.doi.org/10.1371/journal.pone.0047165.

Sibani, S., & LaBaer, J. (2011). Immunoprofiling using NAPPA protein microarrays. *Methods in Molecular Biology*, *723*, 149–161. http://dx.doi.org/10.1007/978-1-61779-043-0_10.

Skrabanek, L., Saini, H. K., Bader, G. D., & Enright, A. J. (2008). Computational prediction of protein-protein interactions. *Molecular Biotechnology*, *38*(1), 1–17. http://dx.doi.org/10.1007/s12033-007-0069-2.

Smialowski, P., Pagel, P., Wong, P., Brauner, B., Dunger, I., Fobo, G., et al. (2010). The Negatome database: A reference set of non-interacting protein pairs. *Nucleic Acids Research*, *38*, D540–D544. http://dx.doi.org/10.1093/nar/gkp1026.

Spirin, S., Titov, M., Karyagina, A., & Alexeevski, A. (2007). NPIDB: A database of nucleic acids-protein interactions. *Bioinformatics*, *23*(23), 3247–3248. http://dx.doi.org/10.1093/bioinformatics/btm519.

Spurrier, B., Ramalingam, S., & Nishizuka, S. (2008). Reverse-phase protein lysate microarrays for cell signaling analysis. *Nature Protocols*, *3*(11), 1796–1808. http://dx.doi.org/10.1038/nprot.2008.179.

Stumpf, M. P., Thorne, T., de Silva, E., Stewart, R., An, H. J., Lappe, M., et al. (2008). Estimating the size of the human interactome. *Proceedings of the National Academy of Sciences of the United States of America*, *105*(19), 6959–6964. http://dx.doi.org/10.1073/pnas.0708078105.

Tao, S. C., & Zhu, H. (2006). Protein chip fabrication by capture of nascent polypeptides. *Nature Biotechnology*, *24*(10), 1253–1254. http://dx.doi.org/10.1038/nbt1249.

Tarcea, V. G., Weymouth, T., Ade, A., Bookvich, A., Gao, J., Mahavisno, V., et al. (2009). Michigan molecular interactions r2: From interacting proteins to pathways. *Nucleic Acids Research*, *37*, D642–D646. http://dx.doi.org/10.1093/nar/gkn722.

Tarrant, M. K., & Cole, P. A. (2009). The chemical biology of protein phosphorylation. *Annual Review of Biochemistry*, *78*, 797–825. http://dx.doi.org/10.1146/annurev.biochem.78.070907.103047.

Thaxton, C. S., Elghanian, R., Thomas, A. D., Stoeva, S. I., Lee, J. S., Smith, N. D., et al. (2009). Nanoparticle-based bio-barcode assay redefines "undetectable" PSA and biochemical recurrence after radical prostatectomy. *Proceedings of the National Academy of Sciences of the United States of America*, *106*(44), 18437–18442. http://dx.doi.org/10.1073/pnas.0904719106.

Tran, N. H., Choi, K. P., & Zhang, L. (2013). Counting motifs in the human interactome. *Nature Communications*, *4*, 2241. http://dx.doi.org/10.1038/ncomms3241.

Tyagi, M., Hashimoto, K., Shoemaker, B. A., Wuchty, S., & Panchenko, A. R. (2012). Large-scale mapping of human protein interactome using structural complexes. *EMBO Reports*, *13*(3), 266–271. http://dx.doi.org/10.1038/embor.2011.261.

Vaquerizas, J. M., Kummerfeld, S. K., Teichmann, S. A., & Luscombe, N. M. (2009). A census of human transcription factors: Function, expression and evolution. *Nature Reviews. Genetics*, *10*(4), 252–263. http://dx.doi.org/10.1038/nrg2538.

Venkatesan, K., Rual, J. F., Vazquez, A., Stelzl, U., Lemmens, I., Hirozane-Kishikawa, T., et al. (2009). An empirical framework for binary interactome mapping. *Nature Methods*, *6*(1), 83–90. http://dx.doi.org/10.1038/nmeth.1280.

Vigil, A., Chen, C., Jain, A., Nakajima-Sasaki, R., Jasinskas, A., Pablo, J., et al. (2011). Profiling the humoral immune response of acute and chronic Q fever by protein microarray. *Molecular & Cellular Proteomics*, *10*(10), M110.006304. http://dx.doi.org/10.1074/mcp.M110.006304.

Vigil, A., Ortega, R., Nakajima-Sasaki, R., Pablo, J., Molina, D. M., Chao, C. C., et al. (2010). Genome-wide profiling of humoral immune response to Coxiella burnetii infection by protein microarray. *Proteomics*, *10*(12), 2259–2269. http://dx.doi.org/10.1002/pmic.201000064.

Virok, D. P., Simon, D., Bozso, Z., Rajko, R., Datki, Z., Balint, E., et al. (2011). Protein array based interactome analysis of amyloid-beta indicates an inhibition of protein translation. *Journal of Proteome Research*, *10*(4), 1538–1547. http://dx.doi.org/10.1021/pr1009096.

von Eichborn, J., Murgueitio, M. S., Dunkel, M., Koerner, S., Bourne, P. E., & Preissner, R. (2011). PROMISCUOUS: A database for network-based drug-repositioning. *Nucleic Acids Research*, *39*, D1060–D1066. http://dx.doi.org/10.1093/nar/gkq1037.

Wang, L., Cole, K. D., He, H. J., Hancock, D. K., Gaigalas, A. K., & Zong, Y. (2006). Comparison of ovalbumin quantification using forward-phase protein microarrays and suspension arrays. *Journal of Proteome Research*, *5*(7), 1770–1775. http://dx.doi.org/10.1021/pr060074v.

Warren, C. L., Kratochvil, N. C., Hauschild, K. E., Foister, S., Brezinski, M. L., Dervan, P. B., et al. (2006). Defining the sequence-recognition profile of DNA-binding molecules. *Proceedings of the National Academy of Sciences of the United States of America*, *103*(4), 867–872. http://dx.doi.org/10.1073/pnas.0509843102.

Wilson, B., Liotta, L. A., & Petricoin, E. F. (2010). Monitoring proteins and protein networks using reverse phase protein arrays. *Disease Markers*, *28*(4), 225–232. http://dx.doi.org/10.3233/DMA-2010-0705.

Wingender, E. (2008). The TRANSFAC project as an example of framework technology that supports the analysis of genomic regulation. *Briefings in Bioinformatics*, *9*(4), 326–332. http://dx.doi.org/10.1093/bib/bbn016.

Wingender, E., Chen, X., Fricke, E., Geffers, R., Hehl, R., Liebich, I., et al. (2001). The TRANSFAC system on gene expression regulation. *Nucleic Acids Research*, *29*(1), 281–283.

Wingren, C., & Borrebaeck, C. A. (2009). Antibody-based microarrays. *Methods in Molecular Biology*, *509*, 57–84. http://dx.doi.org/10.1007/978-1-59745-372-1_5.

Woodbury, C. P., Jr., & von Hippel, P. H. (1983). On the determination of deoxyribonucleic acid-protein interaction parameters using the nitrocellulose filter-binding assay. *Biochemistry*, *22*(20), 4730–4737. http://dx.doi.org/10.1021/bi00289a018.

Xu, X., Song, Y., Li, Y., Chang, J., Zhang, H., & An, L. (2010). The tandem affinity purification method: An efficient system for protein complex purification and protein interaction identification. *Protein Expression and Purification*, *72*(2), 149–156. http://dx.doi.org/10.1016/j.pep.2010.04.009.

Yang, J., Roy, A., & Zhang, Y. (2013). BioLiP: A semi-manually curated database for biologically relevant ligand-protein interactions. *Nucleic Acids Research*, *41*, D1096–D1103. http://dx.doi.org/10.1093/nar/gks966.

Yarden, Y., & Sliwkowski, M. X. (2001). Untangling the ErbB signalling network. *Nature Reviews. Molecular Cell Biology*, *2*(2), 127–137. http://dx.doi.org/10.1038/35052073.

Zhang, Q. C., Petrey, D., Garzon, J. I., Deng, L., & Honig, B. (2013). PrePPI: A structure-informed database of protein-protein interactions. *Nucleic Acids Research*, *41*, D828–D833. http://dx.doi.org/10.1093/nar/gks1231.

Zhu, C., Byers, K. J., McCord, R. P., Shi, Z., Berger, M. F., Newburger, D. E., et al. (2009). High-resolution DNA-binding specificity analysis of yeast transcription factors. *Genome Research*, *19*(4), 556–566. http://dx.doi.org/10.1101/gr.090233.108.

CHAPTER NINE

Application of Proteomics in Diagnosis of ADHD, Schizophrenia, Major Depression, and Suicidal Behavior

Khaled Alawam[1]
Forensic Medicine Department, Ministry of Interior, Kuwait City, Kuwait
[1]Corresponding author: e-mail address: alawam2@hotmail.com

Contents

Advances in Protein Chemistry and Structural Biology, Volume 95
ISSN 1876-1623
http://dx.doi.org/10.1016/B978-0-12-800453-1.00009-9

Abstract

This report focuses on the application of different proteomic techniques in diagnosis and treatment of psychiatric disorders such as major depression, suicidal behavior, schizophrenia, and attention deficit/hyperactivity disorder (ADHD). Firstly, we briefly describe different analytic approaches that can be applied for the discovery of specific biomarkers for diagnosing the above disorders, as well as for monitoring the effect of their treatment. Secondly, we discussed the types of biomarkers in general used in biomedicine for characterizing different disorders and diseases. Next, the potential applications of these biomarkers for diagnosing and managing major depression, suicidal behavior, schizophrenia, and ADHD are discussed in details. Forensic aspects of these biomarkers for the above disorders are also considered. Finally, we discuss the potential of specific biomarkers for distinguishing between comorbid psychiatric disorders in clinical setup as well as their potential for understanding mechanisms underlying the disorders and in discovery of new treatment strategies.

1. INTRODUCTION

The primary focus of this review is the application of proteomic and metabolomic protocols to the study of serum-based biomarkers of depression and associated suicidal tendencies. Major depression is a widespread and devastating psychiatric condition with the majority of cases exhibiting suicidal behavior after a time suffering from depression. Furthermore, other psychiatric disorders may also exhibit suicidal behavior in severe states. While biomarkers have been discovered via the application of modern proteomic and metabolomics techniques for a wide range of clinical disorders, it is only recently that such investigations have included different neuropsychiatric disorders. As a result of this, compared to other clinical conditions, such as cancer, the application of metabolomics and proteomics is still relatively new in relation to clinical psychiatry disorders.

Proteomic technologies, based primarily on mass spectrometric analysis, are increasingly being used as a valuable tool in psychiatric research. This modern methodology allows for a high-throughput protein profiling and identification revealing expression levels, posttranslational modifications, and complex protein-interaction networks. In contrast to other methods such as molecular genetics, this approach provides the opportunity to determine modifications at the protein level. Therefore, it can be argued that the resulting protein levels relate more closely to the pathophysiological processes underlying the clinical phenomenology of specific psychiatric conditions. In the field of biomarker research, proteomic tools can be used to identify quantitative and qualitative protein patterns in easily accessible biological fluids, such as serum and urine, and also in other biofluids, such as cerebrospinal fluid (CSF). Alongside proteomic analysis, mass spectrometry (MS) (and allied techniques in spectroscopy) can be used to obtain profiles of metabolites within biofluids and combining different types of extraction and analysis can allow for a large dataset of biological compounds to be monitored in a single sample. As a change in the proteome would be expected to cause a change in the balance of metabolites within a given biological sample, proteomic and metabolomic analysis can be thought of as complimentary methods of studying the same question. The idea that psychiatric disorders can be linked (and partially diagnosed at least) to altered protein/metabolite levels is not without precedence. For example, there is existing knowledge that neuropsychiatric conditions, such as depression, are linked with altered levels of folate and its' associated metabolism products.

It is well recognized that many psychiatric patients with conditions such as depression also exhibit comorbid psychiatric disorders as well as depression. Therefore, many of the analyses undertaken in studying biomarkers for depression need to be also applied to patients with these certain comorbid disorders in order to determine whether the biomarkers are selective for depression. The individual value of biomarkers specific for the other, often comorbid conditions, within the context of forensic psychiatry will also be considered outside of the consideration of their relevance to depression.

2. ANALYTICAL APPROACHES APPLIED

2.1. Background

Mental health disorders are complicated regarding their affected genetic and biological pathways and thus modern approaches such as proteomic analyses are given a great hope to reveal a global view of the pathways involved in such disorders. The proteomic tools are a technology-driven approach to attempt to study the whole proteome data in a given sample. Comparative genomic analyses study the genome alterations that reflect the pathophysiology of psychiatric disorders. Therefore, proteomic studies are complementary with the comparative genomic approaches supporting genetic analysis at protein level. Metabolomic approaches determine changes that appear as a result of the proteomic changes. Thus, all three methodologies act to gain a global insight into the effect of changes within any given biological system. Application of these techniques in clinic has been used to identify proteins involved in pathological processes and to understand how illnesses can lead to altered protein expression. The primary proteomic research approach with respect to biomarker discovery focuses on the comparison of the protein and/or peptide profile between patients and healthy control individuals. Significant success has been detected among various clinical research disorders. The most obvious success in proteomic biomarker discovery has been achieved among disorders which have a clearer biological cause, such as cancer studies and regarding tumor metabolic activity (Cohen et al., 2011; Liu, Pan, Shen, Wang, & Yong, 2011; Zhao, Liu, Sun, Peng, & Ding, 2011). Recently, proteomic techniques have begun to be applied to psychiatric disorders which have a known biological cause such as Alzheimer's disease and Parkinson disease (Li et al., 2011; Lopez et al., 2005). Many psychiatric disorders, such as attention deficit/hyperactivity disorder (ADHD) and autism spectrum disorder, are multifactorial in biological nature and therefore the identification of primary pathophysiological changes is much more challenging. However, recent research efforts working to screen for potential biomarkers which may be of value in diagnosing such disorders have suggested some promise (Corbett et al., 2007; Taurines et al., 2010).

2.2. Proteomic analysis

Proteomics can be considered to represent the change in all proteins expressed by a genome. The application of this technique in clinic has been used to identify proteins involved in pathological processes and to understand how illness can lead to altered protein expression. Clinical proteomics can offer the opportunity and the potential to develop new diagnostic and prognostic tests, to identify new therapeutic targets, and eventually to allow the design of individualized patient treatment. Proteomic technologies are predominantly based on MS and can assist in studying global protein expression levels and identification of potential biomarker related to psychiatric pathophysiology.

The proteomic approach has four aspects:

- Expression proteomic
- Structural proteomic
- Interaction proteomic
- Functional proteomics

Of these, expression proteomics has been most commonly applied to medical areas of investigation, studying the disregulation of certain proteins in different disorders. Proteomics can be considered to be part of a set of studies representing alterations in biological activity at different levels. These represent the flow of information from gene to RNA transcript, transcript to protein, and the final metabolic change brought about by proteomic changes as represented by the metabolome.

2.3. Genomic and transcriptomic analysis

Genomic and transcriptomic analysis seeks to study the changes associated with gene profiles, while genomic studies allow analysis at the level of DNA (sequencing, mutation, or single nucleotide polymorphism analysis), the transcriptomic approach provides details of changes at the RNA level. Psychiatry is generally considered as representing polygenic disorders, meaning that multiple genes might contribute to the pictures of disorders. Furthermore, symptom expression might not occur until much later in life as seen in Alzheimer's disorders (Karch et al., 2012). Recently, genomic studies have been shown to have real implications in the identification of susceptibility genes related to autism showing a greater promise in furthering our understanding of etiology, diagnosis, and successful therapeutic intervention of such psychiatric disorders.

2.4. Metabolomics

Metabolomic analysis encompasses the study of metabolite profiles, which represent the end products of cellular processes in biological samples.

Therefore, proteomic, transcriptomic, and metabolomic analyses can provide a complete physiological picture of a living organism. Metabolomics is a new approach toward identifying metabolite biomarkers related to human disorders. The topic of "functional genomics" has been developed to determine the complex dynamic aspects in physiology/pathophysiology such as gene transcription, translation, protein–protein interactions, and metabolite state analysis in order to understand the overall phenotype of disorders. MS approaches (and allied spectroscopy techniques) have been developed to study metabolites, offering a potential impact on such understanding and the potential to reveal the relationship between the genomic and phenotype pictures (Villas-Bôas, Mas, Akesson, Smedsgaard, & Nielsen, 2005). Proteomic and genomic analyses are newly diagnostic methods beginning to emerge in psychiatry, while metabolomics too offers further promising in diagnostic tools but as yet has not been as commonly applied in the field. Further assessment of these methods, however, is required to fully demonstrate their value and accuracy (Smith, Renshaw, & Bilello, 2013).

3. ANALYTICAL TECHNIQUES

3.1. Protein profiling using matrix-assisted laser desorption-ionization time-of-flight mass spectrometry

Matrix-assisted laser desorption is an ionization method having been described first by Karas (Costello, 1997; Karas & Hillenkamp, 1988). Their work demonstrated that this techniques and results obtained can be considered a revolution in the way that MS experiments allowing the analysis of biomolecules (biopolymers such as proteins, peptides, and sugars) and large organic molecules (such as polymers, dendrimers, and other macromolecules). Franz Hillenkamp and colleagues by 1987 discovered that ions can be produced from large proteins by laser desorption if these molecules are mixed with small organic compounds working as matrices that strongly absorbance the laser wavelength aimed at them during ionization. The samples if ionized can be analyzed based on their differing masses and relative abundance. Matrix-assisted laser desorption-ionization time-of-flight mass spectrometry (MALDI-TOF MS) has become an important tool in the analysis of biomolecules.

The MALDI ionization process consists of a number of variables which are optimized based upon the type of biomolecule being analyzed and sample source, as discussed below:

- Matrix: A solution of crystallized molecules (commonly used matrices include sinapinic acid and alpha cyanohydroxycinnaminic acid for proteins and peptides). The matrix is mixed with highly purified water and organic solvent (normally acetonitrile or ethanol). Trifluoroacetic acid may also be added. The matrix solution is then mixed with the analyte (protein sample) and then spotted onto a stainless steel MALDI plate and allowed to dry before it is then analyzed inside the MALDI-TOF analyzer, where the acidic matrix act as a proton source for the analyte.
- Laser: Laser energy is pulsed onto the matrix-sample mixture to be ionized and causes localized vaporization. Molecules become positively charged and travel along the time-of-flight (TOF) tube. At the time of each laser pulse, a voltage is applied to accelerate the sample toward the detector.
- Electric field: A strong electric field in a flight tube is applied so the ions enter a vacuum can be accelerated toward the detector.
- Analyzer: The analyzer measures the TOF taken for particular ions to hit the detector where smaller ions take a shorter time than larger ions. The flight time of an ion is related to its mass-to-charge (m/z) ratio.
- Recorder: Ion fractions arriving at the end of the flight tube are detected by a recorder that produces a signal for each ion group, mass spectrum data are then generated from simple time measurement (the flight time of an ion is related to its m/z ratio); thus, mass spectra can be generated.

MALDI-TOF MS has the ability to determine molecular weight information accurately and is also suitable for the analysis of high molecular weight compounds, including the determination of molecular masses of biopolymers such as peptides and proteins, carbohydrates, oligonucleotides, and synthetic polymers with relative masses up to several hundred kiloDalton (Karas & Hillenkamp, 1988). This technique has been considered for the future development of biomarkers among neuropsychiatric disorders (Dudley, Hässler, & Thome, 2011). Such biomarkers might be used as indicators regarding progress or severity of diseases/disorders.

3.2. Gas chromatography–mass spectrometry

The hyphenation of gas chromatography with mass spectrometry combined with the complementary technique the Fourier transform infrared spectrometry (discussed in Section 3.3) have made a substantial contribution in the clinical and forensic analysis. Clinical application can be seen in recent reports which allow researchers to distinguish free fatty acids as potential biomarkers between healthy controls and Alzheimer's patients (Wang et al., 2012). Gas chromatography allows for the separation of volatile compounds in the gas phase, briefly, a gaseous sample or small volume of a liquid sample is injected into a flow of inert gas (usually helium), which carries the compounds in gas form across a capillary column. The compounds separate by partitioning between the helium (carrier gas) and the chemistry of the column (the stationary phase). Compounds which are strongly retained on the column stationary phase are eluted by increasing the temperature in which the column is held, thereby favoring release of the compounds and elution. A temperature gradient is applied to gradually elute compounds individually therefore separating them before release into the mass spectrometer. Once released from the GC, the gaseous compounds are transferred to the ion source of the mass spectrometer (usually an electron impact (EI) or chemical ionization (CI) source). During EI ionization compounds are bombarded with electrons causing fragmentation and release of electron from the compounds themselves giving rise to ionized species, while in CI, compounds are reacted with gas phase ions such as methane to form ions. The ions so formed are separated by a quadrupole mass spectrometer and the m/z ratio of the ions determined. There is a large database (produced by the National Institute of Standards and Technology (NIST)) of the EI spectra of known compounds against which spectra for each compound can be compared in order to identify each compound analyzed.

3.3. Fourier transform infrared spectroscopy

Fourier transform infrared spectroscopy (FTIR) has traditionally been used in material analysis within the chemical industry; however, more recently, it has been applied to biological and biomedical analyses. FTIR is an adsorption spectroscopy that uses an incandescent source of light to emit a bright ray in the IR wavelength range. This beam arrives at a half-silvered mirror, which divides it into 50% which is directed toward a fixed mirror and 50% to movable mirror. The reflected beams return to the half-silvered mirror, which rejoins them and sends them to the detector. Even if the two beams arrive at the detector simultaneously, they have done so after covering a different optical path: depending on the differences of optical paths of two rays, a constructive or destructive interference has therefore been created and these interferences are converted into a signal at the detector. The signal detected in this manner is proportional to the difference in the optical path of the two beams and to the position of movable mirror during analysis. Each organic compound has a specific infrared spectrum depending on its molecular structure. The absorption of infrared radiation generates characteristic vibrational movements in molecules, defined as stretching and bending due to a changing of electric dipole; in fact, the molecule changes its vibrational state as it passes from fundamental vibrational state to excited vibrational state.

FTIR has found increasing use in metabolomics analysis as a complimentary method of analysis alongside MS. Furthermore, the FTIR technique has been used to monitor drug toxicology (Pandey, Pandey, Tiwari, Tiwari, & Rai, 2012) and offers a viable, low cost test of bones in forensic analysis (Fredericks, Bennett, Williams, & Rogers, 2012) besides many other clinical applications.

4. BIOMARKERS

4.1. Background

Psychiatric research commonly focuses on genomic pathways to reveal genes abnormally expressed in patients and their products thus understanding the pathophysiology of different neuropsychiatric disorders. Protein biomarker discovery can also be employed in neuropsychiatric disorders in addition to genetic studies. There is also a growing need to identify various biomarkers associated with psychiatric disorders to support the theories regarding the initial development of the disorders and identify new target sites for therapy development.

Biomarker assessment might be undertaken by studying either bodily fluid or body tissue as initial sources of biological material. Body fluids available for study include serum, saliva, urine, and CSF, whereas body tissue relies on biopsy (which is impractical and dangerous to the patient if at all reliable in neuropsychiatric disorders), human postmortem samples, or cell lines. Of these options, biological fluid collection is the easiest to access and also the least invasive to the patient and serum has been a predominant focus for previous biomarker studies. Biomarkers may also facilitate the identification of treatment effects on disease progression and so aid the therapeutic development related to the disorder and personalized treatments. Furthermore, specific biomarkers might also be considered not only in diagnostic criteria but also as an indicator of prognosis.

4.2. Types of biomarker

Proteomic research in psychiatry disorders is focused on comparing protein expression levels between patients and healthy controls. Biomarkers which are related to specific psychiatry disorders are divided into trait or state biomarkers. The trait biomarkers are enduring and their levels reflect the property of behavior and biological functioning which potentially play a role in the etiopathology of disorders. Therefore, trait biomarkers can be present in clinically unaffected individuals prior to any development of the disorder. Trait biomarkers are representative for the origin of the disorder rather than its expression in individuals. The state biomarkers are transient and characterize the clinical manifestation of disorders rather than any disposition toward them. Unlike trait biomarkers, state biomarkers would return to "normal" levels with successful therapy and so can act as a prognostic tool (Chen, Bidwell, & Norton, 2006).

5. DEPRESSION

5.1. Definition of depression

Depression is considered in part to be a completely normal stress reaction to a number of factors that may be experienced by an individual. The depression is only considered as being abnormal if it is disproportionate to the triggering event, and when the individuals are determined to be unable to relieve themselves of the depressive state. Depression is a very common psychiatric disease worldwide and is estimated to become the second most common cause of disability, after heart disease by 2020 (Peveler, Carson, & Rodin, 2002). People suffering with depression may be subjected to discrimination and are more likely to be unemployed compared to other people in the general population and other groups with mental disorders. The effect of the disorder itself and the stigma attached to the disorder leads to a greater potential for people with depression becoming socially isolated, which in turn reemphasizes the individuals' feelings of worthlessness and hopelessness. Depressive disorder as a psychiatric condition is characterized by psychological, behavioral, and physical symptoms (Greden, 2003).

5.2. Symptoms

The symptoms of depression develop over an extended time period and for the purpose of a clinical diagnosis it is considered that five symptoms should be present and persist for a period of 2 weeks. Such diagnosis is based upon patients' self-reported or relative-reported behavior of the patient. Diagnosis criteria for depression are detailed by the Diagnostic and Statistical Manual-fourth edition (DSM-IV) of the American Psychiatric Association and the International Classification of Diseases and Related Health Problems

(ICD-10) of the World Health Organization, with continual updating. The Diagnostic and Statistical Manual of Mental Disorders outlines diagnostic criteria for psychiatric disorders published by the American Psychiatry Association, whereas the International Statistical Classification of Disease outlines diagnostic criteria for human disease (published by the World Health Organization). Emotional symptoms: sadness and depression, continual feeling of hopeless and misery, excessive crying and even suicidal thoughts are common symptoms exhibited by depression patients. Further symptoms include a lack of ability to find satisfaction and/or enjoyment in activities, a loss of interest, motivation or initiative in day-to-day activities, negative thoughts, and low self-esteem. Physical symptoms commonly include a loss of appetite, insomnia, and fatigue. Furthermore, the depressive state magnifies the effect of the physical symptoms felt by the patient, resulting in an increase in the patients concern regarding their health. These symptoms must be present during the same 2-week period episode for a clinical depression diagnosis to be confirmed and last most of the day, occurring daily. Manic episodes also typically occur, resulting in sudden mood elevation over a few days and persisting for at least 1 week with three or more of the following symptoms: more talking than usually, reduced need for sleeping, difficulties in concentration, inflated self-esteem, racing of ideas, and engagement in activity which appear pleasurable (Peveler et al., 2002).

5.3. Types of depression

Depression covers a wide range of forms of illness from unipolar, bipolar forms, and shorter. The unipolar form (major depression) involves persistent low moods and can be classified into different forms. (a) Major depression (MDD) is diagnosed if the patient has five depression symptoms or more over a 2-week period. MDD includes further subclasses such as (b) postnatal depression after childbirth, (c) seasonal affective disorder depression associated with winter, and (d) dysthymia. Bipolar depression (also referred to as manic depression) involves transient low moods separated by periods of mania. Shorter forms of stress reaction in everyday life can result in depression-like symptoms; however, overall clinical manifestations of psychiatric depressive disorder must be considered in diagnosis.

5.4. Prevalence of major depression

The most common mental illness prevalence of the depressive states is major depressive disorder (51.3%) followed by psychosis (38%) (Mekonnen &

Kebede, 2011). The prevalence of major depression (MDD) is observed as being about twice as common in women than in men suggesting that gender may play a role. This is possibly reflected by changes in female endocrine hormones that might trigger the stress reactions symptomatic of depression and influence-associated neuropsychological processes (Kuehner, 2003). Other factors, such as exercise activity levels have also been implicated in prevalence (Reichert, Diogo, Vieira, & Dalacorte, 2011).

5.5. Major depression theories

Like many other neuropsychiatric disorders, depression is thought to be caused (at least partially) by some form of neurobiochemical disturbances which might influence the severity of symptoms and is reflected on the phenomenological level by the typical clinical characteristics. For example, the monoamine hypothesis postulates that a deficiency of certain neurotransmitters is responsible for the corresponding features of depression, although in the past two decades research has revealed multiple limitations of the hypothesis (Hirschfeld, 2000; Meyer et al., 2006). An abnormality of the dopaminergic system has been reported in depressive disorders and suicide-comorbid major depression (Arango, Underwood, & Mann, 1996; Dailly, Chenu, Renard, & Bourin, 2004), while another clinical study of depression concluded that neural plasticity plays a major role, as atrophy and death of neurons in specific brain regions (hippocampus, prefrontal cortex) could contribute to the pathophysiology of depression (Duman, Malberg, & Thome, 1999; Thome, Henn, & Duman, 2002). A further theory implicated anatomical structure abnormalities that might increase risk of MDD, concluding that volume reductions in the hippocampus seen during aging might be associated with a greater risk for MDD (Amico et al., 2011; Halbach, 2010). A review study hypothesized that brain-derived neurotrophic factor (BDNF) is critically involved in the regulation of age-related processes in the hippocampus. Therefore, the study suggested that disturbances in the BDNF system have such a role in hippocampal dysfunctions, as seen in major depression and in Alzheimer disease. Essential dietary levels of some bioactive compounds, such as folic acid and vitamin B12, have also been previously implicated as possibly playing a role in depression outcome (Coppen & Bolander-Gouaille, 2005). Hormonal theory, such as the use of estrogen, has also been suggested as having potential for the treatment of depression patients, especially perimenopausal women (Soares, Almeida, Joffe, & Cohen, 2001). Although depression can be caused by a variety of these potential factors among others, it may also be possible that other factors including genetic, psychological, and environmental factors might also be involved in the development of the condition.

5.6. Comorbidity of other conditions with major depression

Anxiety symptoms and ADHD can have a major impact on the course of a depressive illness and commonly co-occur together in patients presenting in the primary care setting (Hirschfeld, 2001; McIntosh et al., 2009). Furthermore, major depression is one of the most frequent comorbid conditions in posttraumatic stress disorder (PTSD) individuals with the incidence of first-onset depression being elevated among individuals who develop PTSD following trauma compared to the trauma-exposed individuals who do not develop PTSD (Breslau, Davis, Peterson, & Schultz, 2000). More frequently depression is the most common psychiatric comorbid condition in epilepsy disorders (LaFrance, Kanner, & Hermann, 2008). Pain complaints are commonly accompanied by major depressive disorder (Husain et al., 2007) and high rates of comorbidity have been found among alcohol use disorder patients and individuals with a number of mood and anxiety disorders (Dawson, Grant, Stinson, & Chou, 2005). Depression is a serious disorder and so misdiagnosis or lack of treatment in such an individual can have fatal consequences due to prevalence of suicidal tendencies in such patients.

5.7. The potential benefit of biomarkers in depression

Psychiatric assessment of depression patients is usually based on doctor–patient interviews carried out within the health services, for example, hospital- or home-based sitting assessments. The majority of mental diagnostic criteria for most psychiatric conditions are based on observation data rather than biological data. Biological biomarker studies would have the benefit of providing potential biological indicators for diagnosis and prognosis of depressive disorders. Therefore, such biomarkers might be able to distinguish depressed patients from healthy people and minimizing the differences between self-reported assessment and clinical observation assessment for different psychiatry scale assessments. Furthermore, biomarkers would act as objective physical measurements that might guide the clinical prognosis of depressive disorders and monitor therapeutic intervention. A previous study monitored the treatment of depressive disorder by selective serotonin reuptake inhibitors (SSRI) and showed that emotion and behavior control was affected if SSRI treatment was discontinued (Scoppetta, Di Gennaro, &

Scoppetta, 2005), potentially biomarker profiles might be changed before and after discontinuity. Any proteomic and/or metabolomics biomarker represents physical evidence concerning disorders and so would require validation and qualification support to have value in diagnostic purposes. In addition, these biomarkers could be used to stratify patients into subgroups thereby allowing patients to be diagnosed and treated as more homogenous groups of patients dependent on a biological basis. These biomarkers may also have legal implications in cases in which forensic reports should include physical evidence (beyond mere interview-based diagnosis), especially for those convicted of criminal offenses linked to depressive behavior. Also, civil reports in medico-legal assessments can be verified by biomarkers in cases where compensation is required or where the mental state might be responsible for some criminal behavior. Furthermore, health insurance agencies require qualified assessments for individuals as a routine check-up, current and past medical history, family history, and treatment history and these clinical histories could be verified through biomarker data history further elucidating an accurate picture of the candidate health. Therefore, the potential benefits and applications of biological, objective biomarkers with statistical credibility are wide ranging and may affect multiple disciplines.

5.8. Forensic aspects of depressive disorders

Mental health professionals (psychiatrists, psychologists, and social workers) are frequently asked for mental health evaluations for offenders in a forensic setting. These evaluations are utilized for documenting the role (if any) of mental illness in any recorded antisocial or criminal behavior (Ackerman, 2006). Furthermore, depending on the underlying mental illnesses related to the defined criminal activity, an accused individual can be transferred to highly secure health services if judged to be not guilty by reason of insanity as part of the legal process. The criminal-mental evaluation reports actually cover two questions for the court. Firstly, the report defines whether a defendant is deemed to be of sufficient mental state to be fit to plead; secondly, the report provides details as to whether the defendants' mental state may have contributed, totally or in part, to their offense. Therefore, the purpose of the forensic psychiatry report as a clinical evidence document (based on an individual's mental disease) aims to assist the court in reaching their conclusions.

Forensic reports should cover whether the defendant has a psychiatric disorder and proof of this and also whether that disorder could contribute to their offense.

Therefore, it is more convincing for such a report to include objective biological/physical evidence compared to information gained via clinical interview assessment as explained for the forensic situation. In this way, any validated biomarker would assist in such cases. Tendency to criminal behavior can also be associated with a psychiatric disorder background and so violence committed by such patients could be reduced if screening is performed on those at risk. Recently, criminal behaviors linked to psychiatry history have been investigated regarding criminal responsibility prediction, prevention, and treatment. To date, the only biological information that has been obtained that can be used when preparing such forensic reports has been based on genetic research regarding key genes involved in various psychiatric disorders. To further improve the use of biological factors in assessing individuals in forensic reports, the overall research required would need to identify psychopathological predictor factors, for example, biomarkers for future violent for optimal use in any criminal behavior prevention program. The role of forensic psychiatric examinations is important as part of offender assessment, in particular in relation to the prediction of criminal linked-behavior. A recent study reported that 50% of offenders have depressive disorders (de Andrade, Assumpção, & Teixeira, 2011), however, less exhibit violent behavior (Graz, Etschel, Schoech, & Soyka, 2009). Among inmates with major depression there is evidence that suicide risk is highly elevated (Baillargeon et al., 2009) and such findings highlight the importance of improving forensic prediction tools especially within this group of depression individuals. Furthermore, as inmates may represent depression symptoms, it is important to consider not only the risk of suicide behavior as a result of any untreated depressive illness as well as treating them purely as a criminal and only considering their criminal behavior.

6. SUICIDAL BEHAVIOR

6.1. Background

Suicidal behavior includes complete suicide, attempted suicide, and suicide ideation and such behavior may be considered to be an indicator of inadequate methods or application of screening of patients by the mental health services within any specific region. Suicide behavior is always deliberate, conscious, and represents emotional feelings with a strong passion for strong, aggressive self-destructive acts. The suicide attempt is generally considered to be a "call for help" or a way to communicate with others with the aim of changing the behavior of others, the primary motive is not therefore the death of the individual.

6.2. Definition

The overall definition of suicide has been widely debated and varies depending upon the mental illness, cultural, and religion. Generally, suicide is defined by a person intentionally attempting to take his or her own life and the World Health Organization defines suicide as the act of deliberately killing oneself. Many mental disorders are considered to be risk factors for suicide behavior; thus, collaboration between psychiatric and social services may improve suicide prevention strategies, initiating care of the patient when suicidal thoughts first occur. Although suicide is a serious problem for patients with psychiatric distress, especially for those with depression, it is also considered as a component of the psychiatric symptoms and may be related to other disorders associated with depression and therefore can be easily misdiagnosed. For example, suicidal behavior may also be a component of disorders such as anxiety and alcohol abuse. Generally, individuals who make serious suicide attempts have high rates of mental disorders (Beautrais et al., 1996); therefore, suicide is considered to the misdiagnosis of a mental disorder. Improving physician education in depression disorder recognition and treatment and restricting access of those with such risk factors to lethal methods has been shown to be a positive method of reducing suicide rates (Mann et al., 2005).

6.3. Risk factors

Mental disorders (such as depression, personality disorder, alcohol dependence, schizophrenia) and some physical illnesses, such as neurological disorders, cancer, and HIV infection are risk factors for suicidal behavior. Furthermore, gender and geographical differences have also been previously implicated as strong association factors (Arsenault-Lapierre, Kim, & Turecki, 2004). The annual suicide rates have been reported as being higher in males compared to females (Oquendo, Lizardi, Greenwald, Weissman, & Mann, 2004).

Gender seems to play various roles as in the case of men, low social and family support and depersonalization is related to suicide, while in the case of women depressive mood and anxiety is more strongly associated (Yoshimasu et al., 2006). Awareness of risk factors is therefore also a valuable tool for clinicians in estimating the suicide risk among psychiatry disorder patients.

Risk factors in the case of depressed patients include agitation, depressive mixed states (pseudo-unipolar depression), higher number of depressive episodes, comorbid anxiety, personality disorders, and alcohol dependence, as well as sociodemographic and psychosocial factors (Rihmer & Akiskal, 2006). An investigation in the New York Prison Service found that 90% of inmates who committed suicide had a history of substance abuse and that heavy smoking also represented a risk factor of suicidal thoughts (Iwasaki, Akechi, Uchitomi, & Tsugane, 2005; Way, Miraglia, Sawyer, Beer, & Eddy, 2005). Suicide thoughts can also be strongly triggered by other socioeconomic factors such as unemployment, poverty, homelessness, and discrimination (Qin, Agerbo, & Mortensen, 2003) and it is now recognized that genetic factors may also induce suicide behavior in subjects with other suicide risk factors (Bondy, Buettner, & Zill, 2006). Suicide studies in 2004 reported that in America there was a frequent association of mental health diagnoses with suicidal behavior: depression (85.2%), bipolar disorder (7.4%), and schizophrenia (3.3%). Furthermore, 16.5% had alcohol dependence problems and nearly 19.0% of suicide victims had made previous attempts (Centers for Disease Control and Prevention (CDC), 2006). In the majority of patients with mood disorders, the suicidal behavior seemed to be predictable and even preventable (Rihmer, 2007). Approximately 87.3% of individuals had one or more psychiatric disorders at the time they committed suicide (Arsenault-Lapierre et al., 2004).

6.4. Prevalence

Among ethnic groups suicide behavior is more recognized in association with depression (Oquendo et al., 2004). Several reasons for the prevalence of suicide behavior need to be considered as it is recognized as a multicausal behavior with biological factors, psychopathological factors, and social and cultural environment all playing important roles in the determination and manifestation of suicidal behavior. Results from a study in Denmark have shown that, across the entire population, prevalence of suicidal ideation is 6.9% and of suicidal attempts 3.4% (Kjøller & Helweg-Larsen, 2000). Being female is associated with a higher prevalence of suicide ideation and attempts by females who are divorced or widowed at a young age show a further increase in prevalence. Although, there were significant differences in gender as a Swedish study showed a higher suicide rate toward men (71%) compared to women (29%) (Holmgren & Jones, 2010). Good mental health services might increase life expectancy because females with major depression have a much increased rate (28%) of suicidal attempts (Bernal et al., 2007). Therefore, as suicide is considered to be a sign of such mental disorders, much emphasis has been placed on reducing the suicide prevalence and suicide attempts associated with these disorders. In relation to individuals with depression, biomedical approaches to identify biomarker profiles which may represent predictive tools can contribute to the reduction in suicide prevalence rates.

6.5. Comorbidity

As mentioned previously, among cases of suicide, approximately 90% of individuals had one or more psychiatric disorders at the time they committed suicide (Arsenault-Lapierre et al., 2004). The most prevalent disorders were depressive syndromes (66%) and alcohol dependence/abuse (43%) (Lönnqvist et al., 1995). The study of suicide victims from the nationwide "suicide population" showed that at least one psychiatric diagnosis was present in 93% of the victims. Other results among the subjects who had committed suicide demonstrated that 38.7% were afflicted by major depression, 24% by alcohol dependency, and 28.7% were dependent on drugs (Grunberg et al., 1994). Therefore, mental disorders have been found in most people making serious suicide attempts with suicide risk being particularly high in those with unobserved suicide intent, depression patients, patients with chronic alcohol and drug misuse, social isolation, and current physical illnesses (Mitchell & Dennis, 2006).

6.6. Biomarkers and suicide

Suicide acts represent a challenging behavior to trace, prevent, and predict and so there is a clear rationale to explore any biology which may be related to its manifestation. Generally, serotogenic system dysfunction is often investigated in neurobiological research of suicide behavior and associated comorbid disorders at the level of transporter systems of neurotransmitters, receptors, relevant enzymes, and genes. The genetic role regarding suicide behavior remains unknown and much effort is directed toward examining genetic variants which may play roles in the development of suicide behavior. Recently, more effort has focused on elucidating biomarkers which might be associated with suicide behavior, primarily in relation to their predictive value and application in prevention programs. Biomarker measurements might also be useful for evaluation of and in combination with the suicide risk scale currently used as a diagnostic tool to provide a more accurate diagnosis. The study of the risk of suicide among hospitalized patients represents an efficient way in which to study biomarkers profile due to the ease of access to the patients. For example, the suicide risk among elderly patients with a history of hospital-treated psychiatric disorders (such alcohol related disorder and depression) were reported as a high risk (Karvonen, Hakko, Koponen, Meyer-Rochow, & Räsänen, 2009) and so as well as a more controlled cohort of individuals being available, the prevalence of suicide behavior is present in such a population. Therapy provision of the disorders regarding suicide risk may be assisted through any accurate biomarker profile. Such a biomarker might also be used alongside the recording of other prominent suicide risk behavior. For example, a previous study reported that most depressive patients exhibiting suicide behavior are those who have not taken their medication at time of death (Isacsson, Holmgren, Wasserman, & Bergman, 1994) and so the combination of any biological biomarker profile with such observational diagnoses would seek to reduce suicide prevalence.

6.7. Forensic aspects of suicide behavior

Suicide victims are usually examined, postmortem, by a forensic practitioner. A Coroner is required to examine all circumstances surrounding the suicide and report their findings, elaborating the cause of death, via the death certificate. Some countries require physical evidence such as notes written by the victims as proof of suicide, while in many countries this depends on a different range of factors associated with the circumstances of death. Psychiatry history is an important part in reporting suicide on a death certificate, and so clinical findings, medical records, and details of previous suicide attempts are helpful in such reporting. Biomarker profile comparison before and/or after death might reveal victims tendency to commit suicide and suicide behavior itself may be considered as a predictor of later mental disorders such as depression (MacDonald, Taylor, & Clarke, 2009). Individuals with mental disorders showed increased violence to others detained in a secure hospital setting and also a higher risk to suicide behavior after discharge (Jones, Hales, Butwell, Ferriter, & Taylor, 2010). Former psychiatric conditions are reported as a diagnostic predictor for suicide and therefore such individuals are usually considered for more intensive care supervision (Powell, Geddes, Deeks, Goldacre, & Hawton, 2000; Tidemalm, Elofsson, Stefansson, Waern, & Runeson, 2005). Any biological biomarkers could further contribute to this care and biomarker levels can be recorded and shared for future use. The health records in forensic psychiatry for reporting the clinical evidence regarding suicide victims are based on an interviewed assessment which might not be sufficient, or even possible, in some cases and so adding objective biomarkers may assist in helping reach a forensic decision. Furthermore, biomarkers may exhibit a predictive value and serve as screening tools in the prevention program.

7. SCHIZOPHRENIA

7.1. Background

Schizophrenia is a psychiatric disorder characterized by disruption in cognition and emotion which affects many patients' attributes including thoughts, perception, and speech. Diagnosis of schizophrenia is usually made with reference to the criteria in DSM-IV and ICD-10 which states that for a definitive diagnosis at least 1 month duration of two or more positive symptoms which reflect a function distortion should be identified. The negative symptoms of schizophrenia are more difficult to evaluate and persist in schizophrenic patients during periods in which positive symptoms are absent. The disorder has several neurobiological abnormalities with a previous etiology study suggesting that the disorder developed due to a gene–environment interaction (van Os, Rutten, & Poulton, 2008). Biological abnormalities of the disorder are mostly detected through image-based biomarkers such as abnormally large brain ventricles due to CSF, decrease in grey matter, and alteration of white matter (Davatzikos et al., 2005). However, the study of the dopamine system has demonstrated abnormal dopamine concentrations, especially in the acute psychotic state of the disorder, which has represented a major focus of schizophrenia research (Howes & Kapur, 2009). Moreover, other studies have focused on the increase in dopamine synthesis, dopamine release, and resting-state synaptic dopamine levels. This too has dictated much research in respect of schizophrenia treatment strategy, with most treatments mediating the blocking of dopamine receptors especially in the psychosis state (Kapur, Agid, Mizrahi, & Li, 2006). In depression patients with schizophrenia, the SSRI treatments are the most effective therapy (Kasckow, Felmet, & Zisook, 2011). Gender variation also has been suggested to have a role as the disorder is significantly more common in males (Castle, Wessely, & Murray, 1993; McGrath, Saha, Chant, & Welham, 2008).

7.2. Symptoms

The recognized symptoms of schizophrenia are listed by DSM-IV and are grouped into three groupings. Positive symptoms commonly include delusion (false beliefs), hallucination (including visual, auditory or tactile, olfactory, and gustatory hallucinations), disorganized speech, and disorganized behavior (catatonic behavior, abnormal and/or extreme rigidity/flexibility of limbs). Negative symptoms include alogia (lack of additional normal speech), avolition (lack of motivation), affective flattening (lack of emotion), and finally, cognitive symptoms such as slow/disorganized thinking, poor concentration, poor memory, difficulty in understanding, and abnormal feeling behavior can be observed.

7.3. Types of schizophrenia

Various forms of schizophrenia are documented depending on the characteristics of the disorder. The types of schizophrenia are based on symptoms subclassification. According to the DSM-IV-TR subclassification of schizophrenia consists of the following:

- Paranoid: delusion (false beliefs) and auditory hallucination are present but other symptoms such as thinking, speech and emotion disorders are not.
- Disorganized: disorganized behavior to perform normal activities, confused, thought disorders, and inappropriate emotion disorders are present.
- Catatonic: immobile, unresponsive, and unwilling to move and speech affected.
- Undifferentiated: if diagnosed but symptoms do not clearly fall into one of other types.
- Residual: if positive symptoms present with low severity when originally diagnosed.

7.4. Biomarkers and schizophrenia

There are no physical tests that confirm the presence of schizophrenia and the disorder shares a significant number of symptoms with other disorders making misdiagnosis a common problem. Current physical diagnosis tests (such as brain imaging, eye-tracking, and IQ tests) are either not sufficiently specific or sensitive to allow for an accurate diagnosis. Therefore, a correct diagnosis may be delayed beyond the first onset of symptoms illustrating the importance in developing more specific diagnostic methods in order for proper treatment, identification of partial, or complete recovery and prevention of further complications caused by the disease symptoms. Biomarkers may therefore have the potential to play a crucial role in assisting clinicians' schizophrenia assessments and diagnostic decisions based on patients' biomarkers profile. Although current physical tests cannot effectively diagnose schizophrenic disorders, further studies have tried to develop validated disease-specific biomarkers regarding schizophrenia that enable better and earlier diagnosis and effective monitoring of the treatment during disease stages. At present, imaging biomarkers are widely used (Rathi et al., 2010). Unlike molecular psychiatry, most diagnostic tests relate to physical characteristics such as electroencephalogram, functional magnetic resonance, and olfactory and eye movement tests. These are used to measure the abnormal performance manifestations that might help in diagnosis. It is envisaged that molecular research alternatively might develop disease-specific diagnostic tests through the identification of specific biomarkers related to disorders that reveal the mysteries, pathophysiology boundaries, and even define a biological basis for subtype division.

7.5. Forensic aspect of schizophrenic disorders

With respect to schizophrenic individuals who are arrested for criminal behavior, positive symptoms of psychotic disorder (hallucinations and delusions) are reported to be present at the time of the offense in most cases (Yee, Large, Kemp, & Nielssen, 2011). Other misdiagnosed disorders, such as adult ADHD, are also associated with criminal behavior (de Andrade et al., 2011) and thought to underline the schizophrenia and so forensic examiners are required to improve their awareness in relation to undiagnosed disorders leading to persistence of antisocial behavior if not referred and treated by expert. Moreover, screening at custody services may have potential value in predicting other family members who may be at risk of not being diagnosed as also having symptoms, misdiagnosis, and the impact of this on potential future criminal behavior. Schizophrenia is reported to be associated with criminal behavior and most patients present the paranoid subtype (and usually demonstrate violence behavior) compared to the other subtypes (Teixeira & Dalgalarrondo, 2009; Ural, Öncü, Belli, & Soysal, 2013). Individuals suffering from schizophrenia who are arrested for criminal behavior usually present positive symptoms of psychotic disorders (hallucinations and delusions) at the time of the offense (Yee et al., 2011). Moreover, schizophrenia can be associated with comorbid depression symptoms which result in a higher rate of death via suicide and suicide behavior compared to patients with schizophrenia alone (Caldwell & Gottesman, 1990; Inskip, Harris, & Barraclough, 1998; Isjanovski, Vukovic, Hadjihamza, Pejoska-Gerazova, & Raleva, 2010; Kim et al., 2010; Palmer, Pankratz, & Bostwick, 2005; Saha, Chant, & McGrath, 2007). Therefore, suicide victims may encompass a wide range of underlying mental disorders; moreover, transient depressive mood pictures represent an increased risk stage for suicide behavior (Fenton, 2000; Hawton, Sutton, Haw, Sinclair, & Deeks, 2005).

Other misdiagnosed disorders, such as adult ADHD, are also associated with criminal behavior (de Andrade et al., 2011) and thought to underline any schizophrenia disorder present. Therefore, forensic examiners are required to improve their awareness in relation to undiagnosed disorders leading to persistence of antisocial behavior if not referred and treated by expert. Moreover, screening at custody services may have potential value in predicting other family members who may be at risk of not being diagnosed as also having symptoms, misdiagnosis, and the impact of this on potential future criminal behavior.

8. ATTENTION DEFICIT/HYPERACTIVITY DISORDER

8.1. Background

ADHD is a high-prevalence neurodevelopmental disorder which manifests itself in childhood and can often persist into adulthood (Barkley, Fischer, Smallish, & Fletcher, 2002; Kessler et al., 2006; Rubia, 2002; Voeller, 2004). While ADHD in children and adolescents has been well researched in recent years, ADHD in adult patients has been under-researched for a long time, and only recently has attention been given to the fact that adults still may exhibit considerable ADHD symptoms. The behavior symptoms can be diagnosed by the age of seven and can last to adulthood (Mayes, Bagwell, & Erkulwater, 2008). Prevalence in adults has been reported to be around 2.5% and declines with age (Simon, Czobor, Bálint, Mészáros, & Bitter, 2009). The worldwide prevalence rate of ADHD has been variously estimated as 5.8% (using DSM-IV criteria) and 1.5% (ICD-10 criteria used) (Rohde et al., 2005) with gender studies concluding that boys are more often diagnosed with ADHD compared to girls (Müller et al., 2011). ADHD is a complex heterogeneous disorder with unknown specific causes; therefore, several research centers have focused efforts in many areas including genetic variation and environmental factors which might reveal the etiology of disorder. Furthermore, social factors and artificial food additives have also been considered as contributing to the etiology of ADHD (McCann et al., 2007). Such research has the potential to identify positive family history factors and help tracing trait biomarkers which underline the disorder. The pathophysiology of ADHD is unclear, however, imaging analysis has previously shown several abnormalities of brain volume in childhood patients with a total reduction of cortical and cerebral areas being reported alongside a decrease in cortical folding (Wolosin, Richardson, Hennessey, Denckla, & Mostofsky, 2009). There is also some evidence that the dopaminergic neurotransmitter system may be involved in the pathophysiology of ADHD (Tripp & Wickens, 2009). Additionally, a plethora of other factors such as neurotrophins and CLOCK genes have been implicated in the path-cascade ultimately leading to ADHD (Conner et al., 2008; Kent et al., 2005; Kissling et al., 2008).

8.2. Symptoms

ADHD symptoms are listed in the DSM-IV and International Statistical Classification of Disease-10 Revision (ICD-10) with core symptoms being evident in early childhood. These symptoms are predominantly: increased levels of hyperactivity, impulsivity, and inattention. Adult ADHD patients typically also exhibit these same tendencies (Wender, Wolf, & Wasserstein, 2001).

8.3. ADHD subtypes

ADHD patients can be further differentiated into subtypes of ADHD, dependent on the individual symptoms/elevated levels of behavior. The main subtypes are hyperactive/impulsive, inattentive, and combined.

The three main types of ADHD are characterized as follows:

- Inattentive: marked by impaired attention and concentration.
- Hyperactive-impulsive: marked by hyperactivity without inattentiveness.
- Combined: most common type includes all of the symptoms.

8.4. Comorbidity

ADHD may be associated with other disorders which can complicate the diagnosis and treatment of the patient. The most commonly identified comorbid disorders are conduct disorder, mood disorder, depression, anxiety disorder, and learning disability disorder (Biederman, Newcorn, &

Sprich, 1991). Furthermore, adult ADHD patients often exhibit several comorbid conditions such as mood swings, addiction disorders, and personality problems (Kessler et al., 2006; Wolf & Wasserstein, 2001).

8.5. Biomarkers and ADHD

Protein biomarkers from human biological fluid can be studied in neuropsychiatry disorders, especially those with relatively well-established etiology disorders such as Alzheimer's disease and schizophrenia; however, ADHD has had much less previous study regarding biomarkers linked to the neurodevelopmental disorder (Dudley et al., 2011). Therefore, it is more of a research challenge to identify specific diagnostic biomarkers in biological fluids; however, some genotype biomarkers previously explored have been linked to ADHD (Wallis, 2010). It has been suggested that ADHD-like features might also be an early vulnerability marker for schizophrenia (Keshavan, Sujata, Mehra, Montrose, & Sweeney, 2003) and a so-called "attention deficit disorder psychosis" has been previously described in some patients (Bellak, Kay, & Opler, 1987). However, very little scientific work has been published concerning the comorbidity of ADHD with schizophrenia in adult patients. ADHD screening may have a crucial impact on diagnosis and treatment and thereby greatly assist in the reduction of symptoms in misdiagnosed and untreated individuals. This may be considered to be especially important among young offenders whom require treatment despite a severe lack of recognition of the contribution of ADHD and similar disorders to continued offending in some individuals. The majority of the effort in studying ADHD has been focused on the genetic etiology of the disorder and so RNA products of suspected genes have been analyzed to study how they might be applied as diagnostic biomarkers.

8.6. Forensic aspects of ADHD

Psychiatric disorders are often linked to criminal behavior and therefore most offenders should, if possible, be screened for such disorders. ADHD was the most important predictor of violent offending and so the Offender Assessment System might develop identification tools based on any reliable biomarker profile in offenders which may provide information regarding offender's behavior and potential treatment requirements. A previous study reported that ADHD was the most commonly identified such disorder among offenders 33.3% (de Andrade et al., 2011). A further study suggested that childhood ADHD might increase the risk of developing antisocial behavior and substance use disorder in adulthood (Mannuzza, Klein, & Moulton, 2008). A previous report utilized genetic information regarding specific gene polymorphisms to confirm the role of ADHD in antisocial behavior in children (Caspi et al., 2008). Adults who meet the criteria of ADHD disorders are more likely to exhibit criminal and antisocial behavior (Langley et al., 2010) and so ADHD can be an important risk factor for the development of later antisocial behavior. While ADHD-like symptoms might be present in much of the population at any given period of time, ADHD individuals present such behavior more severely and this behavior leads to clinically significant problems. Therefore, forensic reports regarding the medical assessment of criminal offenders are required in some countries, including data detailing clinical assessments and physical evidence confirming the diagnosis. Improving the health services in prisons toward individuals with criminal behaviors, especially with regard to ADHD is important alongside assisting the future prevention and prediction programs. However, left untreated such disorders might maintain the criminal behavior and reduce the ability of an individual to change such behavior patterns. A meeting held in November 2009 discussed the awareness of ADHD in respective health services throughout the United Kingdom regarding recognition, assessment, and management. The meeting discussed the needs of ADHD offenders at different stages of the offender pathway which might lead to crime reduction (Young et al., 2011). The forensic services of UK prison studies demonstrated that 24% adult male prisoners were positive for disorder childhood history and a variety of symptoms (fully symptomatic, partial remission, and full remission). Thus, it was considered important to screen and identify prisoners who remain symptomatic for ADHD and refer such individuals to an expert for further diagnosis and treatment, as most prison inmates are young men with complex healthcare needs (Young et al., 2009).

9. CONCLUDING REMARKS

The psychiatry service has a long way to go to develop standard criteria based on laboratory-derived evidence, alongside the usually performed psychometric results which are considered as a psychiatric (although potentially subjective) opinion concerning the mental health of individuals. Moreover, the development of a linkage between biomarker evidence and different psychometric tools would be envisaged to optimize such psychiatric-standard screening, diagnosis, and management strategies. Therefore, the establishment of novel biomarker information resources for certain psychiatric disorders utilizing proteomic techniques could provide a novel framework of biological-based criteria as a parallel to the interview-based criteria. Such a resource would further improve good mental healthcare services and also potentially contribute to the forensic assessment of individuals by overcoming substandard screening tests and the limited identification of certain disorders commonly linked to criminal behavior or self harm. Beyond individual evaluation, such biological evidence would also have a potential role as a screening tool for larger populations of selected individuals, such as criminal detainees or mental health patients with other disorders or disabilities. The application of proteomic and metabolomic analyses in psychiatry and their utilization in searching for prospective biomarkers among psychiatry disorders is currently increasing (Dudley et al., 2011; Taurines et al., 2010). The study of biological differences in psychiatric patients requires cooperation between many groups to validate different biological, imaging, and genetic biomarker data (both state and trait) that may help in categorizing psychiatry diagnosis criteria using physical evidence. Such an improvement in our understanding of biological differences will also provide further insights into the biological and pathogenesis variations in psychiatric disorders. However, any central resource of data would require continuous monitoring and updating as new evidences arise that add to or contradict previous evidences.

REFERENCES

Ackerman, M. J. (2006). Forensic report writing. *Journal of Clinical Psychology*, *62*, 59–72.

Amico, F., Meisenzahl, E., Koutsouleris, N., Reiser, M., Möller, H. J., & Frodl, T. (2011). Structural MRI correlates for vulnerability and resilience to major depressive disorder. *Psychiatry and Neuroscience*, *36*, 15–22.

Arango, V., Underwood, M. D., & Mann, J. J. (1996). Fewer pigmented locus coeruleus neurons in suicide victims: Preliminary results. *Biological Psychiatry*, *39*, 112–120.

Arsenault-Lapierre, G., Kim, C., & Turecki, G. (2004). Psychiatric diagnosis in 3275 suicides: A meta-analysis. *BMC Psychiatry*, *4*, 37.

Baillargeon, J., Penn, J. V., Thomas, C. R., Temple, J. R., Baillargeon, G., & Murray, O. J. (2009). Psychiatric disorders and suicide in the nation's largest state prison system. *Journal of the American Academy of Psychiatry and the Law*, *37*, 188–193.

Barkley, R. A., Fischer, M., Smallish, L., & Fletcher, K. (2002). The persistence of attention-deficit/hyperactivity disorder into young adulthood as a function of reporting source and definition of disorder. *Journal of Abnormal Psychology*, *111*, 279–289.

Beautrais, A. L., Joyce, P. R., Mulder, R. T., Fergusson, D. M., Deavoll, B. J., & Nightingale, S. K. (1996). Prevalence and co-morbidity of mental disorders in persons making serious suicide attempts: A case control study. *American Journal of Psychiatry*, *153*, 1009–1014.

Bellak, L., Kay, S. R., & Opler, L. A. (1987). Attention deficit disorder psychosis as a diagnostic category. *Psychiatric Developments*, *5*, 239–263.

Bernal, M., Haro, J. M., Bernert, S., Brugha, T., de Graaf, R., Bruffaerts, R., et al. (2007). Risk factors for suicidality in Europe: Results from the ESEMED study. *Journal of Affective Disorders, 101*, 27–34.

Biederman, J., Newcorn, J., & Sprich, S. (1991). Comorbidity of attention deficit hyperactivity disorder with conduct, depressive, anxiety, and other disorders. *American Journal of Psychiatry, 148*, 564–577.

Bondy, B., Buettner, A., & Zill, P. (2006). Genetics of suicide. *Molecular Psychiatry, 11*, 336–351.

Breslau, N., Davis, G. C., Peterson, E. L., & Schultz, L. R. (2000). A second look at comorbidity in victims of trauma: The posttraumatic stress disorder-major depression connection. *Biological Psychiatry, 48*, 902–909.

Caldwell, C. B., & Gottesman, I. I. (1990). Schizophrenics kill themselves too; our view of risk factors for suicide. *Schizophrenia Bulletin, 16*, 571–590.

Caspi, A., Langley, K., Milne, B., Moffitt, T. E., O'Donovan, M., Owen, M. J., et al. (2008). A replicated molecular genetic basis for subtyping antisocial behavior in children with attention-deficit/hyperactivity disorder. *Archives of General Psychiatry, 65*, 203–210.

Castle, D. J., Wessely, S., & Murray, R. M. (1993). Sex and schizophrenia: Effects of diagnostic stringency, and associations with and premorbid variables. *British Journal of Psychiatry, 162*, 658–664.

Centers for Disease Control and Prevention (CDC). (2006). Homicides and suicides—National Violent Death Reporting System, United States, 2003–2004. *Morbidity and Mortality Weekly Report, 55*, 721–724.

Chen, Y., Bidwell, L. C., & Norton, D. (2006). Trait vs. state markers for schizophrenia: Identification and characterization through visual processes. *Current Psychiatry Reviews, 2*, 431–438.

Cohen, M., Yossef, R., Erez, T., Kugel, A., Welt, M., Karpasas, M. M., et al. (2011). Serum apolipoproteins C-I and C-III are reduced in stomach cancer patients: Results from MALDI-based peptidome and immuno-based clinical assays. *PLoS One, 6*, e14540.

Conner, A. C., Kissling, C., Hodges, E., Hünnerkopf, R., Clement, R. M., Dudley, E., et al. (2008). Neurotrophic factor-related gene polymorphisms and adult attention deficit hyperactivity disorder (ADHD) score in a high-risk male population. *American Journal of Medical Genetics Part B: Neuropsychiatric Genetics, 147B*, 1476–1480.

Coppen, A., & Bolander-Gouaille, C. (2005). Treatment of depression: Time to consider folic acid and vitamin B12. *Journal of Psychopharmacology, 19*, 59–65.

Corbett, B. A., Kantor, A. B., Schulman, H., Walker, W. L., Lit, L., Ashwood, P., et al. (2007). A proteomic study of serum from children with autism showing differential expression of apolipoproteins and complement proteins. *Molecular Psychiatry, 12*, 292–306.

Costello, C. E. (1997). Time, life... and mass spectrometry. New techniques to address biological questions. *Biophysical Chemistry, 68*, 173–188.

Dailly, E., Chenu, F., Renard, C. E., & Bourin, M. (2004). Dopamine, depression and antidepressants. *Fundamental & Clinical Pharmacology, 18*, 601–607.

Davatzikos, C., Shen, D., Gur, R. C., Wu, X., Liu, D., Fan, Y., et al. (2005). Whole brain morphometric study of schizophrenia revealing a spatially complex set of focal abnormalities. *Archives of General Psychiatry, 62*, 1218–1227.

Dawson, D. A., Grant, B. F., Stinson, F. S., & Chou, P. S. (2005). Psychopathology associated with drinking and alcohol use disorders in the college and general adult populations. *Drug and Alcohol Dependence, 77*, 139–150.

de Andrade, R. C., Assumpção, F., Jr., & Teixeira, I. A. (2011). Prevalence of psychiatric disorders in juvenile offenders in the city of Rio de Janeiro (RJ, Brazil). *Ciência & Saúde Coletiva, 16*, 2179–2188.

Dudley, E., Hässler, F., & Thome, J. (2011). Profiling for novel proteomics biomarkers in neurodevelopmental disorders. *Expert Review of Proteomics, 8*, 127–136.

Duman, R. S., Malberg, J., & Thome, J. (1999). Neural plasticity to stress and antidepressant treatment. *Biological Psychiatry, 46*, 1181–1191.

Fenton, W. S. (2000). Depression, suicide, and suicide prevention in schizophrenia. *Suicide & Life-Threatening Behavior, 30*, 34–49.

Fredericks, J. D., Bennett, P., Williams, A., & Rogers, K. D. (2012). FTIR spectroscopy: A new diagnostic tool to aid DNA analysis from heated bone. *Forensic Science International. Genetics, 6*, 375–380.

Graz, C., Etschel, E., Schoech, H., & Soyka, M. (2009). Criminal behaviour and violent crimes in former inpatients with affective disorder. *Journal of Affective Disorders, 117*, 98–103.

Greden, J. F. (2003). Physical symptoms of depression: Unmet needs. *Journal of Clinical Psychiatry, 64*, 5–11.

Grunberg, F., Lesage, A. D., Boyer, R., Vanier, C., Morissette, R., Buteau, C. M., et al. (1994). Suicide among young male adults in Quebec: Psychopathology and utilization of medical services. *Santé Mentale au Québec, 19*, 25–39.

Halbach, O. B. (2010). Involvement of BDNF in age-dependent alterations in the hippocampus. *Frontiers in Aging Neuroscience, 13*, 36.

Hawton, K., Sutton, L., Haw, C., Sinclair, J., & Deeks, J. J. (2005). Schizophrenia and suicide: Systematic review of risk factors. *British Journal of Psychiatry, 187*, 9–20.

Hirschfeld, R. M. (2000). History and evolution of monoamine hypothesis of depression. *Journal of Clinical Psychiatry, 61*, 4–6.

Hirschfeld, R. M. (2001). The comorbidity of major depression and anxiety disorders: Recognition and management in primary care. *Primary Care Companion to the Journal of Clinical Psychiatry, 3*, 244–254.

Holmgren, A., & Jones, A. W. (2010). Demographics of suicide victims in Sweden in relation to their blood-alcohol concentration and the circumstances and manner of death. *Forensic Science International, 198*, 17–22.

Howes, O. D., & Kapur, S. (2009). The dopamine hypothesis of schizophrenia: Version III—The final common pathway. *Schizophrenia Bulletin, 35*, 549–562.

Husain, M. M., Rush, A. J., Trivedi, M. H., McClintock, S. M., Wisniewski, S. R., Davis, L., et al. (2007). Pain in depression: STAR*D study findings. *Journal of Psychosomatic Research, 63*, 113–122.

Inskip, H. M., Harris, E. C., & Barraclough, B. (1998). Lifetime risk of suicide for affective disorder, alcoholism and schizophrenia. *British Journal of Psychiatry, 172*, 35–37.

Isacsson, G., Holmgren, P., Wasserman, D., & Bergman, U. (1994). Use of antidepressants among people committing suicide in Sweden. *British Medical Journal, 308*, 506–509.

Isjanovski, V., Vukovic, V., Hadjihamza, K., Pejoska-Gerazova, V., & Raleva, M. (2010). Some characteristics of patients with schizophrenia who attempted suicide. *Prilozi, 31*, 183–193.

Iwasaki, M., Akechi, T., Uchitomi, Y., & Tsugane, S. (2005). Cigarette smoking and completed suicide among middle-aged men: A population-based cohort study in Japan. *Annals of Epidemiology, 15*, 286–292.

Jones, R. M., Hales, H., Butwell, M., Ferriter, M., & Taylor, P. J. (2010). Suicide in high security hospital patients. *Social Psychiatry and Psychiatric Epidemiology, 46*, 723–731.

Kapur, S., Agid, O., Mizrahi, R., & Li, M. (2006). How antipsychotics work-from receptors to reality. *NeuroRx, 3*, 10–21.

Karas, M., & Hillenkamp, F. (1988). Laser desorption ionization of proteins with molecular masses exceeding 10,000 daltons. *Analytical Chemistry, 60*, 2299–2301.

Karch, C. M., Jeng, A. T., Nowotny, P., Cady, J., Cruchaga, C., & Goate, A. M. (2012). Expression of novel Alzheimer's disease risk genes in control and Alzheimer's disease brains. *PLoS One*, *7*, e 50976.

Karvonen, K., Hakko, H., Koponen, H., Meyer-Rochow, V. B., & Räsänen, P. (2009). Suicides among older persons in Finland and time since hospitalization discharge. *Psychiatric Services*, *60*, 390–393.

Kasckow, J., Felmet, K., & Zisook, S. (2011). Managing suicide risk in patients with schizophrenia. *CNS Drugs*, *25*, 129–143.

Kent, L., Green, E., Hawi, Z., Kirley, A., Dudbridge, F., Lowe, N., et al. (2005). Association of the paternally transmitted copy of common Valine allele of the Val66Met polymorphism of the brain-derived neurotrophic factor (BDNF) gene with susceptibility to ADHD. *Molecular Psychiatry*, *10*, 939–943.

Keshavan, M. S., Sujata, M., Mehra, A., Montrose, D. M., & Sweeney, J. A. (2003). Psychosis proneness and ADHD in young relatives of schizophrenia patients. *Schizophrenia Research*, *59*, 85–92.

Kessler, R. C., Adler, L., Barkley, R., Biederman, J., Conners, C. K., Demler, O., et al. (2006). The prevalence and correlates of adult ADHD in the United States: Results from the National Comorbidity Survey Replication. *American Journal of Psychiatry*, *163*, 716–723.

Kim, S. W., Kim, S. J., Mun, J. W., Bae, K. Y., Kim, J. M., Kim, S. Y., et al. (2010). Psychosocial factors contributing to suicidal ideation in hospitalized schizophrenia patients in Korea. *Psychiatry Investigation*, *7*, 79–85.

Kissling, C., Retz, W., Wiemann, S., Coogan, A. N., Clement, R. M., Hünnerkopf, R., et al. (2008). A polymorphism at the 3'-untranslated region of the CLOCK gene is associated with adult attention-deficit hyperactivity disorder. *American Journal of Medical Genetics Part B: Neuropsychiatric Genetics*, *147*, 333–338.

Kjøller, M., & Helweg-Larsen, M. (2000). Suicidal ideation and suicide attempts among adult Danes. *Scandinavian Journal of Public Health*, *28*, 54–61.

Kuehner, C. (2003). Gender differences in unipolar depression: An update of epidemiological findings and possible explanations. *Acta Psychiatrica Scandinavica*, *108*, 163–174.

LaFrance, W. C., Jr., Kanner, A. M., & Hermann, B. (2008). Psychiatric comorbidities in epilepsy. *International Review of Neurobiology*, *83*, 347–383.

Langley, K., Fowler, T., Ford, T., Thapar, A. K., van den Bree, M., Harold, G., et al. (2010). Adolescent clinical outcomes for young people with attention-deficit hyperactivity disorder. *British Journal of Psychiatry*, *196*, 235–240.

Li, Y. H., Wang, J., Zheng, X. L., Zhang, Y. L., Li, X., Yu, S., et al. (2011). Matrix-assisted laser desorption/ionization time-of-flight mass spectrometry combined with magnetic beads for detecting serum protein biomarkers in Parkinson's disease. *European Neurology*, *65*, 105–111.

Liu, C., Pan, C., Shen, J., Wang, H., & Yong, L. (2011). MALDI-TOF MS combined with magnetic beads for detecting serum protein biomarkers and establishment of boosting decision tree model for diagnosis of colorectal cancer. *International Journal of Medical Sciences*, *8*, 39–47.

Lönnqvist, J. K., Henriksson, M. M., Isometsä, E. T., Marttunen, M. J., Heikkinen, M. E., Aro, H. M., et al. (1995). Mental disorders and suicide prevention. *Psychiatry and Clinical Neurosciences*, *49*, S111–S116.

Lopez, M. F., Mikulskis, A., Kuzdzal, S., Bennett, D. A., Kelly, J., Golenko, E., et al. (2005). High-resolution serum proteomic profiling of Alzheimer disease samples reveals disease-specific, carrier-protein-bound mass signatures. *Clinical Chemistry*, *51*, 1946–1954.

MacDonald, R., Taylor, J., & Clarke, D. (2009). The relationship between early suicide behaviors and mental health: Results from a nine-year panel study. *Journal of Adolescence*, *32*, 1159–1172.

Mann, J. J., Apter, A., Bertolote, J., Beautrais, A., Currier, D., Haas, A., et al. (2005). Suicide prevention strategies: A systematic review. *JAMA, 294*, 2064–2074.

Mannuzza, S., Klein, R. G., & Moulton, J. L., 3rd. (2008). Lifetime criminality among boys with attention deficit hyperactivity disorder: A prospective follow-up study into adulthood using official arrest records. *Psychiatry Research, 160*, 237–246.

Mayes, R., Bagwell, C., & Erkulwater, J. (2008). ADHD and the rise in stimulant use among children. *Harvard Review of Psychiatry, 16*, 151–166.

McCann, D., Barrett, A., Cooper, A., Crumpler, D., Dalen, L., Grimshaw, K., et al. (2007). Food additives and hyperactive behaviour in 3-year-old and 8/9-year-old children in the community: A randomised, double-blinded, placebo-controlled trial. *Lancet, 370*, 1560–1567.

McGrath, J., Saha, S., Chant, D., & Welham, J. (2008). Schizophrenia: A concise overview of incidence, prevalence, and mortality. *Epidemiologic Reviews, 30*, 67–76.

McIntosh, D., Kutcher, S., Binder, C., Levitt, A., Fallu, A., & Rosenbluth, M. (2009). Adult ADHD and comorbid depression: A consensus-derived diagnostic algorithm for ADHD. *Neuropsychiatric Disease and Treatment, 5*, 137–150.

Mekonnen, D., & Kebede, Y. (2011). The prevalence of suicidal ideation and attempts among individuals attending an adult psychiatry out-patient clinic in Gondar, Ethiopia. *African Health Sciences, 11*, 103–107.

Meyer, J. H., Ginovart, N., Boovariwala, A., Sagrati, S., Hussey, D., Garcia, A., et al. (2006). Elevated monoamine oxidase a levels in the brain: An explanation for the monoamine imbalance of major depression. *Archives of General Psychiatry, 63*, 1209–1216.

Mitchell, A. J., & Dennis, M. (2006). Self-harm and attempted suicide in adults: 10 practical questions and answers for emergency department staff. *Emergency Medicine Journal, 23*, 251–255.

Müller, U. C., Asherson, P., Banaschewski, T., Buitelaar, J. K., Ebstein, R. P., Eisenberg, J., et al. (2011). The impact of study design and diagnostic approach in a large multi-centre ADHD study. Part 1: ADHD symptom patterns. *BMC Psychiatry, 11*, 54.

Oquendo, M. A., Lizardi, D., Greenwald, S., Weissman, M. M., & Mann, J. J. (2004). Rates of lifetime suicide attempt and rates of lifetime major depression in different ethnic groups in the United States. *Acta Psychiatrica Scandinavica, 110*, 446–451.

Palmer, B. A., Pankratz, V. S., & Bostwick, J. M. (2005). The lifetime risk of suicide in schizophrenia: A reexamination. *Archives of General Psychiatry, 62*, 247–253.

Pandey, S., Pandey, P., Tiwari, G., Tiwari, R., & Rai, A. K. (2012). FTIR spectroscopy: A tool for quantitative analysis of ciprofloxacin in tablets. *Indian Journal of Pharmaceutical Sciences, 74*, 86–90.

Peveler, R., Carson, A., & Rodin, G. (2002). ABC of psychological medicine. Depression in medical patients. *British Medical Journal, 325*, 149–152.

Powell, J., Geddes, J., Deeks, J., Goldacre, M., & Hawton, K. (2000). Suicide in psychiatric hospital in-patients. Risk factors and their predictive power. *British Journal of Psychiatry, 176*, 266–272.

Qin, P., Agerbo, E., & Mortensen, P. B. (2003). Suicide risk in relation to socioeconomic, demographic, psychiatric, and familial factors: A national register-based study of all suicides in Denmark, 1981–1997. *American Journal of Psychiatry, 160*, 765–772.

Rathi, Y., Malcolm, J., Michailovich, O., Goldstein, J., Seidman, L., McCarley, R. W., et al. (2010). Biomarkers for identifying first-episode schizophrenia patients using diffusion weighted imaging. *Medical Image Computing and Computer-Assisted Intervention, 13*, 657–665.

Reichert, C. L., Diogo, C. L., Vieira, J. L., & Dalacorte, R. R. (2011). Physical activity and depressive symptoms in community-dwelling elders from southern Brazil. *Revista Brasileira de Psiquiatria, 33*, 165–170.

Rihmer, Z. (2007). Suicide risk in mood disorders. *Current Opinion in Psychiatry, 20*, 17–22.

Rihmer, Z., & Akiskal, H. (2006). Do antidepressants t(h)reat(en) depressives? Toward a clinically judicious formulation of the antidepressant-suicidality FDA advisory in light of declining national suicide statistics from many countries. *Journal of Affective Disorders, 94*, 3–13.

Rohde, L. A., Szobot, C., Polanczyk, G., Schmitz, M., Martins, S., & Tramontina, S. (2005). Attention-deficit/hyperactivity disorder in a diverse culture: Do research and clinical findings support the notion of a cultural construct for the disorder? *Biological Psychiatry, 57*, 1436–1441.

Rubia, K. (2002). The dynamic approach to neurodevelopmental psychiatric disorders: Use of fMRI combined with neuropsychology to elucidate the dynamics of psychiatric disorders, exemplified in ADHD and schizophrenia. *Behavioural Brain Research, 130*, 47–56.

Saha, S., Chant, D., & McGrath, J. (2007). A systematic review of mortality in schizophrenia: Is the differential mortality gap worsening over time? *Archives of General Psychiatry, 64*, 1123–1131.

Scoppetta, M., Di Gennaro, G., & Scoppetta, C. (2005). Selective serotonin reuptake inhibitors prevents emotional lability in healthy subjects. *European Review for Medical and Pharmacological Sciences, 9*, 343–348.

Simon, V., Czobor, P., Bálint, S., Mészáros, A., & Bitter, I. (2009). Prevalence and correlates of adult attention-deficit hyperactivity disorder: Meta-analysis. *British Journal of Psychiatry, 194*, 204–211.

Smith, K. M., Renshaw, P. F., & Bilello, J. (2013). The diagnosis of depression: Current and emerging methods. *Comprehensive Psychiatry, 54*, 1–6.

Soares, C. N., Almeida, O. P., Joffe, H., & Cohen, L. S. (2001). Efficacy of estradiol for the treatment of depressive disorders in perimenopausal women: A double-blind, randomized, placebo-controlled trial. *Archives of General Psychiatry, 58*, 529–534.

Taurines, R., Dudley, E., Conner, A. C., Grassl, J., Jans, T., Guderian, F., et al. (2010). Serum protein profiling and proteomics in autistic spectrum disorder using magnetic bead-assisted mass spectrometry. *European Archives of Psychiatry and Clinical Neuroscience, 260*, 249–255.

Teixeira, E. H., & Dalgalarrondo, P. (2009). Violent crime and dimensions of delusion: A comparative study of criminal and noncriminal delusional patients. *Journal of the American Academy of Psychiatry and the Law, 37*, 225–231.

Thome, J., Henn, F. A., & Duman, R. S. (2002). Cyclic AMP response element-binding protein and depression. *Expert Review of Neurotherapeutics, 2*, 347–354.

Tidemalm, D., Elofsson, S., Stefansson, C. G., Waern, M., & Runeson, B. (2005). Predictors of suicide in a community-based cohort of individuals with severe mental disorder. *Social Psychiatry and Psychiatric Epidemiology, 40*, 595–600.

Tripp, G., & Wickens, J. R. (2009). Neurobiology of ADHD. *Neuropharmacology, 57*, 579–589.

Ural, C., Öncü, F., Belli, H., & Soysal, H. (2013). Violent behavior variables among patients with schizophrenia under forensic psychiatric process: A case-control study. *Türk Psikiyatri Dergisi, 24*, 17–24.

van Os, J., Rutten, B. P., & Poulton, R. (2008). Gene-environment interactions in schizophrenia: Review of epidemiological findings and future directions. *Schizophrenia Bulletin, 34*, 1066–1082.

Villas-Bôas, S. G., Mas, S., Akesson, M., Smedsgaard, J., & Nielsen, J. (2005). Mass spectrometry in metabolome analysis. *Mass Spectrometry Reviews, 24*, 613–646.

Voeller, K. K. (2004). Attention-deficit hyperactivity disorder (ADHD). *Journal of Child Neurology, 19*, 798–814.

Wallis, D. (2010). The search for biomarkers for attention deficit/hyperactivity disorder. *Drug News & Perspectives, 23*, 438–449.

Wang, D. C., Sun, C. H., Liu, L. Y., Sun, X. H., Jin, X. W., Song, W. L., et al. (2012). Serum fatty acid profiles using GC-MS and multivariate statistical analysis: Potential biomarkers of Alzheimer's disease. *Neurobiology of Aging*, *33*, 1057–1066.

Way, B. B., Miraglia, R., Sawyer, D. A., Beer, R., & Eddy, J. (2005). Factors related to suicide in New York state prisons. *International Journal of Law and Psychiatry*, *28*, 207–221.

Wender, P. H., Wolf, L. E., & Wasserstein, J. (2001). Adults with ADHD. An overview. *Annals of the New York Academy of Sciences*, *931*, 1–16.

Wolf, L. E., & Wasserstein, J. (2001). Adult ADHD. Concluding thoughts. *Annals of the New York Academy of Sciences*, *931*, 396–408.

Wolosin, S. M., Richardson, M. E., Hennessey, J. G., Denckla, M. B., & Mostofsky, S. H. (2009). Abnormal cerebral cortex structure in children with ADHD. *Human Brain Mapping*, *30*, 175–184.

Yee, N. Y., Large, M. M., Kemp, R. I., & Nielssen, O. B. (2011). Severe non-lethal violence during psychotic illness. *The Australian and New Zealand Journal of Psychiatry*, *45*, 466–472.

Yoshimasu, K., Sugahara, H., Tokunaga, S., Akamine, M., Kondo, T., Fujisawa, K., et al. (2006). Gender differences in psychiatric symptoms related to suicidal ideation in Japanese patients with depression. *Psychiatry and Clinical Neurosciences*, *60*, 563–569.

Young, S. J., Adamou, M., Bolea, B., Gudjonsson, G., Müller, U., Pitts, M., et al. (2011). The identification and management of ADHD offenders within the criminal justice system: A consensus statement from the UK Adult ADHD Network and criminal justice agencies. *BMC Psychiatry*, *11*, 32.

Young, S., Gudjonsson, G. H., Wells, J., Asherson, P., Theobald, D., Oliver, B., et al. (2009). Attention deficit hyperactivity disorder and critical incidents in a Scottish prison population. *Personality and Individual Differences*, *46*, 265–269.

Zhao, L., Liu, Y., Sun, X., Peng, K., & Ding, Y. (2011). Serum proteome analysis for profiling protein markers associated with lymph node metastasis in colorectal carcinoma. *Journal of Comparative Pathology*, *144*, 187–194.

AUTHOR INDEX

Note: Page numbers followed by "*f*" indicate figures, "*t*" indicate tables and "*np*" indicate footnotes.

C

G

H

I

K

L

M

N

Q

R

S

U

V

W

SUBJECT INDEX

Note: Page numbers followed by "*f*" indicate figures and "*t*" indicate tables.

D

E

F

Kolibri 360 GmbH
Ferdinand-Jühlke-Straße 7
99095 Erfurt, Deutschland
produktsicherheit@kolibri360.de